# About Island Press

Since 1984, the nonprofit organization Island Press has been stimulating, shaping, and communicating ideas that are essential for solving environmental problems worldwide. With more than 1,000 titles in print and some 30 new releases each year, we are the nation's leading publisher on environmental issues. We identify innovative thinkers and emerging trends in the environmental field. We work with world-renowned experts and authors to develop cross-disciplinary solutions to environmental challenges.

Island Press designs and executes educational campaigns, in conjunction with our authors, to communicate their critical messages in print, in person, and online using the latest technologies, innovative programs, and the media. Our goal is to reach targeted audiences—scientists, policy makers, environmental advocates, urban planners, the media, and concerned citizens—with information that can be used to create the framework for long-term ecological health and human well-being.

Island Press gratefully acknowledges major support from The Bobolink Foundation, Caldera Foundation, The Curtis and Edith Munson Foundation, The Forrest C. and Frances H. Lattner Foundation, The JPB Foundation, The Kresge Foundation, The Summit Charitable Foundation, Inc., and many other generous organizations and individuals.

The opinions expressed in this book are those of the author(s) and do not necessarily reflect the views of our supporters.

# HOLISTIC
# MANAGEMENT

# HOLISTIC MANAGEMENT

*A Commonsense Revolution to Restore Our Environment*

Third Edition

*Allan Savory*
with Jody Butterfield

**ISLAND**PRESS  Washington | Covelo | London

Island Press is a trademark of The Center for Resource Economics.

Island Press would like to thank the Regenerative Rangelands Fund for generously supporting the publication of this book.

Holistic Management is a registered trademark of Holistic Management International.

Library of Congress Control Number: 2016941253

Printed on recycled, acid-free paper ♺

Manufactured in the United States of America
10  9  8  7

*Keywords:* climate change, desertification, carbon sequestration, biodiversity, soil carbon, soil health, livestock, grasslands, sound policy, regenerative agriculture, regenerative grazing, land restoration, wildlife habitat, holistic management, natural resource management, ecosystem management, environmental refugees, paradigm shift

*To Bradey, Hugh, Ian, Luke, and Mika,*
*and your generation of team humanity—*
*our future is in your hands now*

# CONTENTS

## Part 5:  The Tools We Use to Manage Our Ecosystem

## Part 6:  Holistic Decision Making

## Part 7:  Guidelines for Using the Management Tools

## Part 8:  Procedures and Processes Unique to Holistic Management

## Part 9:  Completing the Feedback Loop

## Part 10:  Conclusion

# PREFACE

*As a youngster, my only aim in life* was to live in the wildest African bush forever. Though I eventually did have that opportunity, I ended up forsaking it in order to work toward saving the wildlife that was my reason for being in the bush. Even in the wildest areas, the land was deteriorating, in fact gradually turning to desert, rendering it ever less able to support life of any kind. I was determined to find a way to reverse this process.

That quest took me in a direction I would never have anticipated, compelling me to work first with people who for generations had been caretakers of the land and whose livestock I believed were responsible for initiating the deterioration, then with those who were advising them, and eventually with many others as a member of Parliament attempting to deal with environmental degradation and its effects at the policy level.

What I learned from these experiences was that the remorseless spread of deserts and the human impoverishment that always resulted *were* related to the presence of livestock in many cases, but it was how those livestock were managed that was the problem, not their presence, and it was the *management* that had to change. This also had implications for climate change and our ability to tackle it, because desertification causes soils to release carbon and renders them less able to store it.

This book describes the way forward that emerged. It involves a new framework for management that enables people to make decisions which satisfy immediate needs without jeopardizing their future well-being or that of others. In turn, of course, the actions ensuing from any decision must also enhance the well-being of the environment that sustains us now and will have to sustain future generations. The greatest strength of this new management framework is that it leads us to see that we serve our own interests best when we account for the environmental, as well as the social and economic consequences, of our actions.

*Holistic Management: A Commonsense Revolution to Restore Our Environment* is the revised and updated version of *Holistic Management: A New Framework for Decision Making* (1999), which was a second-edition attempt to explain the development of the framework underlying Holistic Management and how it could be used in decision making. The new subtitle is a reflection of the urgency with which we must restore our environment. This requires a revolution in management, agriculture, policy making, and institutional behavior that can only be initiated at the grassroots level. My hope is that this book inspires the revolution by identifying the challenges we have to address and how we can address them successfully.

The change of subtitle is only the most obvious difference. The book's ideas have been clarified and strengthened through the contributions of those I have worked with over the last two decades on six continents. These people include many thousands who make their living from the land and are learning to restore it profitably through practices that mimic nature.

This edition also reflects the contributions made by my wife and colleague, Jody Butterfield. It would not have appeared if she hadn't coauthored it with me. Because so much of the book is built on previous editions and my own history, she insisted it continue to be told in my voice, rather than "our" voice. But hers runs subtly throughout.

The key differences between the second edition and this one are woven throughout the book as critical themes and will become apparent to anyone who has read both volumes. This edition includes fewer pages as we've been able to simplify a number of concepts, eliminate detail that clouded understanding, and jettison management guidelines that were not unique to Holistic Management and that others have written about far more competently. With the creation of the Savory Institute in 2009, which Jody and I cofounded with a group of my early students to continue this work, and following a TED Talk I gave in 2013, since viewed by millions of people, our primary readership has greatly expanded. Readers now also include health care providers, nutritionists, fitness aficionados, students of all ages, scientists from multiple disciplines, and many thousands of parents and others increasingly concerned about where their food comes from, how it is raised, and how well the land that provided it is cared for. Many of them are also climate change activists who know that we cannot successfully tackle climate change by reducing fossil fuel emissions alone; we also have to come to grips with agriculture's enormous contribution to climate change. This will require a shift to agricultural practices that produce healthy, *living* soils that can once again serve as the world's largest carbon sink.

Our fate as a civilization is tied to the land and its health, and millions of ordinary people in making their living from the land control that fate to a large degree. Unless land managers have the support of the millions of people living in cities who depend on their efforts, and whose opinions influence the rate at which institutions and governments can change, those on the land cannot succeed. Given greater public awareness of the need for a shift to Holistic Management and policy making, in what we increasingly recognize as a holistic world, I see exciting times ahead and real hope for generations to come.

*Allan Savory*
Victoria Falls, Zimbabwe
April 2016

# ACKNOWLEDGMENTS

*Anything we do in science* is built on the work of those who have gone before us. Both from their successes and from their failures we learn and thus advance. Without the theory of holism provided by South African statesman Jan Smuts, Holistic Management could not have been developed. I am deeply indebted to the many who established the science and ecological principles I could use and build upon, as well as to the great philosophers and theorists who strove to find better ways for us to live in harmony with each other and our environment. Had I not had the shoulders of these greater minds to stand on, I could not have seen nearly so far, nor found the way forward that I did.

From the time I departed from the conventional thinking of my training to become an independent scientist, I have been supported and helped by many people, and I welcome this opportunity to thank them. I am particularly indebted to the many farmers and ranchers in southern Africa, and later in North America and Australia, who recognized that their land continued to deteriorate no matter what they did, and who were prepared to work with me in the search for answers. I am no less indebted to those working in government agencies and universities who supported our efforts, despite the criticism of their peers, and in some cases the opposition of their institutions. Without the courage and enthusiasm of all these people we would never have succeeded in finding a better way.

A number of friends and colleagues read portions of the manuscript, providing invaluable criticism and correcting embarrassing errors. They are Tre' Cates, Hannah Gosnell, Daniela Howell, Archie Mossman, Joel Salatin,

Kate Sherren, Keith Weber, and Kristina Wolf. In addition, the following persons read the entire manuscript (or most of it) helping to improve the overall structure, challenging points that were weak or vague, and greatly reducing the number of incomprehensible passages: Brandon Dalton, Christopher Cooke, Andrea Malmberg, Matt Raven, Jason Rowntree, Byron Shelton, and Abbey Smith. To all these people Jody and I give our most sincere thanks.

# PART 1

# INTRODUCTION

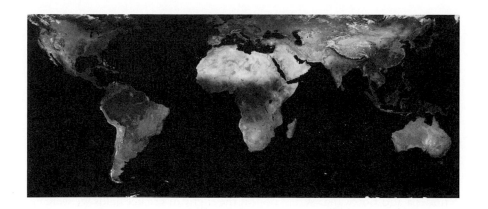

# 1

## Managing Holistically

*In 1948 I entered Plumtree School,* a boarding school in the British tradition set in the African bushveld on the border of what was then Southern Rhodesia and Botswana. When not on the rugby or cricket fields we were encouraged to get out into the bush, a gesture of liberality that offset all my adolescent frustration with formal education. I became fanatic about the bush and its big game, and a passion to return to it drove me through a university education that qualified me for a Northern Rhodesian Game Department post at the age of twenty.

Once in the Game Department I began to realize that all I loved was doomed. Not for the commonly talked of reasons—poaching and overexploitation—but rather because of our own ignorance as professional bureaucrats. My colleagues did not want to admit to ignorance or to raise the questions I did about the environmental deterioration I was seeing everywhere—massive amounts of bare ground, deep gullies, dead vegetation, and dried-up rivers. It was not only destroying the wildlife we were meant to protect but would ultimately threaten all other life on our planet. I took on a new post as a research officer in the Southern Rhodesia Game Department but again faced the same challenges and eventually resigned to become an independent scientist free to seek knowledge and solutions from any field in any country.

I supported my growing family through a variety of additional occupations—farmer, game rancher, cattle rancher, management consultant—while also becoming a soldier during a long and bitter civil war, and a member of Parliament leading the opposition to the racist government of Ian Smith. The latter got me into hot water and I was forced into exile, where I continued my consulting work in the Americas. No matter what I was doing over these years the problems I encountered every day in every place stemmed ultimately from a deteriorating environment. I had quickly learned that poor land leads to poor people, social breakdown, political upheaval, and war. This was at its worst where humidity and rainfall were seasonal and livestock production was the chief occupation.

I had long believed, like most of my peers, that livestock were responsible for the destruction I was seeing in these areas, but new insights (described in chaps. 3–6), enabled me to see that the problem was the way we were managing livestock, not the animals themselves. Properly managed, livestock could be part of the solution. Flowing from this knowledge I was able to develop an entirely new approach to livestock management using a planning process that improved the land for wildlife, livestock, and people. But rather than exciting most of my peers, or even many of the ranchers who stood to benefit, the counterintuitive logic of using livestock herds to restore degraded land caused a ruckus. It has taken close to five decades to work through what started as vigorous opposition from many quarters to growing support for the ideas. This is due in no small part to the hundreds of people who worked with me initially, demonstrating their own successes and providing support and insights. Although some belonged to institutions opposed to the new ideas, they found ways to collaborate as individuals.

## The Agriculture Problem

Opposition to the idea that properly managed livestock could restore degraded land led to a delay in the widespread application of Holistic Management that has been costly, as the amount of land turning to desert has only accelerated. Over these same decades agriculture as a whole has transitioned from a soil-maintaining enterprise to a soil-depleting enterprise based on chemical inputs, with the result that we are losing our ability to feed a growing population of nearly nine billion people.

Farmers are increasingly dependent on synthetic fertilizers and pesticides, which kill soil organisms and poison waterways. And anytime soil is exposed—through plowing, or through harvesting crops and clearing or burning the residue—soil organisms die and thus soils do too, releasing carbon dioxide into the atmosphere. When combined with the unsustainable techniques used for factory farming pigs, poultry, and cattle, it becomes apparent that modern agriculture is a major contributor to both desertification and climate change.

If we do not address the agricultural problem realistically and rapidly, irreparable climate change could continue long after we replace fossil fuels with environmentally benign energy sources. Each year, the earth loses seventy-five billion tons of soil to erosion, mostly from agricultural land.[1] That's more than ten tons per human alive, or twenty times as much eroding soil as food required per human each year. Seventy percent of the grasslands—broadly defined as any environment where grasses play a critical role in stabilizing soil—are now considered degraded, or turning to desert. This has led to increasing hunger, poverty, violence, and tens of millions of "environmental refugees." As I will show in many of the following chapters, the land degradation figures I've cited are almost certainly much too conservative.

---

Grasslands, broadly defined, are those environments in which grasses play a critical role in stabilizing soil—from dry deciduous forests to savannas or open grasslands to arid and semiarid rangelands.

---

The appalling amount of soil destruction is silting up once highly productive coastal fisheries. The annual burning of billions of hectares of crop residues, grasslands, and forests is adding to the atmospheric pollution contributing to climate change. Soil destruction now accounts for thirty percent of the carbon dioxide emissions entering the atmosphere[2] and biomass (vegetation) burning eighteen percent[3]—nearly equaling the emissions from fossil fuels.

## Addressing Climate Change

Healthy, *living* soils are key to reversing climate change because once we reduce the carbon dioxide coming from agriculture and fossil-fuel emissions, there will still be many billions of tons of carbon dioxide in the atmosphere that need to be drawn down to Earth and safely stored if we are to maintain a livable climate.

The oceans have long played a role in drawing down atmospheric carbon dioxide, but when carbon dioxide dissolves in the ocean carbonic acid forms. So much carbon dioxide has seeped into the oceans that they are now becoming increasingly acidic and inhospitable to a variety of sea life, especially shell-growing animals. Equally worrisome, the oceans' capacity to store carbon dioxide could diminish.[4]

Planting trees is not a solution for desertification or climate change because only a few environments receive sufficient rainfall to sustain tree plantations or a full soil-covering canopy of leaves. And, using fossil-fuel-powered earth-moving techniques to bring water to them is not commercially viable or scalable. Trees do store carbon, just as all living things do, but then release it as carbon dioxide when they die. Soils, however, can hold carbon for millennia in the form of organic matter.[5] The vast grassland soils, with the help of the grazing animals that evolved with them, can store the greatest amounts of carbon, which is why so many of the world's primary grain-growing regions, with their once deep, carbon-rich soils, are former grasslands.

We don't have time to waste in reforming agriculture and regenerating our soils to draw down the "legacy load" of carbon dioxide from the atmosphere: in 2014 atmospheric carbon dioxide levels reached 400 parts per million—50 parts per million higher than scientists believe is safe for human life.[6] Fortunately, a growing number of farmers working human-scale, rather than industrial-scale, farms are showing us the way, and ranchers and pastoralists are demonstrating what is possible on the world's grasslands.[7]

## A Sustainable Economy

Setting aside the urgency of climate change for a moment, consider the economic importance of establishing a sound and sustainable agriculture. Agriculture made civilization possible. The domestication of crop and livestock species enabled farmers to create surpluses. This freed people to pursue activities that led to the development of cities and all their amenities. Without agriculture we could not have an orchestra, museum, university—or even a city. Agriculture was once the cornerstone of every city's economy.

Although we've lost sight of the fact today, the only basis for an economy that can sustain a community or nation is derived from photosynthesis—the process through which green, growing plants convert sunlight and carbon dioxide into the carbohydrates and sugars that feed all terrestrial and most aquatic life. Healthy, regenerating soils can grow more plants that can convert more sunlight to food, and keep on doing so. Soils rendered lifeless by synthetic fertilizers and pesticides, and practices that keep soils exposed, will at some point no longer be able to grow plants, nor store the water they depend upon. The mineral resources we so prize—coal, oil, gold, and diamonds—are nonrenewable and cannot feed and clothe people; they could never become the basis of a self-sustaining economy.

## Two Management Frameworks

In the 1970s, as farmers and ranchers began to demonstrate just how effective livestock could be at restoring degraded land, I realized, as chapter 3 explains, that if we focused only on land restoration, we would not achieve lasting change. We had to keep a steady eye as well on the financial soundness of our efforts, and the well-being of the people involved. And this was no simple task. It led me to see the need for a basic framework to help guide us through the complex situations we were attempting to manage, and I enlisted many others in its development—clients, students, fans, and detractors.

Only after developing what became the Holistic Management framework, did I realize that we already were using a framework, one that appears to be genetically embedded in all tool-using animals but is not holistic in nature, nor successful at guiding the management of the environment that sustains us. It's helpful to look at this embedded framework first because the holistic framework builds on it.

### The Genetically Embedded Framework

There was a common denominator in our management failures. This was tied to how we decided what actions to take. Something was faulty, and it had been faulty for a very long time. But where was it at fault, and how were we to find out? The answer doesn't become apparent until you first examine *how* we make the decisions that inform our actions.

Fundamentally, we use a process common to all tool-using animals:

- We have an objective (or goal, vision, mission).
- To achieve that objective we apply one or more of the tools available to us.
- We decide which tool to use and how to use it, based on whether or not we think it can do the job and meet our objective.

For example, a hungry otter has an objective: break open a clamshell; he uses a simple tool—technology, in the form of a stone—to do so, based on past experience, or what he learned from his mother. Or, the president of the United States declares an objective: to put a man on the moon within a decade (before the Soviets achieve it). He and his team use the same tool—technology—but various and more sophisticated forms of it, and base their choices on research and expert advice, past experience, cost, and so on.

It's the same process, or framework, in both cases; only the degree of sophistication varies. For humans, who, unlike other tool-using animals, can create visions beyond the simplest objectives, the process has been wildly successful: we have indeed put a man on the moon. But this framework has also led to big trouble: we're destroying life on our own planet at an alarming rate.

So, there are a few things we've added to this genetically embedded management framework, as shown (in bold) in table 1-1. With these additions, the Holistic Management framework helps ensure that we succeed in our aims while beginning to restore our ailing planet.

**Table 1-1.** The genetically embedded framework versus the holistic framework

| Genetically Embedded Framework | Holistic Framework |
| --- | --- |
| • Simple context | • **Whole under management** |
| • Objectives (goals, mission, or vision) | • **Holistic context** |
| • Tools: technology, fire, resting environment | • Objectives (goals, mission, or vision) |
| • Actions based on one or more of many factors | • Tools: technology, fire, resting environment, **living organisms** |
| | • Actions based on one or more of many factors and **checked to ensure in context** |

*The Context*

As used in table 1-1, *context* refers to the reason we want to do something, and there always is one, even if we're not conscious of it. In the previous examples, the context for the otter was "hunger"; for the U.S. president the context was "competition with the USSR for national prestige." In both cases the context was simple, as it has always been for our objectives and the actions we take to achieve them. Commonly, the context is related to a need or desire—we want a profit, to create something, to achieve more than someone else—or to a problem that needs fixing—brush is invading our pasture, there's a gas leak in the house, we're running short of cash. But when we are attempting to manage anything, and especially when managing land, people, and finances together, a simple context is too narrow, and we tend to overlook vital aspects that a larger context would encompass. So we now create a *holistic* context (covered in chap. 9) that describes how we want our lives to be in the whole we manage and the environment and behaviors that will sustain that quality of life for future generations. There is no mention of problems; the holistic context is a reflection of what lies beyond them.

*The Tools Available*

All the creativity, money, or labor we expend to influence our environment has to be applied through one or more *tools*, which is why tool using defines us as a species. For millennia we used sticks and stones—*technology*—as our first tool, and we barely impacted our environment. Once we also added *fire* to the toolbox that changed. Fire enabled us to dramatically impact whole landscapes and to develop increasingly sophisticated technology as we moved through the copper, bronze, and iron ages and into today's society, driven by advanced technology. Other than technology and fire, the only other tool we've applied to managing our environment at large is *rest*, or nondisturbance, to restore biodiversity.

None of these tools in and of themselves can be relied upon to regenerate the world's soils, which has to be done through biological rather than chemical means, not only because it is a biological issue but also due to the nature and scale of the challenge. Thus we now add a fourth tool—*living organisms*, covered in chapters 21 through 23. In perennially humid environments, the cycle of life—birth, growth, death, decay—functions well in the absence of large herding animals. In seasonally humid environments, which experience prolonged periods of little or no growth during the year due to dryness or cold, the vegetation life cycle is impeded in the absence of significant numbers of large herding animals. So in these environments we utilize large animals to help restore or enhance ecosystem functioning. The seasonally humid, or brittle, environments, described in chapter 4, encompass nearly two-thirds of Earth's landmass and include most of the world's grasslands. They evolved with large herds of wild grazers whose behavior in the presence of pack-hunting predators had a dramatic effect on soils and soil life, described in chapter 5. It is this behavior that I observed in my early days in some of the wildest areas left in Africa, and that I realized livestock could be managed to mimic.

### Deciding How We'll Use the Tools

In weighing up the actions we might take to achieve our objectives, we consider all the usual criteria—past experience, research results, cost, and so forth—but we now also check for their social, economic, and ecological soundness by asking a few questions (described in chaps. 24–31) to help filter out actions that are not in context and could lead us in the wrong direction—short- or long-term.

### *The Holistic Management Framework*

The need for a holistic management framework has long been obscured because of the successes we have achieved without one. We've been able to develop increasingly sophisticated forms of technology with which to exploit Earth's resources, and to make life genuinely more comfortable, but we haven't been able to do so without damaging our environment at the same time.

Now, more than ever, we require the ability to make decisions that *simultaneously* consider economic, social, and environmental realities, both short- and long-term. And, the holistic framework for organizing management and decision making helps us do this. I believe it will be key to creating an agriculture that produces more food than eroding soil, which is far more critical than financing bank bailouts or developing ever more lethal weapons.

Creating the framework has been a driving force in the development of Holistic Management, but as the next five chapters will show, we had much to learn before it took shape. Four key insights discovered over the last century, when taken together, proved critical. The first insight made the argument for why such a framework was needed and the form it should take. The next three insights enabled us to understand why some environments rapidly deteriorate under practices that benefit others, and added pieces to the new framework that proved vital for completing it. The Holistic Management framework is summarized in chapter 7 and described at length in the remaining chapters of this book.

This book should prove enlightening to those working within agriculture, or to those working to address the problems arising from environmental degradation and desertification—loss of biodiversity, increasing droughts, floods, hunger, poverty and social breakdown, and mass emigration of rural populations to cities.

If you are one of many others who feel disconnected from the land, my hope is that this book helps reestablish that connection. I can guarantee that after reading it you will never view land the same way again.

# PART 2

# FOUR KEY INSIGHTS

# 2

## Introduction
### *The Power of*
### *Paradigms*

HAD WE CONCLUDED A HUNDRED YEARS AGO that we needed to change the way we make management decisions, we could not have done so successfully. Our knowledge still lacked some vital pieces. Four new insights proved key to removing the obstacles in our path. They were all discovered separately over the last century but were either ignored, forgotten, or bitterly opposed because they represented new knowledge that contradicted the beliefs held by most people—by no means a new problem.

We could draw parallels to early innovators, such as Copernicus or Galileo, or later ones, such as Hungarian physician Ignaz Semmelweis, and have faith that one day the world will accept new knowledge readily. The fact is that, although we would like to believe otherwise, even as trained scientists people still approach new knowledge in much the same way they did in Galileo's time. They will always judge new ideas in the light of prevailing beliefs, or paradigms, according to Thomas Kuhn, in *The Structure of Scientific Revolutions*. Rarely can people be objective about new information.

If a new idea is in line with what we believe, said Kuhn, we accept it readily. A good example is the belief throughout society that technology can improve our lives and solve most of our problems. Thus technological advances, or "solutions" are readily accepted and spread rapidly.

But when a new idea goes against our experience, knowledge, and prejudices—what we *know* rather than what we *think*—our mind blocks it out, distorts it, or rebels against it. A good example is the new idea introduced by nineteenth-century physician Ignaz Semmelweis, who demonstrated that physicians could cut maternal deaths in the delivery room dramatically if they washed their hands in a chlorine solution between autopsy work and examining patients. He couldn't prove why it worked, but the results were compelling. Medical institutions rejected the idea not only because they knew disease was caused by an imbalance of the "four humors" and that it spread through "bad air," but also because physicians were offended by the idea of having to wash their hands (as they were gentlemen, it was impossible for their hands to be unclean). It was several decades after Semmelweis's death that hand washing with an antiseptic became standard medical practice.

As this example shows, it is not just individuals who are slow to accept new ideas outside prevailing beliefs; it is also the institutions protecting the body of knowledge tied to those beliefs. And this is a major reason it has taken us nearly fifty years to work through the institutional opposition to the notion that livestock can be used to restore deteriorating land. The prevailing belief of society, and thus of our institutions, was just the opposite.

This inability to accept new ideas outside prevailing beliefs is called the *paradigm effect*, and none of us can escape it. To demonstrate this to yourself take a few seconds to read the following sentence, and as you do count the number of times the letter *f* appears.

> Finished files are the result of years of scientific study combined with
> the experience of many years of experts.

Chances are, you probably counted two, three, or four. Few people count more.

Now, take a few seconds to read the following sentence, and again count the number of *f*s.

> Strepxe fo sraey ynam fo ecneirepxe eht htiw denibmoc yduts cifitneics
> fo sraey fo tluser eht era selif dehsinif.

Chances are you counted six or seven. There were seven in both cases. You probably realize that the second sentence was the same as the first—only typed backward to prevent your mind seeing the words. The way you were taught to read made you see words more easily than letters in the first sentence. When there were no recognizable words in the second sentence you could easily see the *fs*. I doubt you hold any deep beliefs about the existence of *fs*, or that you have a PhD in that field, or that your self-esteem is tied to *fs* in any way. Had you been somewhat emotional, even subconsciously, about *fs*, there would have been an even greater disparity in the results.

When it came to understanding the underlying causes of much of the grassland degradation I was seeing in Africa and elsewhere, many scientists already *knew* the answer, and their institutions backed them up. Their assurance that, given enough money, technology would put things right differed little from the conviction of Renaissance theologians that God caused the sun to circle the earth and not vice versa, as Copernicus had suggested.

With the benefit of hindsight we can easily see what smaller revelations had to occur before people accepted Copernicus's theory. After people became truly comfortable with the notion of a round Earth, the theory of gravity, certain advances in mathematics, and the moons of Jupiter, the movement of the planets became a simple matter too. In the meantime, a number of people went to the stake.

It was no different in the case of Semmelweis several centuries later. After Louis Pasteur confirmed the germ theory of disease, and physician Joseph Lister built on that knowledge to introduce antiseptic surgery in Scottish hospitals, antiseptic practices became standard in hospitals everywhere and were strongly advocated by all medical institutions. Semmelweis wasn't burned at the stake, but he was committed to an asylum, where he died following a severe beating at age forty-seven.

Four such bottlenecks of understanding impeded the development of Holistic Management. The insights that enabled us to move forward came late and painfully, however, because, although they were each rather simple to grasp, they only became obvious when taken together. Thus it had been difficult to discover any one of these concepts in isolation.

The first insight overturned the notion that the world could be viewed as a machine made up of parts that could be isolated for study or management. In reality the world is composed of patterns—of matter, energy, and life— that function as *wholes* whose qualities cannot be predicted by studying any aspect in isolation. We would know very little about water, for instance, by making an exhaustive study of hydrogen or oxygen, even though every molecule of water is composed of both. Likewise, we could never manage a piece of land in isolation from the people who work it or the economy in which both the land and the people are enmeshed. As chapter 3 explains, this insight led to the development of a holistic framework for management and decision making. Defining the whole we were dealing with became the starting point in Holistic Management.

The next three insights contradicted long-held beliefs about the causes of the environmental deterioration I had first witnessed in Africa and later found in America. As chapter 4 explains, there were two broad categories of environment we had not recognized before—brittle and nonbrittle—that had evolved in different ways and responded differently when the same actions were applied to them. The types of animals associated with the two categories of environment also differed. As chapters 5 and 6 show, much of the land deterioration that has occurred in the world was initiated by the severing of a vital relationship between herding animals and their pack-hunting predators. Armed with this new knowledge we could more accurately predict how any piece of land might respond to our management. And this in turn would influence the decisions we made in determining which actions to take.

The four key insights are as follows:

1. A holistic perspective is essential in management. If we base management decisions on any other perspective, we are likely to experience results different from those intended because only the whole is reality.

2. Environments may be classified on a continuum from nonbrittle to very brittle according to how well humidity is distributed throughout the year and how quickly dead vegetation breaks

down. At either end of the scale, environments respond differently to the same influences. For instance, allowing land to rest restores it in nonbrittle environments but damages it in very brittle environments.

3. In brittle environments, relatively high numbers of large, herding animals, concentrated and moving as they naturally do in the presence of pack-hunting predators, are vital to maintaining the health of the lands we thought they destroyed.

4. In any environment, overgrazing and damage from trampling bear little relationship to the number of animals, but rather to the amount of *time* plants and soils are exposed to the animals.

The next four chapters will introduce these four key insights one at a time, but an understanding of all four is essential to see why, despite all our efforts, the environments that sustain us continue to deteriorate. No doubt many other insights await discovery, but at this stage we know that these four represent a major advance.

One last thought before moving on. It can take a century or more for prevailing beliefs to change so that new ideas gain acceptance. Given the seriousness of the twin threats of desertification and climate change, we don't have that kind of time. Yet, thanks to the rapid advances in communications technology and the now widespread use of social media, there is hope. A TED Talk on desertification and climate change that I presented in 2013 was available online in thirty-one languages within two months. Millions watched it, shared it with friends, colleagues, students, and then millions more watched it too. That talk, and the ability to share it so rapidly and widely, did more in twenty minutes to increase public awareness of the four key insights than fifty years of struggle to do so, and at a scale sufficient to begin the paradigm change that precedes acceptance.

# 3

# Nature Functions in Wholes and Patterns

*THE DISCOVERY THAT A HOLISTIC PERSPECTIVE* is essential in management is the most vital of the four new insights. We now realize that no whole, be it a family, a business, a community, or a nation, can be managed without looking inward to the lesser wholes that combine to form it, and *outward to the greater wholes of which it is a member*. Each day we put the utmost concentration and energy into our chosen tasks, seldom reflecting that we work within a greater whole that our actions will affect, slowly, cumulatively, and often dramatically.

The need for a new approach to the challenge of making a living without damaging our environment goes back to prehistoric times, when humans acquired language and organizing ability, plus fire, spear, and axe, and with them the ability to alter our environment in ways other animals could not. The sheer bounty of Earth's resources, however, has enabled us to keep the prehistoric attitude to any challenge: if you have a problem, get a rock and smash it.

In the last four hundred years our knowledge and the technological power to respond to any challenge have increased more rapidly than in all of the two hundred thousand or so years of modern human existence. Over the same few centuries the health of our planet has entered a breathtaking decline. The parallel is no coincidence, as table 3-1 helps to illustrate. The first column shows areas of technological success, while the second shows areas of failure, although a few might contain apparent short-term successes. It takes no special insight to generalize about these two realms of endeavor. Every item on the left is of a mechanical nature, and involves something we *make*. Each one on the right involves the nonmechanical world of relationships and wholes with diffuse boundaries—all of the things we *manage*.

Systems science helps explain why these differences are important. Everything humans make, which always involves some form of technology, lies in the realm of "hard systems." The things we make do what they are designed to do: the watch tells us the time, the computer computes. They do not do anything unplanned or unexpected, and they are not self-organizing—they generally won't work if a part is missing, a battery goes flat, or the fuel runs out. They are complicated, rather than complex, and the problems associated with them are referred to as tame because they can generally be solved, given enough time and money.

The things we manage, on the other hand, involve people and human organizations, which systems scientists refer to as *soft systems*, or natural resources, which they term *natural systems*. The things we manage in both soft and natural systems often produce unplanned or unexpected results, and they are self-organizing: if a person or a species dies, the organization or the biological community adjusts and continues, albeit in changed form. These systems are complex, as opposed to complicated, and the problems associated with them are referred to as wicked because they are difficult or impossible to solve. It is this complexity that has made soft and natural systems so hard to manage, leading so often to disappointing results and unintended consequences.

The modern scientific approach to the areas in both columns goes back to the thirteenth-century work of Roger Bacon, who first distinguished experimental science from the unqualified belief in scripture and tradition. This idea developed into the formal scientific method, wherein one seeks to test a hypothesis by controlling all variables of a phenomenon and manipulating them one at a time. By the seventeenth century, scientists began to view the whole world as a machine made up of parts that could be isolated and studied

**Table 3-1.** Mechanical versus nonmechanical areas of endeavor

| Mechanical (Things We Make) | Nonmechanical (Things We Manage) |
|---|---|
| Planes, cars, trains, ships | Croplands |
| Radio, television, telephone, satellites | Grasslands |
| Weapons: conventional, nuclear, laser | Forests |
| Electricity: solar, wind, fossil | Air quality |
| Spacecraft | Fisheries |
| Computers, robots | Water supplies and quality |
| Buildings: homes, churches, factories | Economics and finance |
| Roads, railway lines, bridges | Wildlife |
| Home appliances, swimming pools | Institutions/organizations |
| Clothing | Human relationships |
| Dams and power stations | Human health |
| Medical: brain scanners, artificial limbs, eyeglasses/contact lenses, medicines | |
| Chemicals, fertilizers, pesticides | |
| *(Ever-increasing success stories testifying to the marvels of technology—as long as we ignore effects on the environment)* | *(Ever-increasing problems, testifying to our inability to successfully manage anything complex)* |

by the scientific method, and their success in areas that are in fact mechanical seemed to confirm this as a fundamental truth.

However, in studying our planet and the many creatures inhabiting it we cannot meaningfully isolate anything, let alone control the variables. Earth's atmosphere; its plant, animal, and human inhabitants; its oceans, plains, and forests; its ecological stability; and its promise for humankind can only be grasped when they are viewed in their entirety. Isolate any part, and neither what you have taken nor what you have left behind remains what it was when all were one.

## The Holistic World View

In the 1920s this new worldview was given a name, *holism* (from the Greek *holos*), and a theoretical base by the legendary South African statesman-scholar Jan Christian Smuts (1870–1950); (fig. 3-1). In the years since, others have further elaborated on Smuts's original theory. However, it is Smuts who most influenced my own thinking.

In *Holism and Evolution* (1926), Smuts challenged the old mechanical viewpoint of science. Like modern-day physicists, Smuts came to see that the world is not made up of substance, but of flexible, changing patterns. "If you take patterns as the ultimate structure of the world, if it is arrangements and not stuff that make up the world," said Smuts, "the new concept leads you to the concept of wholes. Wholes have no stuff, they are arrangements. Science

**Figure 3-1.** Jan Christian Smuts (courtesy *The Star,* Johannesburg).

has come round to the view that the world consists of patterns, and I construe that to be that the world consists of wholes."[1]

Individual parts do not exist in nature, only wholes, and these form and shape each other. The new science of Smuts's day, ecology, was simply a recognition of the fact that all organisms feel the force and molding effect of the environment as a whole. "We are indeed one with Nature," he wrote. "Her genetic fibers run through all our being; our physical organs connect us with millions of years of her history; our minds are full of immemorial paths of pre-human experience."[2]

Without realizing it, American biologist Robert Paine provided dramatic evidence of the holistic nature of communities in a study he did in a seashore environment. When he removed the main predator, a certain species of starfish, from a population of fifteen observable species, things quickly changed. Within a year, the area was occupied by only eight of the original fifteen species. Numbers within the prey species boomed and in the resulting competition for space, reasoned Paine, those species that could move left the area; those that could not simply died out. Paine speculated that in time even more species would be lost. Over the same time period, his control area, which still contained the predatory starfish, remained a complex community where all species thrived.[3]

Viewed from the old paradigm that nature can be viewed as a machine made up of parts, the results of this study were interesting but not surprising. When a critical part (the starfish) was removed, the food chain was dramatically affected because all those species (or parts) were interconnected. Looked at in Smuts's terms, Paine's findings are more dramatic. Although there were fifteen observable species in the environment studied, they were more than a collection of interconnected species. They were a whole, just as algae and fungi that cling so closely to one another have become lichens, or hydrogen and oxygen have become water—and just as billions of bacterial, nerve, muscle, skin, blood, and bone cells have become you. You do not see yourself, or your parents, or your children as communities of interconnected cells, you see them as whole persons. Removing one element in the whole, as Paine did, severely disrupted the whole community. Given that there can be up to a billion organisms in a teaspoon of water, we really have no idea how much more deeply the community in Paine's study was affected.

I witnessed a similar disruption in two much larger communities in Africa. For a period in the 1950s I worked as a game department biologist in the Luangwa Valley in Northern Rhodesia (today Zambia) and the lower Zambezi Valley of Southern Rhodesia (today Zimbabwe). Both areas contained large wildlife populations—elephant, buffalo, zebra, more than a dozen antelope species, hippo, crocodiles, and numerous other predators. On more than one occasion I saw more than forty lions in a day's walk, which gives a good indication of just how large the populations were that they preyed upon. Buffalo herds were so thick that one day when a friend shot one buffalo with a light rifle twenty-seven adult buffalo were trampled to death in the resulting stampede. Yet, despite these numbers, the riverbanks were stable and well vegetated (color plate 1). People had lived in these areas since time immemorial in clusters of huts away from the main rivers because of the mosquitos and wet season flooding. Near their huts they kept gardens that they protected from elephants and other raiders by beating drums throughout much of the night or firing muzzle-loading guns to frighten them off. The people hunted and trapped animals throughout the year as well.

But the governments of both countries wanted to make these areas national parks. It would not do to have all this hunting going on, and all the drum beating, singing, and general disturbance, so the government removed the people. Like Paine, we, in effect, removed the starfish. But in our case we put a different type of starfish back in. We replaced drum beating, gun firing, gardening, and farming people with ecologists, naturalists, and tourists, under strict control to ensure they did not disturb the animals or vegetation.

Just as in Paine's study, the results were quick and dramatic. Within a few decades, miles of riverbank in both valleys were devoid of reeds, fig thickets, and most other vegetation (color plate 2). With nothing but the change in behavior of one species these areas became terribly impoverished and are still deteriorating seriously as I write. Why this resulted will become clearer in the following chapters. For now, let me just say that the change in human behavior changed the behavior of the animals that had naturally feared them, which in turn led to the damage to soils and vegetation. Had I better understood what Smuts was saying, I might have seen the danger in what we had done before it was too late. It was years, however, before his message finally made sense to me.

## Learning the Hard Way

Like many young boys growing up in Africa during the Second World War, I had idolized Jan Smuts for his exploits as a field marshal in the British Commonwealth forces, but his philosophy had lain far beyond my grasp. Even though I used the word *holistic* for years, I had to go through a long and intellectually unsophisticated school of hard knocks before I could even read his book, let alone understand holism well enough to put it to practical use. That experience is nevertheless probably worth relating for what it shows about the biases that must be overcome in our culture because of the paradigms we hold.

I received my scientific training in the conventional approach that viewed events in isolation. My professors discouraged any attempt to combine what we learned in one discipline with what we covered in another, and the sanctity of research was held inviolate, even when it offended common sense.

Once a visiting lecturer from Cambridge informed us that research had shown no use for the flap of flesh behind a crocodile's ear and that, despite having the musculature to move the flap, the croc never did so. As I had kept a tame croc myself I knew I could make him move his earflaps any day just by teasing him. When I ventured that crocs raised their flaps in response to a threat, the lecturer quickly put me down. My observations just could not tip the scales against years of experiments in a controlled environment. Small as this incident was, it fueled a growing disillusionment with the artificiality of any approach that isolated parts of nature for study. The crocodile isolated from his environment and behaving differently was not the same animal.

Later, as an adviser to private ranchers on land management, I encountered a similar dilemma. They listened well to my opinion on techniques that would improve their land, but several, while making great progress toward that end, still went bankrupt after committing scarce funds to government-sponsored irrigation schemes. From my own farming days I knew the risk of tying up capital in government incentive programs to build irrigation dams, as these ranchers had done, but as I was an ecologist, not a financial adviser, they ignored my friendly warnings.

Thus I learned that, to help ranchers manage their land, livestock, and wildlife, I had to become involved in the financial planning to some extent.

Such involvement helped, but other troubles frequently arose that expertise in financial planning did not address. Sound operations could also flounder because of communication problems among the people involved and conflict over differing objectives or lack of any unifying long-range goals. This led me to add organizational management to the specialties I studied.

By this time I was calling my work holistic ranch management, but in reality it was anything but holistic. I still had not read Smuts and saw no need, assuming myself to have advanced beyond anything he might have written in 1926. I had a record of good results on the land to support this opinion and had spent two five-year periods in partnership with consultants from whom I had learned much about the latest economic thinking. Nevertheless I still got sporadic results and clearly something was missing.

Other scientists, encountering similar frustrations, had concluded that we often aggravated management problems by approaching them from the perspective of narrow disciplines. No animal nutritionist, soil scientist, economist, or any other specialist alone had meaningful answers. Where I had accumulated knowledge in several fields and had even teamed up with other experts, others formed interdisciplinary teams of various kinds but fared no better.

Why these teams, my own included, did not work deserves a close look because increasing numbers of people are calling their work holistic, as I once did, when perhaps it is not. Their fundamental weakness was described in the book *Landscape Ecology* by Zev Naveh and Arthur Lieberman (1983):

> In a computerized simulation game, Dörner [a researcher] asked 12 professionals from different relevant disciplines to propose an integrated development plan for the overall improvement of an imaginary African country, Tana. The results achieved were very disappointing: if these proposals were carried out they would worsen the lot of the people, destroy the agricultural-economic base, and create new, even more severe problems.[4]

My only criticism of this work was the need to invent an imaginary country. "Tana" could have been any one of my clients or any state in America or any developed country in the world using integrated planning teams. What Dorner simulated with a computer was what I had seen and experienced repeatedly in practice from a farm level to a policy level and as president of a political party.

First of all, specialists often communicate poorly, not only because they have different perspectives but also because they speak different languages. Often the same words in one jargon mean something else in another. Even where team members have training in several disciplines, as I myself did to some extent, the tendency is to simply swap hats and keep talking without ever being able to stand back and see the whole.

I, however, did not see any of this until many years later when I undertook to explain my approach to management in training courses I was asked to provide for professionals in U.S. government agencies. It is a great credit to American openness that this happened because my ideas departed radically from those given in the training most of these people had embraced for years. Nonetheless, the courses were very stressful events, and in struggling to teach what I knew of holism to an audience of very skeptical peers, I came to realize I didn't understand it myself.

## Where the Holistic Worldview Differs

Finally, I read Smuts and realized wherein the holistic worldview differed from everything prior. It became obvious to me that not only are there no parts in nature, there are no boundaries either. Your skin could be viewed as the boundary between the community of cells that compose you as a person and the outside world. Yet skin is permeable, and the traffic passing through it in both directions is heavy. Viewed at the molecular level, skin is more space than substance.

Anytime we talk about interconnectedness we are implying that boundaries exist between whatever is being connected. To more accurately view the world, one has to accept that, in reality, there are no boundaries, only wholes within wholes in a variety of patterns. And to understand the world, according to Smuts, we must first seek to understand the greater whole, which has qualities and characteristics not present in any of the lesser wholes that form it.

The design in figure 3-2 shows this well. Take a close look at it and think of it as depicting our world. This is the sort of confusing picture we saw

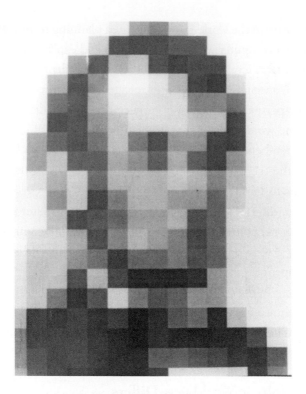

**Figure 3-2.** The sort of confusing picture we saw when first trying to understand ourselves and our environment (courtesy Alcatel-Lucent USA, Inc.).

when first trying to understand our environment and ourselves. According to scientific custom we isolated the individual squares for study, believing that if we could learn enough about each of them we would understand the whole. However, in the case of nature, as in this pattern of squares, this leads nowhere.

Now stand well back from the pattern, squint your eyes so the squares blur into each other and the picture appears as a whole. It is a face, and a familiar one to many of us (Abraham Lincoln's). If you had set out to somehow manage this design by paying attention only to the individual squares, anything you might have learned would have been useless, for no square has any meaning in isolation from all the rest.

Of course, once you see the whole in the pattern, detailed knowledge of the squares does become useful. You would need a great deal of such knowledge to reproduce, enlarge, preserve, market, or modify the work in any way, but only having first seen the whole could you even ask the right questions about the details.

I personally had tremendous difficulty in seeing why the fact that wholes have qualities not present in their parts causes the interdisciplinary approach to flounder, until I actually could work it out with my own hands as a young child might.

I took four balls of kindergarten-type modeling clay in red, green, yellow, and blue, and began kneading them together until they slowly blended into a fifth color, gray. Mentally I let gray represent the world we originally set out to understand. Close inspection of my gray ball revealed traces of the four colors I had begun with. So to understand this world of gray I would study the colors I knew to be in some way involved in it.

Today we have thousands of disciplines, but for simplicity I used four in my clay ball exercise, represented by the four colors, to make my point, as shown in color plate 3 (top left). Although a few hundred years of intense effort have greatly increased our knowledge of the four colors, we still had no knowledge of gray itself.

Next I pulled the four colors together as shown in plate 3 (top right) to represent a multidisciplinary team. Immediately I could see that the problem was a lack of knowledge of gray, not a lack of communication between disciplines as we previously thought.

Next I mixed, or *integrated*, the four colors until I had four balls, each of which contained equal divisions of green, red, yellow, and blue to represent interdisciplinary teams with knowledge of each color, as illustrated in plate 3 (lower left). Still no knowledge of gray! Now I could see why the interdisciplinary approach could not succeed.

In practice, I realized, all management decisions had to be made from the perspective of the whole under management. If we based our decisions on any other perspective, we could expect to experience results different from those intended because only the whole is reality. First, however, the whole had to be defined, bearing in mind that it always influenced, and was influenced by, both greater and lesser wholes. To know which management actions were appropriate we needed to identify a *holistic* context for our actions. Last, we needed a means of weighing up the many ramifications stemming from our actions to ensure they were in context and unlikely to produce adverse consequences. Thus the beginnings of a framework for management and decision making had emerged. In using that framework we can now take the perspective of the whole by reversing the arrows and weighing up management decisions within a holistic context, as depicted in plate 3 (lower right).

The rest of this book will deal with what this means in terms of what you will do after breakfast this morning, but, before moving on, there is a final point. A number of people have suggested that, given the complexity of our world, computer software may be more capable than we of weighing up the consequences stemming from our actions. But there is a strong argument to the contrary: the human mind can see patterns and make decisions out of a deep, even unconscious, sense of the whole, and given an awareness of the necessity, and the mental crutch of a framework for management that keeps our focus on the whole, far-reaching changes can be brought about by the likes of you and me.

## Conclusion

Successfully addressing the complexity inherent in anything we manage is, I believe, our greatest challenge, and one we have to meet if we are to regenerate the world's soils so that we reverse desertification, address climate change, and successfully feed nine billion or more people.

In *The Watchman's Rattle: A Radical New Theory of Collapse*, American sociobiologist Rebecca Costa concludes that many past civilizations failed when societies proved unable to address the complex problem of rising populations within deteriorating environments. Unable to devise solutions, she says, these past societies turned to faith and sacrifice to address the symptoms that accompany a failing civilization—poverty, hunger, violence, war, and so on—shelving the underlying problem for future generations to solve. Costa looks to technology for a solution—in her, case technology to rapidly speed the development of that part of the human brain believed to best handle complexity. But I believe the solution is simpler and more readily at hand.

It involves recognizing that the genetically embedded framework we've long used to manage our affairs has flaws that we can correct, which the

Holistic Management framework attempts to do. This framework already enables us to look outward and to choose from all the available knowledge that which is appropriate for the whole we are managing and the environment that sustains it. It also lets us predict results ahead of time. With a bit of practice, almost anyone can use the Holistic Management framework when soliciting advice from specialists and judge when such advice is or is not in context for them—socially, financially, environmentally.

Frequently, advice that appears perfectly sound from an economist's, engineer's, or any other's point of view proves unsound holistically in a particular situation at a particular time. This has spelled disaster, as Dörner's study predicted, for many a foreign aid project and national policy, but also for families, communities, and businesses large and small. As a culture we have acquired such an awe of expertise that we have trained ourselves simply to phone any certified expert whenever adversity arises. This is not a new behavior.

Back in eighteenth-century Europe, during the Age of Enlightenment, Napoleon led an initiative to reform the organizations of the day by requiring their leaders, who had previously either bought or inherited their positions, to now be recruited based on their qualifications. The intent was for these new, professional leaders to use their expertise to better manage the complex challenges organizations faced. In *Voltaire's Bastards*, Enlightenment scholar John Ralston Saul took a look at the results of this initiative, all the way down to our present day, and found that the focus on expertise had led to numerous unintended consequences. The blunders had actually increased and continue to do so: "The reality," says Saul, "is that the division of knowledge into feudal fiefdoms of expertise has made general understanding and coordinated action not simply impossible but despised and distrusted."[5]

I believe Saul's conclusion is right, and that these "fiefdoms of expertise" will continue to challenge us. But I also believe we now have a way to begin managing the complexity inherent in human organizations and Nature based on the developments that have arisen from this first key insight:

> *A holistic perspective is essential in management. If we base management decisions on any other perspective, we are likely to experience results different from those intended because only the whole is reality.*

# 4

## Viewing Environments in a Whole New Way

*THE FIRST KEY INSIGHT ENABLED US* to develop a holistic framework for management and decision making, but we still lacked certain insights that were key to reversing the environmental deterioration that had accompanied the rise of numerous civilizations, including our own, and without which the holistic framework would not be complete.

The second insight overturns the belief that all environments respond in the same manner to the same influences. They don't. The standard classifications of environments by vegetative features—desert, prairie, savanna, dry deciduous or rain forest, and so on—accurately describe major variations within our global ecosystem, as do such climatic categories as arid, semiarid, temperate, and so on. Nevertheless, in looking at why some deteriorate to the extent that deserts form, while others don't, leads to an alternative way of classifying them.

More specifically, then, the second insight is the principle that environments may be classified on a continuum from nonbrittle to very brittle according to how well humidity is distributed throughout the year, and how quickly dead vegetation breaks down. In nonbrittle environments, dead grass leaves and stems are soft and moist, and they crumple when squeezed in your hand. In brittle environments, dead leaves and stems are brittle and dry, and they shatter when squeezed in your hand.

In nonbrittle environments, dead grass leaves and stems are soft and moist, and they crumple when squeezed in your hand. In brittle environments, dead leaves and stems are brittle and dry, and they shatter when squeezed in your hand.

We have long recognized that some environments readily deteriorated under human management. Herodotus, for example, described Libya in the fifth century BC as having deep, rich soils and an abundant supply of springs that provided a highly productive agricultural base for a large population. Today only desert remains. Those who chronicled this sort of deterioration believed that such regions were vulnerable to desertification because they were arid or semiarid.

The bulk of the world's arid and semiarid regions are in fact predominantly grasslands of one form or another where livestock production has long been the chief occupation. When livestock management practices produce bare ground, a critical share of available moisture either evaporates from the exposed surface or runs off it. Springs dry up; silt chokes dams, rivers, and irrigation ditches; and less water remains for croplands, industry, and people in nearby cities.

Going back even beyond Herodotus, common sense has always assumed that, once damaged, the best remedy for such land is to rest it from any form of human disturbance, including livestock. But, despite the application of this wisdom, the croplands and rangelands of Libya and its neighbors have desertified anyway, as much of North America is now doing. The old assumption that resting land will restore it to its former productivity and stability appears logical. Moreover, it does apply to the fairly stable environments of northern Europe and the eastern United States, where modern agricultural science has its roots.

## Faulty Conclusions

I worked under the old belief that resting land restored it in my early game department days, but a unique experience led me to suspect a fundamental flaw in this belief. For many years, Zimbabwe had a practice of eradicating all game animals over vast areas in order to deny the tsetse fly a source of blood—its only source of nourishment. Once the tsetse fly was gone, and the fatal human and livestock diseases it carried, livestock could safely be introduced. As a research officer for both the game department and later the Department of Tsetse Control, I worked often in these areas.

I witnessed environmental damage I could not explain and that did not fit the neat scientific theories I had learned. The land in tsetse fly areas deteriorated seriously once the original game populations were decimated and the incidence of fires, to make the hunting easier, increased. At that time we *knew* that fire helped maintain grassland. The only other influence we *knew* could cause such damage was overgrazing. However, neither game nor domestic animals were present and so there was no overgrazing. It was very puzzling.

Yet another experience added to my confusion. We had a massive buildup of animal numbers in a game reserve on the Botswana border known as the Tuli Circle, and as a result thousands of animals starved to death. I believed, as did the wildlife biologists working with me, that with dramatically fewer animals living there the area would naturally recover, but it continued to deteriorate. Most of our scientists blamed drought, but in the year of the so-called worst drought the records showed one of the best rainy seasons ever, in both volume and distribution.[1] I published a paper at the time in which I concluded that, once land was so badly damaged, it had reached a point of no return and would never recover. Because my peers held the same beliefs, that paper was published, much to my embarrassment today as I now realize how wrong we all were.

Another shock to my conviction that low rainfall and overgrazing inevitably produced desert came out of my first visit to northern Europe. There I saw areas that had as little as 500 millimeters (20 inches) of rain per year that were not desertifying, despite hundreds of years of overgrazing and poor management. Areas in Africa and the Middle East that received as much as 1000 to 1250 millimeters (40–50 inches) of rain annually, however, had desertified under the same practices.

Finally, in the vast and relatively unused lands of North America I discovered what had eluded me in the highly populated and much-used lands of Africa, and in the different environments of Europe—that we had in fact two broad types of environment. At their extremes, these react differently to management. Practices that benefited the one type of environment damaged the other. The terms *brittle* and *nonbrittle* come from that insight.

## The Brittleness Scale

No clear break exists between the extremes that range from nonbrittle to very brittle. The brittleness scale ranges from 1 (evergreen tropical forests) to 10 (true deserts), with all other environments falling somewhere in between. A single vegetative category may cover a wide range on the brittleness scale. Grasslands, for instance, may lie anywhere from 1 to 9 or 10 on the scale; forests from 1 to 7 or 8.

Because the two extremes at 1 and 10 on our brittleness scale show such a clear correlation to total rainfall, it is easy to see in retrospect why we linked an environment's vulnerability to desertification to low rainfall. The degree of brittleness determines this vulnerability, however, more than total precipitation. The closer we get to 10 on the brittleness scale, even with high rainfall (750–2000 millimeters/30–80 inches), the faster the land is likely to deteriorate under modern agricultural practices. This is not to say that nonbrittle environments are invulnerable to deterioration, as the massive clearing of tropical rain forests makes clear.

The features that distinguish any environment's position on the brittleness scale derive mainly from the *distribution* of precipitation and humidity throughout the year. Toward the very brittle end of the scale, environments characteristically experience erratic distribution of both precipitation and humidity during each year. The *distribution* of humidity, *not* the total rainfall, determines the degree of brittleness. An area with 750 to 1250 millimeters

---

### Brittleness versus Fragility

*Brittleness* is not the same as *fragility*, which is why I did not use the latter term, although many of my colleagues refer to desertifying environments as fragile. I believe all environments are fragile and thus damaged by improper management practices, and all environments are remarkably stable and productive under proper management practices.

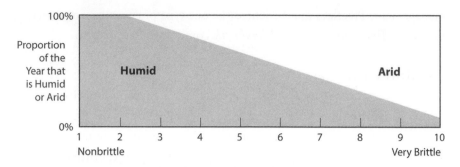

**Figure 4-1.** Humidity and the brittleness scale. Humidity is high over a greater portion of the year the closer an environment is to the nonbrittle end of the scale.

(30–50 inches) of rainfall with prolonged dry periods would be brittle despite good rainfall.

Toward the nonbrittle end of the scale environments characteristically experience increasingly reliable humidity throughout the year. Even though total precipitation may seldom top 500 millimeters (20 inches) a year in some of them, the distribution is such that, throughout the year, atmospheric humidity does not drop severely for any lengthy period. London, England, lies at about 2 to 3 on the brittleness scale, due to its perennial humidity, and is green year-round on only 600 millimeters (24 inches) of average total rainfall. This contrasts greatly with Johannesburg, South Africa, which lies at about 6 or 7 on the scale, due to its seven-month-long dry season, but receives 700 millimeters (28 inches) total rainfall on average.

The distribution of the precipitation as well as the elevation, temperature, and prevailing winds, clearly affects the day-to-day distribution of humidity, and this links very closely to the degree of brittleness. The poorer the distribution of humidity, *particularly in the growing season*, the more brittle the area tends to be, even though total rainfall may be high, as shown in figure 4-1. Very brittle environments commonly have a long period of nongrowth that can be very arid.

## How Vegetation Breaks Down in Nonbrittle Environments

The brittleness scale can perhaps be best understood by looking at how environments at either end of the scale functioned prior to human interference. In the perennially humid, nonbrittle environments, as illustrated in

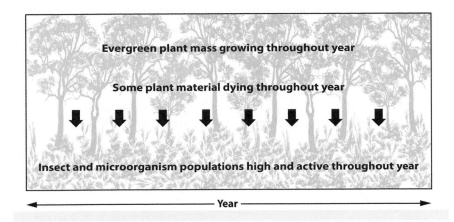

**Figure 4-2.** In the perennially humid, nonbrittle environments, vegetation is evergreen and plants die throughout the year. Decay is rapid due to the presence of insects and microorganisms that are active year-round.

figure 4-2, a mass of evergreen vegetation was produced throughout the year; there was no period of dormancy. Plants died throughout the year as well, but they decomposed quickly due to the high numbers of insects and microorganisms whose populations remained high and active throughout the year due to this constant humidity.

## How Vegetation Breaks Down in Brittle Environments

At the other end of the scale, where humidity was erratic and the environment very brittle, as shown in figure 4-3, vegetation, insect, and microorganism populations would build up during the rainy months of the year. However, when the rains stopped the humidity dropped, and as the soil dried out most of the aboveground leaves and small stems died—only the trunks and branches of trees and brush and the bases of perennial grass plants remained alive. At the same time insect and microorganism activity was drastically reduced as these organisms went into dormancy, died off, or survived only in the egg or pupal form, through the dry, or nongrowth, period. The photo in figure 4-4 was taken in a very brittle environment in the tropics (685-millimeter [27-inch] rainfall) many months after the last rainfall of the growing season. The grass plants have moved energy down to roots and bases and the aboveground stems and leaves are now dead. In fact, the mass of

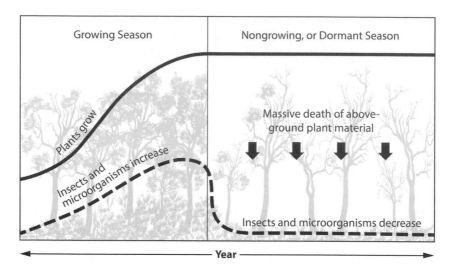

**Figure 4-3.** In the seasonally humid brittle environments, most aboveground vegetation dies over a confined period every year. At the same time, insects and microorganisms decrease or go dormant and decay slows or stops.

**Figure 4-4.** This season's stems and leaves are now dead and will block sunlight from reaching the plants' ground-level growth points. If not removed before the next growing season the plants will produce few new leaves. Zimbabwe.

material remaining on the plants is a liability because it will block sunlight from reaching the growth points at the plants' bases, hampering the growth of new leaves and stems in the next growing season.

So how did all that dead vegetation break down every year over millions of years in the past? Lightning, then as now, would have sparked fires and consumed some of the vegetation, but relatively few areas would have been affected in any given year. There was a much more powerful ecological influence, however—grazing animals, and lots of them. Bison, elk, pronghorn, kangaroo, saiga antelope, the many species of African antelope, elephant, buffalo, zebra all evolved in these environments. So did many other species that have since been lost. (In North America alone, forty-three large mammal species vanished as recently as ten thousand years ago, shortly after the arrival of the first humans. Only eleven species remain today.[2]) The presence of these herds was significant because they consumed a fair amount of plant material while it was still green and growing, *and* they continued consuming it long after all growth had stopped, *and* they trampled what they didn't eat onto the ground. In the moist digestive tracts of these animals, a mass of microorganisms continued to thrive in the nongrowing season and managed to reduce the large volume of material consumed to dung. In the following growing season when insect and external microorganism populations once again became numerous, they would consume the dung, as well as the dead vegetation that had been trampled onto the soil, and thus complete the cycle of decay.

---

In the moist digestive tracts of these animals, a mass of microorganisms continued to thrive in the nongrowing season and managed to reduce the large volume of material consumed to dung.

---

Today, of course, the vast herds have disappeared from these environments (they never did occur in nonbrittle environments), largely due to human interference. In the herds' absence only a small proportion of the vegetation produced is able to decay. Most is left to break down chemically through oxidation—the same process at work on rusting metal, though dead plant material turns gray and then black, rather than reddish-brown—or physically through weathering, where wind, rain, and hail very gradually wear them down. Humans of course also assist the breakdown by burning the dead vegetation, with consequences that will be explored in subsequent chapters.

As the ability of plants to decompose and recycle their nutrients is crucial to the health of the whole environment, determining the degree of brittleness becomes a prime factor in the management of any environment. Nonbrittle and very brittle environments react quite differently to some of the main management practices we engage in daily, yet we have failed to make this distinction.

## Revising Those Faulty Conclusions

At this point let me just say that the old belief that *all* land should be rested or left undisturbed in order to reverse its deterioration has been proven wrong. Environments lying close to the nonbrittle end of the scale do respond in this way. When ancient cities were abandoned in these environments, biological communities recovered and buried the ruins in vegetation, as shown in figure 4-5. In environments leaning toward the very brittle end of the scale, prolonged rest leads to further deterioration and instability. Ancient cities that were abandoned in these environments are now buried under desert sands, as shown in figure 4-6. Since well over half of Earth's landmass is more, rather than less, brittle, it is no surprise that desertification is spreading at the rate it is.

Once we understood the brittle and nonbrittle distinction and had identified the characteristics associated with both types of environment, my puzzling observations of the past became clear. The Tuli Circle, where so much of the game had died off and where so much soil lay exposed, could not recover if left undisturbed because it was a very brittle environment. It required some form of disturbance at the soil surface, similar to what the formerly large herds provided, in order to get more plants growing, as the next chapter explains. In the tsetse fly areas, the increased use of fire had exposed soil, and though old grass plants remained healthy, nothing disturbed the soil sufficiently to allow establishment of new ones. Thus communities declined and soil became unstable.

**Figure 4-5.** When ancient cities in nonbrittle environments were abandoned, biological communities recovered and buried the ruins in vegetation. Palenque, Mexico (courtesy jejim/Shutterstock .com).

**Figure 4-6.** Ancient abandoned cities in very brittle environments are now buried under desert sands. Palmyra, Syria (courtesy Anton Ivanov/Shutterstock.com).

## Conclusion

The second insight, then, is that environments may be classified on a continuum from nonbrittle to very brittle according to how well humidity is distributed throughout the year, and how dead vegetation breaks down—rapidly
through biological decay, or slowly through chemical or physical weathering.
At either end of the scale, environments respond differently to the same influences. Resting land restores biodiversity and land health in nonbrittle environments but leads to desertification in brittle ones.

Taken by itself, this second insight raises the practical question of how
grazing animals might provide the disturbance necessary to the health of an
environment that is brittle to any degree—without overgrazing plants. Fortunately the answer lies in the remaining two insights.

# 5

## The Predator-Prey Connection to Land Health

IN MY UNIVERSITY TRAINING I LEARNED, like all scientists of that era, that large animals such as domestic cattle could damage land. Only keeping numbers low and scattering stock widely would prevent the destructive trampling and intense grazing one could expect from livestock.

Once I left university and went into the field as a biologist, my observations led me to question that dogma. I now defend the exact opposite conclusion. High numbers of heavy herding animals, concentrated and moving as they once did naturally in the presence of predators, support the health of the very lands we thought they destroyed. This revelation came slowly and only after experience in a large variety of situations, because herding animals, like others, have more than one behavior pattern, and the effects on land are often delayed, subtle, and cumulative.

In the mid-1960s, Zimbabwe erupted in civil war, and because of my experience dealing with poachers the army gave me the task of training and commanding a tracker combat unit. Over the next few years I spent thousands of hours tracking people over all sorts of country, day after day. This discipline greatly sharpened my observational skills and also taught me much about the land because, although I was tracking people, my thoughts were constantly on the state of the lands over which we were fighting. I doubt many scientists ever had such an opportunity for learning. I tracked people over game areas, tribal areas, commercial farms and ranches, and many different soil and vegetation types and rainfalls. Often I covered many different areas in a single day as I flew by helicopter from one trouble spot to the next. Everywhere I had to inspect plants and soils for the faintest sign of disturbance by people trying to leave no hint of their passage.

I spent many a long night lying up in the bush with nothing to do but ponder why the tracking had been easier some days than others, knowing that it was somehow linked to how the land was used. Gradually I realized that vast differences distinguished land where wildlife herded naturally, where people herded domestic stock, and where stock were fenced in by people and not herded at all. And compared to areas without any large animals, such as tsetse fly areas in which all large game had been exterminated, the differences were startling indeed.

Most obvious was the fact that, where animals were present, plants were green and growing. In areas without animals they were often gray and dying—even in the growing season—unless they had been burned, in which case the soil between plants was bare and eroding, and tracking was easy. When I compared areas heavily disturbed by animals, where soil was churned up, plants were flattened, and tracking was difficult, it became clear that the degree of disturbance had a proportionately positive impact on the health of plants and soils and thus the whole community.

## How Animal Behavior Affects Soil Condition

I began to pay particular attention to the way animals behaved in different situations, as different behavior and management patterns produced different effects. In tracking large buffalo herds on my own game reserve, for instance, I noted that when feeding they tended to spread out, although not too far for fear of predation, and to walk gently and slowly. They placed their hooves beside coarse plants and not on top of them. They also placed their full weight on their hooves, compacting the soil below the surface but hardly disturbing the surface itself. While thus feeding they had remarkably little impact on the plants and the soil, other than the obvious removal of forage and the soil compaction.

Once feeding was over, however, and the herd began to move, or when predators were about, the animals behaved differently. They bunched together for safety and in their excitement kicked up quite a bit of dust. I noted that while bunched as a herd, animals stepped recklessly, and even very coarse plants, containing much old material that would not be grazed or trampled normally, were trampled down. That provided cover for the soil surface. In addition, the hooves of bunching and milling animals left the soil chipped and broken. In effect, the animals did what any gardener would do in order to get seeds to grow: first loosen the sealed soil surface, then bury the seed slightly, compact the soil around the seed, then cover the surface with a mulch. I also noted that, where the grazing herd had kept off the steep, cutting edges of gullies, the bunched herd now beat down the edges, creating a more gradual slope that could once again support vegetation.

## How Pack-Hunting Predators Affect Herd Behavior

I became convinced that the disturbance created by the hooves of *herding* natural wild animal populations was vital to the health of the land, and that humankind had lost this benefit when we domesticated cattle, horses, sheep, and goats and protected them from predators. Even where people herded livestock, as opposed to merely fencing them in, they did not behave as they would do if naturally herding under the threat of predation.

In the more brittle environments, the large predators differed from those in nonbrittle environments in one important respect: they mostly hunted in packs and ran down their prey. Wolves, lions, cheetahs, African wild dogs, and hyenas are all pack hunters. In the nonbrittle environments the predators, such as tigers and jaguars, were a different type: they hunted singly and ambushed their prey. They had their counterparts (e.g., leopards and mountain lions) in the more brittle environments, but these predators generally did not associate with or follow the large herds. It was the pack-hunting predators who were mainly responsible for producing the behavior of their herding prey. Bunching up tightly in large numbers became the herd's chief form of protection, particularly of females and young, because pack hunters tend to fear large herds and have to isolate animals to kill them.

This relationship between pack hunters, their herding prey, and the soils and plants they trampled and grazed developed over millions of years, and long before humans themselves became predators with the aid of language, organization, fire, and spear. There was no other influence that could realistically have created the necessary soil disturbance to provide a good seedbed for new plants, protected bare soil by trampling down old plant material, or broken down dried-off grass biologically by cycling it through a gut. These functions appear critical to the health of environments at the more brittle end of the scale where grasses provide most of the soil cover.

My understanding of the tremendous significance of these relationships evolved slowly. In Africa I dealt with very large game herds and numerous predators, including enormous prides of lions. It was not until I came to the United States, where predators no longer had significant impact on the wildlife populations, that I realized how much they contributed to creating the kind of soil and vegetation disturbance needed in the more brittle environments.

In the United States, massive destruction of predator populations and wild herds precipitated the decline in the environment we see today throughout most of the western states. We have only exacerbated it by spreading relatively few domestic animals over large areas.

In North America the problem is compounded by the annual freezing and thawing cycle, which creates puffiness under algae- and lichen-encrusted soil surfaces. If the environment is very brittle, this means that not only must there be some agent of disturbance that will remove old, oxidizing material from perennial grasses and break soil surface crusts, but that agent has also to provide soil compaction to increase grass seedling success.

## Hooves and Healthy Land

I was not by any means the first to make the connection between the hooves of animals and the health of land. Many centuries ago shepherds in the less brittle environment of Scotland referred to the "golden hooves" of sheep.[1] In the 1930s, Navajo medicine men in the very brittle environment of the American Southwest warned government officials who were drastically reducing their livestock numbers that a link existed between the hooves of the sheep and the health of the soil:

> Plants come up only where there are sheep and horses. Those are the places where it rains. If the ground is hard, when it rains, the water just flows. It doesn't soak in. It soaks in only when the sheep walk on it. They walk over their dried manure and mix it with the earth. When it rains, it soaks into the ground with the seed.
>
> —George Blueeyes
> (Navajo stockman and medicine man, born 1900–1910)[2]

In southern Africa the old-timers of my childhood had a saying, "Hammer veld to sweeten it." They meant literally hammer the land with herds of livestock to improve forage quality. Unfortunately none of these earlier observations were fully understood because too many of us believed that plants and soils needed protection from the damaging effects of animals. My own early observation of the vital relationship between natural herds, soils, and plants met violent rejection and ridicule from my fellow countrymen and the international scientific community, and still does, though it is at last dwindling.

In the 1980s, thanks to some careful research done over a period of years on the herding wildlife populations in East Africa's fairly brittle grasslands, a relationship between these animals and the plants they feed on was documented independently.[3] Gradually scientists are becoming more comfortable with the idea as support accumulates from multiple fields of study.

## Conclusion

The third key insight then, was that, in brittle environments, relatively high numbers of large herding animals, concentrated and moving as they naturally do in the presence of pack-hunting predators, are vital to maintaining the health of the lands we thought they destroyed. Acceptance of this insight will help to reverse the millennia of damage humankind has inflicted on the land in the more brittle environments by trying to protect it from the effects of grazing and trampling that were perceived as evil. As bare ground increased and the environment deteriorated in response to the lack of *herd effect*, we attributed it to overgrazing, which we in turn blamed on too many animals. As a result, we decreased animal numbers and thus increased the bare ground and the deterioration.

That overgrazing is not in fact due to too many animals is the fourth insight.

# 6

## Timing Is Everything

*A FUNDAMENTAL BELIEF, EMBRACED THROUGHOUT THE WORLD,* holds that the presence of too many animals causes overgrazing, overtrampling, and the resultant destruction of land. Despite massive and sophisticated research on plants, soils, and animals, virtually all land improvement schemes have rested on this very unsophisticated bit of apparent common sense and have called for the reduction or removal of animals.

Until very recently no one truly explored the question of *when* animals are present as opposed to *how many* are present. My own experience illustrates how elusive an obvious principle can be. As a child in Zimbabwe, I too learned of the destruction caused by too many animals when I accompanied my father into the native reserves. He, a civil engineer, had the task of improving water distribution for the people and their stock. Overgrazing and overtrampling had devastated the areas surrounding the few existing water points, and the theory ran that creating more water points would scatter the stock and reduce the damage. The hot and dusty hours spent amid the rabble of cattle, goats, and donkeys on that barren land, together with what others told me, certainly convinced me that my father's work and the government policy that supported it made sense.

I did not question this assumption for a decade or so, until as a young man I encountered historical records that showed what enormous herds of wild animals had existed on the land before people and their domestic stock replaced them. As pioneers made their way into the interior of South Africa they apparently recorded herds of springbok (a pronghorn-sized antelope) so vast that when they migrated through settlements they trampled everything in their path, including yokes of oxen that couldn't be unhitched from wagons fast enough.

Such herds, together with all the millions of other animals on the southern African veld, vastly outnumbered the cattle and sheep herds that came later, yet for millennia they had enjoyed an environment more abundant than anything the descendants of those pioneers can imagine. South African adventurer George Mossop describes the South African plains of the 1860s:

> *The scene which met my eyes the next morning is beyond my power to describe. Game, game everywhere, as far as the eye could see—all on the move, grazing. The game did not appear to be moving; the impression I received was that the earth was doing so, carrying the game with it—they were in such vast numbers . . . hundreds of thousands of blesbok, springbok, wildebeest, and many others were all around us.*[1]

One sign of that former abundance is the animal names borne by villages and towns of today linked to the springs they watered at: Elandsfontein, Springbokfontein, and Buffelsfontein (from the Dutch word *fontein* for fountain or spring). No hint of free-flowing water exists in those places today.

The weather, according to records, did not change, yet the "fountains" disappeared together with the healthy grasslands and the vast herds. As the memory of the wild herds also vanished, people blamed the disappearance of the water and the grasslands on the overgrazing and overtrampling of their own livestock, although fewer in number than the wild herds had been.

This riddle confused me, but I still could only conclude that overgrazing and overtrampling were related to animal numbers. The most obvious deterioration was occurring on the most heavily stocked tribal land, and in certain national parks and game reserves where wildlife numbers were also high, which tended to support the prevailing wisdom. As a research officer in the game department I found myself recommending, despite my questions, a drastic culling of elephant and buffalo to arrest the serious damage done by their overgrazing and overtrampling in areas we were setting aside as future national parks. I made this decision in anguish because by this time I had already noted that the extermination of game in the tsetse fly areas had not in fact brought improvement of the land. Only later did I begin to penetrate the riddle.

## Wild Herds Moved Frequently

I had observed that very large buffalo herds moved constantly and seldom occupied any area for longer than two or three days, after which the land had an opportunity to recover. Could the time they stayed be an important factor? It also began to dawn on me that a lot of game confined to a small area—as happened in some of our newly forming game reserves surrounded by human settlement or tsetse fly areas, where game animals were being destroyed—produced *too many herds in too small an area*. Though each herd moved frequently, plants and soil had little time to recover after being grazed or trampled. I did not yet see *time* as the crucial element, but I was beginning to think that the reduced size of home ranges and territories on which the animals could move lay at the bottom of the problem.

I studied elephant herds. Did they shift location every few days? Did it matter? Did another herd move into areas only recently vacated? I decided to find out, but immediately struck an obstacle—I couldn't tell one herd from another as I was on foot and in dense brush. In those days before methods were refined for drugging and tagging animals, I stalked unsuspecting elephants with homemade paint bombs that I threw from close range. Such work, however, generated minimal enthusiasm among potential helpers, and I simply couldn't paint enough elephants single-handedly. I also lacked sufficient staff to conduct observations over enough area to support any conclusions.

For some time I had possessed a book entitled *Grass Productivity* by French researcher André Voisin. I had bought it because the title interested me, and I thought it might help clarify what was taking place in our wildlife areas. Voisin, however, had worked mainly with cattle on pastures in Europe. After browsing his book I could see no connection between dairy cows on lush French pastures and elephant and buffalo on dry African ranges. The book stayed unread on the shelf.

In the meantime, a loathing for cattle, which I believed were destroying the land, prompted me, along with American Fulbright Scholars Archie Mossman and Ray Dasmann, with whom I was working, to begin promoting a new concept called game ranching. If ranchers could substitute game for livestock, and if we could find ways to effectively market the product and give wild game a financial value, perhaps we could get rid of the cattle and save the game and the land. Ranchers would come to see wildlife as an asset (by custom they considered it vermin), and wildlife would not wreak nearly the damage of domestic stock, or so I thought.

Neither the development of game ranching nor the business of culling the buffalo and elephant proved very popular in the early 1960s, and I was forced out of my game department job by those not so subtle pressures bureaucracies apply to the dissenting. However, I turned to farming, game ranching, and consulting as a livelihood and continued my work with private landowners in Zimbabwe and other countries in southern Africa.

## Domestic Stock and Take Half/Leave Half Management

At the time, governments in southern Africa, my own included, acknowledged that overstocking was probably causing the land to deteriorate, but they ascribed a greater share of the blame to a series of droughts that had occurred. I created somewhat of a controversy when I publicly challenged the latter view by suggesting that our droughts were becoming more frequent *because* our land was deteriorating. To my surprise, several cattle ranchers approached me—a known enemy of their industry—because they believed I was right and wanted to do something about it. I agreed to work with them, but only after I was sure they understood I had no answers myself and that it would be a case of the blind leading the blind.

The many sophisticated schemes for preventing overgrazing on ranches always began by limiting livestock numbers. One of the most common sought to regulate numbers so that animals would not graze off more than half of certain "key indicator" plants in the community. Research indicated that many perennial grasses suffered root damage with the removal of more than fifty percent of their growing leaf area.

The theory, however, had to fail, as wild or domestic severe-grazing animals graze neither half of each plant nor half of the forage available to them. Many plants are grazed severely. Many are not grazed at all, which results in their becoming old, stale, and ever less palatable. Generally the grazed plants are repeatedly grazed and weakened or killed, while those left ungrazed oxidize, choking out new growth and dying prematurely. Both grazed and ungrazed plants become less productive.

As land continued to deteriorate under various attempts at take half/leave half management, government researchers began to doubt that the ranchers had tried seriously. Since my own clients were among them, I paid a visit to the research stations in Zimbabwe and South Africa to learn what I could where things were done "right."

The research stations measured their success by the bulk of forage, the presence of a few species considered desirable, and the general appearance of the land as one glanced at it. By these criteria their ideas had indeed proved successful in practice. By my own criteria, however, even their research plots were desertifying. The soil between the plants was bare and eroding seriously—something not always visible to the person who didn't look for it. Plants were overgrazed severely in some patches, whereas in others they had grown old and excessively fibrous and were smothered with gray to black oxidizing material, and woody plants were establishing in such sites. The researchers had attempted to resolve the accumulation of oxidizing grass by burning about once every four years; the burned areas were quite visibly eroding. But the most shocking characteristic was the almost complete absence of new seedlings, despite massive seed production on parent plants. Desirable species were indeed present as largely old or senile plants, but very few other species.

The bulk of forage produced by the key indicator plant species, and the high individual production, in terms of weight gain and conception rate, on the few animals present, had masked the evidence of degradation. Production per animal was high and increasing, along with supplementation costs, even as production per hectare remained low and declining. For the first time I began to realize the extreme danger in considering short-term high animal production, or, for that matter, species composition, a measure of success. For the ranchers, and my country, blinded by such apparent success, I could only foresee ruination.

## "Overgrazed and Understocked"

Continually seeking answers, I began reading the range management research from various countries, but all appeared to follow the same thinking: excessive animal numbers cause overgrazing. Then one day while assisting a rancher in starting up a game ranching operation, I glanced at a South African farm magazine that lay on his coffee table. In it was an article by a man named John Acocks, who described a grazing system he claimed would heal the land. It made more sense than anything I had read before, so I went to South Africa and tracked him down.

I found Acocks a delightful elderly botanist and very knowledgeable indeed about the extent of the land deterioration that had taken place in South Africa. He believed it came from the selective grazing of the livestock that had replaced the original large and diverse game populations. Livestock, he said, overgrazed the species they preferred until those species disappeared. They then overgrazed another until it, too, disappeared. Gradually only the least desirable species remained. Thus he explained how areas dominated two or three hundred years earlier by perennial grasses had become the domain of desert shrubs. He had plotted the steady movement of these shrubs across South Africa as the desert spread over the years.

Acocks theorized that, as overgrazing weakened or killed plants of a particular species, other species replaced them that appealed less to livestock and thus held an unfair advantage over the grazed plant in the competition for light, water, and nutrients. He concluded that the actual numbers of livestock mattered less than their repeated selection of species that were thus handicapped and eventually replaced in the community. The diverse game species of old, reasoned Acocks, used all plants equally because each species selected a different diet. Thus no plant type had an advantage over another, and many thrived side by side on an even playing field. Based on these observations and interpretations he made the remarkable statement that South Africa was "overgrazed and understocked."[2]

His remedy called for putting the livestock onto a portion of land and holding them there till they had grazed down all the plants evenly. Once they had done this, they could be moved to another area to do the same thing. Each grazed area would then be rested so all the equally grazed plants could recover without unfair competition among them. Acocks's theory did not

answer all my concerns by any means—the deterioration of tsetse fly areas
that had no grazing being one—but it had merit and offered a new direction.
Before leaving he introduced me to a nearby farming couple who were prac-
ticing his ideas.

## The Big Discovery

Len and Denise Howell had a deep concern for the land degradation and were
excited by the results of applying Acocks's idea. So was I. My initial concern at
the amount of bare ground between plants turned to elation when I spotted
a small area where the grass was greener and the soil was completely covered.
The Howells looked on in bewilderment as I fell to my knees and probed my
fingers into the soil, pointing out excitedly what had happened where their
stock had accidentally trodden in very high concentration for a short time in
one corner of a paddock. The surface was broken; litter lay everywhere; water was soaking in rather than running off; aeration had improved; and new
seedlings grew in abundance.

John Acocks, and the accidental bunching of the animals that had occurred on the Howell's farm, had given me a vital piece of knowledge—*that
livestock could simulate the effects of wild herds on the soil.* Here in one area was
the heavy trampling I had seen following game but now done by livestock
and without the damage to the land we had come to expect.

I rushed back to Zimbabwe eager to persuade some of my rancher clients
to concentrate their livestock. The first one to do it rapidly produced the desired effects. Unfortunately, as fast as the land responded, his cattle fell off in
condition. In fact they nearly died. Others who followed my advice reported
the same results.

## Lessons from André Voisin

As disappointing as this was, I still believed we were at last approaching the
answer, and much to their credit a handful of ranchers who loved their land
decided to stick with me until we had it. We had no other allies in the quest.
Under a barrage of criticism and ridicule we confronted the new riddle of
poor livestock performance.

The cattlemen and our government extension officers believed in scattering cattle so they could select the grass species they needed to perform well.

Acocks believed in concentrating the animals and forcing them to graze all plants equally to prevent species from getting selected out of existence. Both arguments had obvious merit and somewhere between them the clues to good livestock and land management had to lie. Now, indulging in the perfect vision of hindsight, I can see that *time* was the factor staring us in the face and always overlooked.

As the problem now involved cattle, I once again dusted off André Voisin's book, and there it was. He had established that overgrazing bore little relationship to the number of animals but rather to the *time* plants were exposed to the animals. If animals remained in any one place for too long or if they returned to it before plants had recovered, they overgrazed plants. The time of exposure was determined by the growth rate of the plants. If plants were growing fast, the animals needed to move on more quickly and could return more quickly. If plants grew slowly, the opposite was true. Suddenly I could see how trampling also could be either good or bad. Time became the determining factor. The disturbance needed for the health of the soil became an evil if prolonged too much or repeated too soon.

## Lessons from the Wild Herds

If this was so, then how had time figured in the grazing and trampling of the vast wild herds of the past? This too could be reasoned out. Animals that bunch closely to ward off predators also dung and urinate in high concentration and thus foul the ground and plants on which they are feeding. No animals normally like to feed on their own feces, as anyone who has kept and observed horses will know. Thus, to be able to feed on fresh plants the herds had to keep moving off the areas they had fouled. And they could not, ideally, return to the fouled area until the dung and urine had weathered and worn off. This meant that plants and soils would have been exposed to massive disturbance in the form of grazing, trampling, dunging, and urinating, but only for a day or so, followed by a period of time that gave the soil and plants an opportunity to recover.

Something similar was at work in Yellowstone National Park following the reintroduction of the wolf in 1995. Elk herds that had lingered along streams and rivers with no fear of predation had overbrowsed willows and aspen to the point they had nearly disappeared, along with many other species that depended on them for food and shelter. Within a few years of the wolf's reappearance the elk herds were once again moving, the willows and aspen returned, and other species, too, including the beaver.[3]

This pattern would have been repeated again and again over millions of years. Even in the age of the dinosaurs, there were pack hunters, and thus presumably herding grazers and browsers. Recovery times for soils and plants would have varied because there were so many different species feeding in an area, and each species (or group of closely related species) was only trying to avoid fouling its own area, not necessarily that of others.

With this realization I now went back to my clients and suggested that we combine the ideas of Acocks and Voisin, concentrating the animals, but not forcing them to graze plants evenly, and timing the exposure and reexposure of plants to the animals according to the rate at which the plants grew. To do this we would follow Voisin's simple *rational* grazing planning technique, which he had devised to solve the problem of poor results from the *rotational* grazing long practiced in Europe. Again, we were very enthusiastic and this time certain of success. We did in fact improve animal performance somewhat, but in many important respects we fell flat on our faces once more.

Voisin had done his work on pastures in a nonbrittle environment, and we had not yet discovered the fundamental differences between nonbrittle and very brittle environments. Thus we did not immediately see how to fit our highly erratic growing conditions and eight months of no rain into his systematic accounting of time and growth rates. Neither, of course, did we know then that the solution in our brittle region also depended on the actual behavior of the cattle—the herding behavior, not merely the concentration of animals, which, unless extreme, does not change behavior. We also encountered greater complexity than Voisin dealt with managing European pastures. Our grasslands had a tremendous variety of grasses, forbs, brush, and trees—all growing at different rates. And they had a variety of soil types that produced varying results. We also had wildlife running on the same land as our stock, seriously affecting our time calculations. Whatever we suspected about time, we could not manage it effectively. But Voisin had provided a vital clue in his use of a planning process to gain control of livestock grazing time.

We would not unravel the whole mystery for a long time, but we knew at least we had discovered the right path. Politics as much as ecology forced us to keep learning. By the early 1970s Zimbabwe, then Rhodesia, had become a pariah among nations because the white-led government, to which I was by then leading the opposition in Parliament, refused to give in to the demands of the black majority. The tragic civil war that had erupted in the mid-1960s had grown increasingly fierce, and to force an end to it the rest of the world raised economic sanctions against us. To survive under the embargo, our ranchers and farmers had to greatly diversify their operations and manage great complexity and constant change on a day-to-day basis.

## A Planning Procedure and What It Taught Us

Livestock operations added new crops, and farmers, who had previously specialized in one or two cash crops, added many others and began to rotate crops and to run cattle as well to maximize income. To handle the difficulties inherent in such diversity, I developed a thorough, but simple, planning procedure based on military planning concepts I already knew (covered in chap. 41). It proved successful, and within a few years it was being applied on over a hundred farms and ranches throughout the country. The experience gathered from such varied situations gave birth to many new insights.

For one it showed us that John Acocks's statement that South Africa was overgrazed but understocked was true, *but not for his reasons*. Selection of the most palatable plant species by livestock and their replacement by less palatable ones did not explain the deterioration in the whole of southern Africa. Nor was the remedy to force livestock to eat off all plants equally, which we found caused stock stress. We realized that livestock do not select species but rather select for a balanced diet, regardless of species. Wild herding animals also select for balanced diets; they don't graze all plants equally, and never has this been necessary.

The overgrazing Acocks observed reflected the *time* plants were exposed to livestock, not the low numbers of animals grazing selectively. The understocking he observed did not damage the land by allowing less palatable plants to escape punishment, but by overresting land and plants that required *herd effect* periodically.

We only discovered all this because the planning procedure allowed us to control the time dimension more subtly than ever before. We found we could minimize overgrazing and overtrampling while still allowing animals to select the plants and nutrients they required. We could also induce adequate disturbance on the soil surface in small areas by briefly attracting the livestock to them, with a bale of hay, for instance, so that new plants could establish. We were able to plan for crops and hay cuttings as well and manage habitat critical to wildlife in particular seasons. All this became possible once we understood the time factor and had a planning procedure that left a good record of what we had done and what result followed.

## Conclusion

Taken together the last three insights provide an explanation for why, despite all our efforts over the centuries, so many environments have continued to deteriorate under human management. The discovery of the brittleness scale and the vital role of herding animals and their predators in maintaining the health of the more brittle environments, showed why these environments were prone to desertification. And the discovery that overgrazing and over-trampling were not a result of too many animals, but rather how long plants and soils were exposed and reexposed to them, led to the development of a grazing planning procedure that enabled us to use concentrated livestock, constantly moving, to provide the same benefits to plants and soils once provided by wild herds.

That we have been unable to stop the deterioration is not surprising, since approximately two-thirds of Earth's landmass is brittle to some degree, and our management has not catered for that fact. Nor until now have we known what to do. With desertification and climate change now threatening our survival as a species, and with the bulk of the world's brittle grasslands and savannas deteriorating so rapidly, we are only left with livestock—managed to mimic the behavior of the once vast wild herds—to begin the urgent process of restoration. While only those actually managing land can implement the changes necessary to reverse the desertification process and regenerate carbon-storing grassland soils, they will need the support of the rest of us, including policy makers, to do this effectively.

# 7

# A New Management Framework

*Until we had arrived at the four key insights* and understood their significance, we could not address what was faulty in our land management and why it so often led to soil destruction and desertification, particularly in the world's grasslands—those environments in which grasses play a critical role in stabilizing soil. The first insight—that nature functions in wholes—enabled us to develop a *holistic* framework to guide management and decision making by modifying the genetically embedded framework we have always used unconsciously. The next three insights—that environments range across a scale from nonbrittle to very brittle, properly managed livestock can improve land health, and time rather than numbers governs overgrazing—only became obvious when the first was understood, and enabled us to complete the holistic framework.

The Holistic Management framework is summarized in figure 7-1. As you can see, it does not illustrate the flow of a process, merely the elements to include. Some of the rows will have little meaning to you at this point, nor seem relevant if you are not engaged in the management of land or natural resources. But we all benefit from understanding basic ecological principles because many of the decisions we make will affect the environment at some point. And, the *tools* of rest, fire, and living organisms may not be required in managing a household, a law office, a bank, or any number of other endeavors, but an understanding of them and how they influence the environment

## Holistic Management Framework

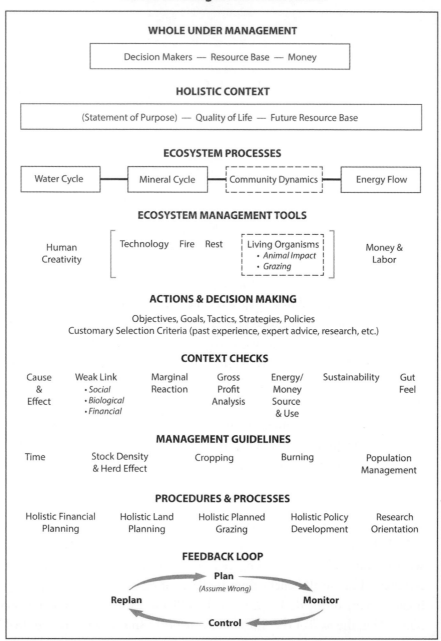

**Figure 7-1.** The Holistic Management framework.

you live in will help you make better decisions about what you eat, buy, and dispose of, or support financially. The management guidelines may not require much attention if you are not engaged in land or resource management but some of the planning procedures will.

We could have developed one framework for those directly managing land and one or more for everyone else, but I have long resisted doing so because the environment is ultimately everyone's concern if we are to sustain our economies, our civilization, and our planet.

## Holistic Management Framework in Brief

A brief review of the framework will give you a fair idea of what to expect in the upcoming chapters:

- **Whole under management.** All management decisions are made from the perspective of the whole under management, bearing in mind that it always influences, and is influenced by, both greater and lesser wholes. Chapter 8 explains that to be manageable the whole has to include, at minimum, the following:
  - **The decision makers.** Those directly involved in its management
  - **The resource base.** The *physical resources*—land, buildings, equipment, and other assets—from which you will generate revenue or derive support, plus the *people* who influence or are influenced by your management
  - **The money.** Available funds or those that you can generate from the resource base
- **Holistic context.** Creating one all-embracing context for management actions in the whole you are managing is the next critical step. You will refer to it often when making day-to-day management decisions, and when developing strategies that have traditionally been framed within a much narrower context. Chapter 9 explains that the holistic context has two aspects:
  - **Quality of life.** An expression of the way people want their lives to be within the whole under management
  - **Future resource base.** A description of the environment and behaviors that will sustain that quality of life for their successors

Ownership in the holistic context is essential, and to gain it the context must have meaning for those who created it. The more ownership they have the more likely everyone will be taking actions within that context and not according to hidden agendas. If you are managing an organization formed for a specific purpose, then prior to creating your holistic context, state that purpose because it will inform the context and your management.

- **Ecosystem processes.** To work with the complexity inherent in the greater ecosystem that sustains us all, we break it into four fundamental processes, each representing vital functions within it: *water cycle, mineral cycle, community dynamics* (the patterns of change and development within communities of living organisms), and solar *energy flow*. We recognize that any action taken to affect one of these processes automatically affects them all, as chapters 10 through 14 explain.

  Where we once viewed our global ecosystem—everything on our planet and in its surrounding atmosphere—mainly as a source of raw materials, we now view it as the foundation on which all human endeavor, all economies, and all life, are built. Even in situations where people may have little power to influence the environment directly, they will, through the cumulative effect of their actions, have an impact on it and must specify in their holistic context what they want that impact to be.

- **Ecosystem management tools.** In conventional management, the tools available for altering any one of the ecosystem processes *on a landscape scale* were limited to three broad categories: *technology, fire*, and *rest* (human creativity and money and labor bracket the other tool headings because none of them can be used on their own to alter ecosystem processes, and one or more is always required in the use of the other tools). In the more brittle environments these tools alone were inadequate to maintain or improve the functioning of the four ecosystem processes. We found a remedy to this shortcoming in the behaviors of the large herding and grazing animals that coevolved with the soils and plants in these environments over eons. Though the value of their dung for increasing soil fertility had long been recognized, most people had rejected the most vital parts of the animals (their hooves and mouths), which can be harnessed as tools for improving ecosystem function on a vast scale under the category of *living organisms* (see chaps. 15–23).

- **Actions and decision making.** As you move forward with management you will create many plans involving a variety of goals and objectives and the tactics, strategies, and policies for achieving them. Just as before, the actions you plan will address your immediate needs or desires for improving your life or your business, and for addressing any problems that stand in the way. But your holistic context reminds you not to lose sight of what is meaningful to you in both the short and the long term. And, as chapter 24 explains, it will lead to the inclusion of what you plan to do to create the quality of life your holistic context describes and the environment and behaviors that will sustain it. This generally necessitates considerable thought, assessment of current actions, and development of plans to achieve yet further goals that will bring about the life you desire and, if an organization, to meet your stated purpose.

  You decide which actions to take and which tools to use to achieve your goals and objectives based on the same commonsense criteria you've always used: past experience, expert advice, research results, cultural norms, peer pressure, expediency, compromise, cost, cash flow, profitability, intuition, laws and regulations, and so on.

- **Context checks.** After selecting which actions are best to take, you now pose a series of questions to filter out actions that might not be in context, and thus not socially, environmentally, and economically sound, both short- and long-term. Chapters 25 through 31 cover these questions, or checks, and the background knowledge that informs them. In brief, the questions are as follows: Does this action address the root cause of the problem? The weakest link in the situation? Does it provide a greater return, in terms of time and money spent, than other possible actions? Which of two or more possible enterprises provides the best gross profit (if choosing among enterprises)? Is the energy or money to be used in this action derived from the most appropriate source, and will it be used in the most appropriate way, based on the holistic context? Will this action lead toward or away from the future resource base described in the holistic context? Finally, based on the picture that has emerged, how do you feel about this action now—how will it affect your quality of life and that of others?

- **Management guidelines.** The five guidelines covered in chapters 32 through 37 are specific to the use of the ecosystem management tools. They reflect years of experience in a variety of situations and will help shape your plans and actions. Some of them depart considerably from guidelines used in conventional management and therefore require some study.

- **Procedures and processes.** The procedures and processes covered in chapters 38 through 43 were developed because Holistic Management enabled us to depart from conventional practice in key ways. *Holistic Financial Planning* includes refinements to the best of conventional financial planning specifically applicable to agriculture. *Holistic Land Planning* enables us to effectively and economically plan (and/or replan) the infrastructure on large tracts of land that livestock will graze. The *Holistic Planned Grazing* procedure is essential for planning livestock moves and behavior to restore degraded land to health while keeping livestock healthy and productive and integrating crop and wildlife production and other land uses. *Holistic Policy Development* enables us to reframe policies created in the context of a problem to ones developed within a holistic context to better ensure their success. And, finally, *Holistic Research Orientation* helps us design projects that meet the needs of management, including the filling in of critical knowledge gaps.

- **Feedback loop.** In Holistic Management the word *plan* has become a twenty-four-letter word: *plan-monitor-control-replan.* In the framework, these words are incorporated into a loop because planning is not an event but a continuous process. Once a plan is made, you need to monitor what happens from the outset because unforeseen circumstances always lie ahead. As chapter 44 explains, when the feedback from your monitoring indicates the plan or action is causing you to deviate from the goal or objective you're trying to reach, you must act to control the deviation. Occasionally events go beyond your control and there is a need to *replan.* If an action, or a plan outlining several actions, attempts to alter ecosystem processes in some way, then, despite having done your best to select the right action or strategy and checked to ensure that it was in context, you assume from the outset that, given nature's complexity, you could be wrong. Then you monitor, *on the assumption you are wrong,* for the earliest possible warnings so you can replan before any damage is done. This keeps management proactive rather than reactive or adaptive.

## Conclusion

To summarize how we can use the new framework to organize our management and decision making, suppose my family and I want to begin managing our farm holistically. The land is currently deteriorating and we are struggling financially, but we ignore this for the moment. We define our whole as those of us making management decisions, plus an uncle who also has a say in them, the farm itself, and the people we do business with and who support us in other ways, and, for money, only what we can generate from the farm. We decision makers now create a holistic context to guide our management going forward: a quality of life statement that reflects what is most meaningful to us; a description of the land not as it is today but as it has to be, in terms of those four ecosystem processes, to sustain our way of life; and our behavior as it will have to be to positively influence those whose support our business needs.

To manage our farm we may choose any of the tools at our disposal. Not being entirely familiar with all our tools, we turn to our set of instruction books, the management guidelines, and planning procedures, to learn more. In most cases we make a plan to get us started that includes goals and objectives related to the business as well as for achieving the larger desires expressed in the holistic context, and the tools we'll use and actions we'll take to achieve those goals and objectives. We base our selection of these tools and actions on the criteria we usually use to make the best choices. But before implementing our plans we check to make sure the choices we've made are socially, environmentally, and economically sound in terms of our holistic context. The context-checking questions help us do this. We implement the plan using the tools and actions selected and monitor our progress to stay on track. When events demand it, we replan, changing the way we're using the tools, substituting different ones, or modifying our actions—whatever it takes to achieve our goals and objectives and remain aligned with our holistic context.

Some years ago I attended a conference sponsored by the International Society for Ecological Economics at the World Bank in Washington, DC. During the plenary session one man was widely applauded for making a statement that reflected the frustration felt by most of those present: "Despite all our attempts to manage ecosystems and economies, we keep finding that we always end up being precisely wrong. Can't someone find a way to be at least approximately right?"

I believe Holistic Management enables us to answer that question in the affirmative. In checking to ensure that our actions are taken within a holistic context we are generally assured of being approximately right, but we still complete those essential feedback loops—planning, monitoring, controlling, and replanning—to make sure.

# PART 3

# THE HOLISTIC CONTEXT

**Holistic Management Framework**

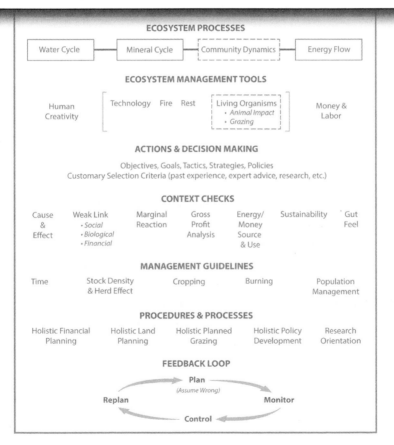

## WHOLE UNDER MANAGEMENT

Decision Makers — Resource Base — Money

### HOLISTIC CONTEXT

(Statement of Purpose) — Quality of Life — Future Resource Base

### ECOSYSTEM PROCESSES

| Water Cycle | — | Mineral Cycle | — | Community Dynamics | — | Energy Flow |

### ECOSYSTEM MANAGEMENT TOOLS

Human Creativity    Technology   Fire   Rest   Living Organisms
• Animal Impact
• Grazing    Money & Labor

### ACTIONS & DECISION MAKING

Objectives, Goals, Tactics, Strategies, Policies
Customary Selection Criteria (past experience, expert advice, research, etc.)

### CONTEXT CHECKS

Cause & Effect    Weak Link
• Social
• Biological
• Financial    Marginal Reaction    Gross Profit Analysis    Energy/ Money Source & Use    Sustainability    Gut Feel

### MANAGEMENT GUIDELINES

Time    Stock Density & Herd Effect    Cropping    Burning    Population Management

### PROCEDURES & PROCESSES

Holistic Financial Planning    Holistic Land Planning    Holistic Planned Grazing    Holistic Policy Development    Research Orientation

### FEEDBACK LOOP

Plan
(Assume Wrong)
Replan              Monitor
Control

# 8

## Defining the Whole
### *What Are You Managing?*

*THE HOLISTIC CONTEXT, THE SUBJECT OF PART 3,* is fundamental to the management of anything complex and will guide the decisions and actions you take to achieve your goals and objectives. That's why the utmost consideration should be given to creating a holistic context. The first step is to define the whole your management encompasses to clarify the boundaries of what you are managing and to identify the people who will create the holistic context.

But how does one define a whole, given Smuts's point that wholes have no defined limits? All wholes, he said, are made up of wholes. In turn, they themselves make up yet greater wholes in a progression that extends from subatomic structures to the universe itself. No whole stands on its own, and, in management especially, many wholes overlap. To isolate a lesser whole by giving it a sharp and arbitrary definition, such as household, business, watershed, national park, would appear to thereafter cripple the management of it.

Yet, for practical purposes, there is a minimum whole at which point Holistic Management becomes possible. In any of the wholes just mentioned (household, business, etc.) this would include the following:

- *The people directly involved in management* and making decisions
- *The resources available to them,* the physical assets, as well as people who can assist, influence, or will be influenced by their management
- *The money on hand or that can be generated*

At the outset you will always start with an arbitrary definition of the entity you want to manage, say a farm or a business. If the entity happens to be a bakery, then the whole might include the owner-manager and her staff; the building itself and the equipment in it; the customers, suppliers, service providers, and advisers; plus the money the business has available as cash on hand, a line of credit at the bank, or cash generated from product sales. This minimum whole is then viewed as one entity for management.

Defining the boundaries of the whole your management encompasses is critical because in doing so you are identifying *who* creates the holistic context and what they will be responsible for managing. You can do this and still acknowledge that any whole you define includes lesser wholes and also lies within greater wholes, both of which will influence your management.

In training sessions people have argued that some wholes are unmanageable. But I have yet to learn of a single case—from an individual household to an entire nation. If a whole is said to be unmanageable, generally it is either one that needs to be broken into smaller wholes, a subject covered later, or, more commonly, one that has not been defined properly—it lacks the minimum criteria. Some people, for example, only loosely define the whole in terms of the land—a national park, a farm, a wilderness area, and so on—without including mention of the people who manage it or derive benefit from it. The fate of the land is so tied to the attitudes and beliefs of these people that only management of people and land together has any hope of success. Likewise, no family, business, or community can be managed in isolation from the land that provides the raw materials for life and the repository for wastes. Now you begin to see where most economists fall short when they advise on the management of a whole economy, which in their eyes is all too often divorced from the land and the attitudes and beliefs of the people in all but the most superficial sense.

A more detailed look at the three components of a minimum manageable whole will enable you to appreciate their importance.

- **The decision makers**. These are the people who will create the holistic context. They should range from those who make the most profound, far-reaching decisions, to those making the most mundane decisions—from the owner of the coffee shop to the person who serves across the counter; from the owner of the ranch to the cowboys handling the cattle; from the trustees and faculty of the college to the secretary in the admissions office; from the senior partners in the legal firm to the person answering the phones. Make a list of them, trying to be inclusive rather than exclusive. *If there are people who, while not making decisions can veto them or in some way alter them, they too should be included in this part of the whole*, a point I will return to. Be prepared to redefine this part of the whole if you later realize that people who should have been included in creating the holistic context were left out. Excluding even minor decision makers can lead to problems down the road.

- **The resource base**. The resource base includes both physical and human resources.
  - *Physical resources:* Next, list the major physical resources from which you can generate revenue or derive support: the land, the factory and its machinery, the office building, your home, or whatever is relevant in your case. These resources need not be owned, merely available to you. You are not after a detailed list of every asset you have, only a very general one. You may later decide to sell some of the assets you own anyway, particularly the liquid or movable ones.
  - *Human resources.* Now make a list of the categories of people who will/can influence or be influenced by the management decisions you make, but won't have the power to veto or alter them—clients and customers, suppliers, advisers, neighbors, family, and so on. As you will learn in the next chapter, these people are no less important than the decision makers included in the first part of your whole. In fact, they are often vitally important to the whole even though they do not make management decisions within it. Be inclusive. Some groups or organizations you consider adversarial may in fact be an important part of your resource base as you may need to work with them to move forward productively. If the entity you are managing includes something as large and complex as state lands or a national park, the people in your

resource base could number in the millions. To overcome the dilemma of how to include them all, list groups of people representative of the larger public—for example, environmental groups, community organizations, and others deeply concerned about how that land or wildlife is managed.

- **Money.** Money will be involved in most wholes under management; for better or worse, today it is the oil that makes the cogs of life go round. Thus, in defining the whole, make a note of the sources of money available to you. This might include cash on hand, money in a savings account, or cash available from relatives, shareholders, or a line of credit at the bank. And it would almost always include money that could be generated from the physical resources listed in your resource base. Don't be sidetracked here by long and involved discussions on the meaning of money and wealth. Just think of money in terms of what you require to live on, or to run the business, institute, agency, or whatever entity you are dealing with.

## Keep Your Focus on the Big Picture

Remember the picture of Lincoln's face in chapter 3 that only became visible when you blurred the detail of the squares? In defining the whole, you are attempting something similar. You don't, at this stage, need to reflect on the detail of the squares because, if you do, you may lose sight of the whole you are dealing with. Just make sure that you identify what resources you actually manage, which of the people involved are decision makers and which are resources to you. Try to keep your lists and notes brief. Great detail is not needed now, only big-picture clarity. The following list provides a selection of examples. In each case a person or group of people have identified an entity they want to manage holistically—their lives, a business, an agency, and so on. Now they're ready to determine what the minimum, manageable, whole would be:

- A single, employed person seeking to manage her life holistically would conclude that there is only one person making decisions—herself. Her resource base includes her home, job and work associates, friends, neighbors, and mentors. For money she has what she earns and can save or invest profitably.

- A family seeking to manage a farm holistically might say that the decision makers include the members of their immediate family and their two employees. Their resource base includes the land they own, plus the 200 hectares (500 acres) they lease, their home, their extended family in town, their customers, their suppliers, the county extension agent, their soil testing laboratory, and their local study group. For money they have only what they can earn from the farm.

- Members of a livestock marketing cooperative in an African community seeking to manage their cooperative, or co-op, holistically might say the decision makers are the co-op members. Their resource base includes the land within the community boundaries, plus the grazing lands they share with other communities, their livestock and sale pens, customers (those who buy their livestock), marketing advisers, veterinary suppliers, government inspectors, and extension agents. They have the money they can earn from livestock sales or that extended family members can loan to the co-op.

- A group of accountants seeking to manage their firm holistically might say that the decision makers include the partners and staff. Their resource base is their office building and equipment, their clients, the people who supply professional and other services to them, and their families. For money they have what they earn from their services and, if needed, a $100,000 line of credit from the bank.

- A town council seeking to manage the local government holistically might say that the decision makers are the council members and their staff. Their resource base includes the physical structures within the town, as well as the parks and other recreational amenities, the businesses and cultural and other public institutions, the tourists and other visitors, and the people living in town and in the surrounding rural communities. They would have the money they raise by taxes, which depends on the money all their people and businesses earn, interest on that money, and grants from state and federal agencies.

I hope you appreciate the level of simplicity in this cross section of examples. Go much beyond this level and you risk clouding your picture of the whole. No amount of detail will help you make holistically sound decisions if you have not got a clear picture of the whole. All the detail imaginable will come into play later.

## Including the Right People in the Right Place

Occasionally, it may be difficult to determine whether some people fit in the first part of the whole, as decision makers who will create the holistic context, or the second part, as resources to you in achieving your goals and aspirations. This is an easy mistake to make when the people concerned aren't involved in day-to-day management yet can veto some of your decisions. These people should be included in the decision maker group and involved in creating the holistic context.

A lesson we learned in this respect involved two eight-year-long "trials" of Holistic Management conducted by the U.S. Forest Service in cooperation with two ranching families permitted to graze livestock on forestlands. In forming the whole we included only the families in the first part of the whole because they were the ones directly involved in management and decision making. We put the Forest Service people in the resource base, along with various environmental groups and a host of others.

Neither of these trials succeeded, in large part because the Forest Service people were not included in the first part of the whole, nor involved in creating the holistic context. Although they saw the ranchers as responsible for decision making, the Forest Service did have veto power over major decisions. Their regulations in fact overrode some crucial decisions that then made it impossible for the management, and thus the trials, to succeed. In subsequent efforts elsewhere, we did include government agency representatives in the first part of the whole and had them create the holistic context with us. When we again made some decisions that could have been vetoed due to their regulations, they worked with us to find a way forward because those regulations were now seen as standing in *their* way too.

Don't be put off defining the whole for fear you won't get it right. You will have plenty of opportunities to further refine it. Initially, those defining the whole—it could be you alone—may have only a limited understanding of Holistic Management and will need time to deepen it. As others become involved, you will likely make changes too. You will make mistakes, but you will do far more good than harm simply by gaining the clarity that defining the whole provides.

In your first attempt, then, do your best to define your minimum whole. The result, no matter how rough, should be adequate to enable you to get on with creating your holistic context. As you work at refining your holistic context you may be obliged to reconsider your whole as well. (For additional online learning resources contact the Savory Institute, http://savory.global.)

## Wholes within Wholes

If the group of people you have included in the first part of your whole is very large, and if the enterprises engaged in are very diverse, or if members are separated from one another geographically, it often becomes impractical to manage the entity as a single whole. In such cases don't hesitate to break management down into smaller, more manageable, wholes. Each of these smaller wholes should ideally meet the minimum whole requirement—that is, include people who are directly responsible for making management decisions at that level, an identifiable resource base, and money available or that can be generated from the resource base. It is also important to include at least one representative from the larger whole as a decision maker to ensure the holistic context created by the smaller whole is consistent with that of the larger whole.

If you have doubts about whether your management would benefit by forming a smaller whole within the whole you currently manage, it helps to remind yourself of what you are trying to accomplish. The primary purpose in creating separate wholes within a greater whole is to give the people within these smaller wholes the opportunity to create a holistic context that relates to their specific management situation and the resources available to them. *The more specific a holistic context can be, the more relevant it will be for the people managing within it.*

Once you have defined your whole and identified the people who need to be involved in creating the holistic context, those whose concerns it needs to address, and the resources at your disposal, you're ready to create your holistic context.

# 9

# Creating the Context for Your Management

FROM TIME IMMEMORIAL, human goals have driven human actions: to make a spear, build a dwelling, buy a car, get an education, reach the moon. Yet when these goals were met, it was often at the long-term expense of factors we failed to consider. Inevitably, each goal was made within a context that was too narrow, given the complexity of situations where human values, livelihoods, and our life-sustaining environment intersect. Defining one all-embracing *holistic* context for your goals in the whole you are managing is a new concept.

The holistic context has two essential components: *quality of life*—how we want our lives to be in the whole we manage; and, *future resource base*—a description of the environment and behaviors that will sustain that quality of life for future generations. If the whole under management includes an organization formed for a specific purpose we create a *statement of purpose* beforehand.

Where the context for our goals and actions may once have been focused on immediate needs, desires, or a problem, the holistic context shifts the focus beyond them to a conscious awareness of the life we want to lead and the life-supporting environment and behaviors that help ensure it. You will find that in framing your management and decision making within a holistic context things rapidly change for the better because many of the problems we face are really symptoms resulting from the unintended consequences flowing from past management.

### From Holistic Goal to Holistic Context

*When in the early 1980s* I first understood that something more than objectives, goals, missions, or visions was needed to guide our actions, I struggled to name the new concept, as well as define it. To minimize jargon I settled for *holistic goal.* But that led people, including myself, to refer to the new concept as something to achieve, which it wasn't exactly, even though it expressed fundamental desires.

After considerable thought about what I had failed to understand well enough to articulate, and discussions with long-time practitioners, the name became obvious. *Holistic context* more accurately reflected my original thinking—we do not strive to achieve a context but to operate within it. And it's working. People more readily grasp the idea that enlarging the context for our goals and the means to achieve them helps us better select the right course of action—one less likely to lead to regrets or unintended consequences.

The change to *holistic context* precipitated yet another change. The holistic *goal* had included *forms of production*—what we had to produce to create the quality of life we were seeking. People, including myself, had found it difficult to articulate forms of production that weren't actions because producing anything requires action. *And actions have no place in the context created to guide them.* But the idea of identifying what we need to produce was still important because unless we take appropriate action to bring about the life we desire, or the future resource base we've described, those desires will remain unfulfilled. So now we undertake the planning that will bring this about, covered in more detail in chapter 24 and in part 8.

## Statement of Purpose

If you are managing an organization or a division within one that was formed for a specific purpose your holistic context needs to reflect that purpose. If the purpose is well known, understood, and recorded in writing you have nothing more to do than include it as a preface to your holistic context.

On the other hand, many an institution formed for a specific purpose later loses sight of it and becomes ineffective or self-serving. In other cases, the purpose may have been worded so vaguely that it was open to interpretation. Or, because times change, and with them the situation that may have prompted the formation of the entity, the original purpose is no longer valid. If any of this is true in your case, take time to create a new statement of purpose.

In stating your organization's purpose, you want to get to the heart of the matter. The statement should reflect, in very few words, what the organization was formed to do. If it takes you more than a sentence or two you have not thought carefully enough, or, more commonly, you have gone beyond a statement of purpose into *how* you see yourselves doing whatever it is you are supposed to do.

The board, faculty, and staff of a private college I once assisted labored for hours over a whole paragraph of flowery words describing the services they would provide and how they would attract students—the sorts of things that would have to be passed through the context checks (see chaps. 25–31) *after* the statement of purpose and the holistic context were created. In frustration, after repeated attempts to steer them away from fussing over words that had little to do with the task at hand, I finally blurted out, "Damn it, I feel like calling you a bunch of bloody academics, but I can't because you are!" They got the message, and soon after came down to a simple statement that expressed what they had been formed to do: "To provide exceptional education that is relevant to the future our students will face."

Revisit your statement of purpose periodically, just to make sure that the purpose for which you were formed continues to be relevant.

Now you're ready to create your holistic context, and it starts with quality of life.

---

**Guidelines for Defining Your Holistic Context**

- *The decision makers identified* in the whole should create the holistic context, and they include anyone making management decisions at any level, or having veto power over them (see chap. 8).

- A trained facilitator can help bring out beliefs and aspirations that might otherwise not be voiced and can be invaluable in keeping the group focused on the task at hand.

- Write out the holistic context so all can see it.

- Keep its length to a page or less so all can remember it.

- Don't worry about getting it perfect; you will refine it over time.

- Make sure the words used mean the same thing to each person.

- Don't prioritize the ideas expressed—it is a *context* in which nothing is more important than anything else.

- Don't include actions or any prejudice against a future action.

- Include what you are for, not what you are against.

- Don't refer to problems, only to what lies beyond them.

A holistic context that has meaning for all the decision makers involved is one they will commit to. *The more ownership they have in it, the more likely everyone will be to act within that context.*

---

## Quality of Life

In creating this first part of your holistic context describe *how you want your life to be* in the whole you have defined, based on what you most value. It should express why you're doing what you're doing, what you are about, and what you want to become. It is a reflection of what motivates and excites you. It is your collective sense of what is important and why.

No one can specify what is or is not appropriate to include in your quality of life statement because what needs to be included is unique to each situation and the people within it. However, there are four areas you may want to consider in thinking about quality of life: economic well-being, relationships, challenge and growth, and purpose and contribution.

### Economic Well-Being

This is essential for meeting basic human needs for food, clothing, shelter, health, and security. The entity as a whole must be prosperous, but so should the individuals within it. How you define economic well-being will always depend on your circumstances.

"Making a lot of money" is rarely as useful in a quality of life statement as naming instead what you gain from having money: security, comfortable surroundings, enough to eat, and the wherewithal to do what you want to do. The same can be said for any material object.

I once knew a young couple who were certain that owning a farm was more important to them than nearly anything else in life. They had used all their working capital, plus a loan, to finance the purchase of one. Now they struggled to make ends meet, and the financial stress was beginning to take its toll on their relationship. Not the sort of life they had envisioned owning a farm would give them.

Had they instead at the outset created a holistic context and discussed what they wanted to gain in their lives by being farmers, they might have realized that owning a farm was not what was important, it was the way of life that was. That would open up other possibilities, such as leasing a farm, and running that idea through the context checks (covered in part 6). In their case, leasing a farm would more likely have left enough working capital to make a comfortable living and to create the way of life they had envisioned. Later, when they could afford the luxury of ownership, they could purchase that or another farm—an action they would again check to ensure it was in context.

### Relationships

Humans are social creatures. When we feel alienated or alone, we rarely function as well as when we feel we belong, that "we're all in this together." As decision makers you might mention qualities such as harmony, clear and open communication, respect for others, freedom to express our ideas, and so on. These qualities, if not already present, will only come about through changes in behavior. Likewise, when defining the whole, you identified people who were important as resources to you, and, as you will see below, when you describe your future resource base your behavior will govern how fully these people support you. For this reason, in your quality of life statement *describe the kind of relationships that result from respectful, caring behavior, rather than the behavior itself.*

### Challenge and Growth

Humans have a need to experience challenge; without it we fail to grow and develop. Where is the challenge in what you do? Try to think in terms of what you find stimulating, what requires all the resourcefulness and creativity you can muster? What sort of working environment stimulates creativity and allows you to be all you can be?

### Purpose and Contribution

People will only give their best to an effort when it has meaning for them. Meaning, in any kind of organization, is not something that can be created by a leader and handed down; it has to be a shared discovery. Ask yourselves three important questions:

1. What are we about?
2. What do we want to be?
3. What do we ultimately want to accomplish?

These same questions, of course, apply just as much to small family businesses and individual endeavors. The answers will help you discover the meaning in what you are doing. If you are working within an organization guided by a statement of purpose, you have an answer to the first question of what you are about. Your answers to the remaining questions help all the decision makers find meaning and personal relevance in the purpose and increase commitment to meeting it.

---

### Crafting Your Quality of Life Statement

*In recording everyone's thoughts initially,* it is important that you capture them in simple phrases, rather than well-worded sentences. You will have plenty of time to edit the results into a unified statement. On the other hand, resist the temptation to break the phrases down into single words. Don't just write: "prosperity, security, family values, and health" when what you really mean is "stable and healthy families where all generations feel secure and are cared for."

In general, people value many of the same things, but the specifics may vary, and in some cases conflict. If that happens work to *accommodate* those differences rather than force a compromise, which can discourage ownership. If differences arise over how an idea is worded keep talking until you find the words that best express your collective meaning.

Don't worry about how well it finally reads; you don't have to show your holistic context to outsiders. What matters most is that the words capture how the decision makers want their lives to be within the whole under management, *and* that the words mean the same thing to each person.

---

Creating a quality of life statement requires a good deal of reflection and numerous conversations before it begins to express what you want it to express. Start with a rough statement that indicates the general idea of who you are and who you want to become and what it is that gets you out of bed in the morning.

## Future Resource Base

Describe your future resource base as it must be if it is to sustain the quality of life you desire. There are two elements to consider: the *people* you included in the resource base when defining your whole and the *land*, or *environment*, even if you did not make reference to it when defining your whole and even when you operate a business that has no direct connection to the land.

## The People

Most businesses and organizations will have clients, customers, suppliers, supporters, members, funders, and others who in one way or another are critical to the health and future of that concern. Individuals managing their lives holistically will have friends, employers, and others upon whom they depend. These groups of people will have been listed in the resource base defined in your whole. They are the people who make no management decisions but can greatly influence them or be influenced by them.

Farmers and ranchers whose products are sold as commodities—milk, beef, grain, and so on—rarely meet either the retailer or the consumer of those products, and thus give little thought to the concerns of either. "Customers and clients" rarely figure in their future resource base. But other people will: buyers, suppliers, extended family, extension educators and other advisers, neighbors, and so on. If, on the other hand, you are a farmer or rancher producing products identified with you and that you are proud to be associated with, then mention in your future resource base the customers who purchase those products, and suppliers who enable you to produce them efficiently. You *want* your customers to know you, like you, respect you, trust you, and promote your products. You *want* your suppliers to be trustworthy, loyal, and committed to serving you well.

How can you describe such people far into the future? You can't. The way we overcome this dilemma is to describe how *we ourselves* must be, not *them*. How will we and our business or organization have to be perceived for these people to remain loyal, respectful, or supportive? In other words, you now describe *your behavior*.

Consider each of the people or groups of people mentioned in your resource base. For your clients, people whose loyalty and patronage you want to maintain, you might describe yourselves as honest, positive, professional, prompt, reliable, caring, producing nothing but the best quality, up-to-date, environmentally and socially responsible, and so on. It is not difficult to develop your list if you just try to think from their point of view. Many of the same behaviors will apply across categories and individuals, but some will be unique, so think through each case carefully. Finally, review your quality of life statement to make sure the behaviors underlying the relationships you desire among yourselves, as decision makers, are covered too.

Remember, the words you finally use are not meant for public consumption, so don't worry whether outsiders might misinterpret them. We are all judged by our behavior and actions, in any case, not by our words, and what you are describing here is how you are going to behave.

### *The Land*

When the whole you have defined does not include land under your management, you may wonder why land should figure at all in the description of your future resource base. It needs to be included simply because, in the long term, the well-being of any family, business, or community is dependent on the stability and productivity of the surrounding land. When I say "land" I am referring to it in the broadest sense, meaning soils, plants, forests, birds, insects, wildlife, lakes, streams, and, ultimately, the oceans as well.

All households and businesses, even those that are service oriented, at some point tie back to the land and its waterways and affect ecological health. For example, most businesses use paper and inks as well as detergents and plastic packaging, and these, in both their production and their final disposal, came from and return to the land with consequences. Banks provide loans for myriad enterprises that impact ecological health in both small and large ways.

Arising from almost every financial transaction there is an effect on the land that is experienced months or years later, and generally far removed from the site of the original transaction. For example, the cotton T-shirt you buy for your child is likely to have been produced from plants grown on deteriorating soils and dyed with chemicals that adversely affect water quality and human health. The pesticide you spray on your lawn can be tracked into your home, and can also end up in the water system, where it accumulates in shellfish, which in turn are eaten by another family many miles away.

The average citizen generally assumes that someone somewhere is going to do something to ameliorate these effects. That *someone*, as it turns out, is going to have to be ordinary folks like us. By including a description of the land around us, as it will have to be far into the future, we give the holistic context a much-needed dimension. When we later check to ensure our decisions and actions are in context we will always be reminded to consider the effects of those decisions on our environment.

Some people might hesitate to describe the land around them as it would have to be far into the future, feeling they don't know enough about it to do so. But one doesn't need a scientific background to be able to express a

---

### Holistic Context Created by a Middle-Aged Couple to Manage Their Lives

*Quality of Life*

- To be engaged in meaningful work for the rest of our lives, and to be excited and enthusiastic about what we have to do and get to do each day
- To be secure financially, physically, and emotionally into old age
- To maintain robust health and physical stamina
- To enjoy an abundance of mutually satisfying relationships
- To explore and experience wild places, and to ensure those places will still be there when our grandchildren's grandchildren seek to find them
- To live simply, and consume sparingly

*Future Resource Base*

- *People:* We are known to be compassionate and thoughtful, well-informed, good listeners, fun to be with, adventurous, and supportive.
- *Land:* The land surrounding and supporting our town will be stable and productive. Wildlife will be plentiful—we'll be able to see animals, or signs of them, anytime we venture out. The river will run clear and be full of life, and eagles will nest in the trees alongside it once again.

## Holistic Context Created by a Government Agency
## Managing a National Forest

*Like most government agencies, this one was formed for a specific purpose that it is obligated to meet.*

### Statement of Purpose

- To ensure this forest is managed sustainably and for the benefit of the nation

### Quality of Life

- To be proud of our work and respected for it

- To live meaningful lives in a caring and collaborative working environment with opportunities to further our learning, and in which our special talents and capabilities are acknowledged and utilized

- To share what we do with our families and people in the surrounding communities

- To manage this forest so that it provides an excellent financial return to the nation, and an even greater return in "biological capital"

### Future Resource Base

- *People:* We will be professional, knowledgeable, open to new ideas and learning, innovative, teamworkers, tolerant, patient, friendly, and helpful, and seen to be serving the public's interest.

- *Land:* Many generations hence, this forest will be healthy and rich in biological diversity, from the trees—in which all age groups are represented—to the abundant birds, mammals, insects, and microorganisms. The soils will be covered throughout the year, and remain where they form. Streams will flow perennially and clear and be healthy enough to drink from. Water and mineral cycles will be maximized and energy flow optimized for all life forms.

---

### A Generic Holistic Context

*I use a generic holistic context* when working in countries with different cultures and reading the papers and research reports related to the problems each country hopes to solve. Using this context for the actions, policies, and development projects proposed to address the problems, I am able to form my own opinion on the likelihood of their success and of unintended consequences. This holistic context is far from perfect, but I believe it is close to what most people desire:

- Stable families living peaceful lives in prosperity and physical security while free to pursue our own spiritual or religious beliefs
- Adequate, nutritious food and clean water
- A good education and good health
- Balanced lives with time for family, friends, and community, and leisure time for cultural and other pursuits

All of this is to be ensured, for many generations to come, by the following:

- *People:* treating each other respectfully
- *Resources:* healthy, regenerating soils and biologically diverse communities on land and in rivers, lakes, and oceans

---

need for surroundings that are stable, productive, and healthy, with clean and clear-running rivers, and covered—rather than bare—soils. And this would be enough of a description for checking to ensure your decisions are in context.

When you are managing land, you need to provide a fairly detailed description of what that land must look like far into the future, and how the four fundamental processes covered in chapters 11 through 14 must function:

- *Water cycle:* A water cycle characterized by flood-causing runoff or excessive soil surface evaporation would generally undermine the productivity of all life on the land sustaining us. Unless you are managing a wetland, you would almost always want the water cycle to be effective.

- *Mineral cycle:* The plants we depend upon will require mineral nutrients from the soil life and air, and to maintain them we must ensure that those nutrients cycle appropriately. If nutrients are trapped in dead vegetation that is not breaking down (unless burned) or if we have to depend on the constant reintroduction of major nutrients, something is wrong.

- *Community dynamics:* Most contexts will require a certain level of complexity in the communities being managed. In the future you will want these levels of complexity to be self-sustaining through the great synergy that exists among diverse species—from minute soil organisms to the greater biological community, be it a grassland, forest, wetland, or cropland.

- *Energy flow:* Most of what we produce from the land will require that the maximum energy be converted, both to maximize productivity and to sustain it.

Your task as a land manager is to describe the environment in terms of these four processes, not as they are now functioning but as they will have to be functioning in the future if you are to sustain the quality of life you have described over many generations. In many cases you will be dealing with several environments—rangelands, croplands, wetlands, riparian areas, forests, and so on. Since each may have different requirements, you can create separate descriptions for each of them.

One of the most common mistakes is describing a future landscape that is not much different from what exists today, when it needs to be. The mistake is understandable in many cases, because people have trouble envisioning something they've only heard about but never seen. I've encountered this in both developed and developing countries.

A Louisiana sugarcane farmer, for instance, had throughout his life seen only enormous fields of sugarcane and found it difficult to envision anything different. Large areas of land with high biodiversity, more effective water and mineral cycling, and far higher energy flow—all possible over time under changed practices—were beyond his ability to describe.

Young people in an African village surrounded by bare ground and starving goats and cattle found it hard to picture grassland with their livestock herded among zebra, sable, impala, and other game species. Having hunted big game as a young man over the same land they now occupy, this seemed simple enough to me. I couldn't understand their struggle until they pointed out that they couldn't picture something that had disappeared before they were born.

If you are faced with a similar situation, visit other areas to help expand your vision, or talk to a few old-timers who have a good memory of what your area was like long ago—though sometimes changes have occurred so slowly they may not have noticed them. As with the rest of your holistic context, how you describe the land in your future resource base may be refined over time as you learn more about that land.

Once you have thought through and described all the elements that make up your future resource base, you will have a holistic context.

## Conclusion

Now that you have a holistic context to guide your actions you can get on with management, and if the whole you manage includes an organization, also meeting the purpose for which it was formed. The world around you is not suddenly going to be different. You are most likely already managing your farm, ranch, business, or other organization, and you may be working to a strategic plan that includes goals, objectives, timelines, and more. However, where the day before the actions you took to achieve your aims were in the context of needs, desires, or problems, the actions you take from today onward will be subjected to the context-checking questions (chaps. 25–31) to ensure they are aligned with your holistic context.

At this point, having put so much consideration into creating your holistic context, you will be aware that, ultimately, almost all we are doing involves improving the quality of our lives. Now is a good time to begin seriously thinking of the many actions you can take specifically geared to improving the quality of your life within the whole you are managing. For example, people commonly want more quality time with colleagues, family, and friends, but this has not been happening. Nor will it in the future unless you take action to bring it about, such as instituting good time management. (Every person has the same amount of time. It is how we manage that time that governs those who achieve much and enjoy leisure time and those who are in constant stress and achieving little.) For people in organizations, this is a good time to seriously question present actions in the light of the organization's statement of purpose. Has it become a self-serving bureaucracy to the detriment of its purpose? If so then different actions, including a possible restructuring, can be investigated to make the organization more effective. If the land you've described in your future resource base is far more healthy and productive than it is today, you will need to consider changes to your current practices and plan how you will bring those changes about.

None of this means that the actions you took yesterday are faulty, or the strategic plan or goals and objectives you are working toward are now irrelevant. Some of your existing plans or actions will survive the context checking; others may be dropped as you create better plans and more effective actions that are in context. None of us do this perfectly, least of all myself, but one thing you can be sure of is that things will begin to improve in many ways as the context checking helps ensure that the actions you take are, to the best of your knowledge, socially, economically, and environmentally sound, both short- and long-term.

Create your holistic context as described in this and the preceding chapter, and even though you might muddle your way through the rest of Holistic Management, you will begin to see a change for the better. It is simply not possible to change from management based on a narrowly defined context to management guided by a holistic context and see the same results.

# PART 4

# THE ECOSYSTEM THAT SUSTAINS US ALL

## Holistic Management Framework

### WHOLE UNDER MANAGEMENT

Decision Makers — Resource Base — Money

### HOLISTIC CONTEXT

(Statement of Purpose) — Quality of Life — Future Resource Base

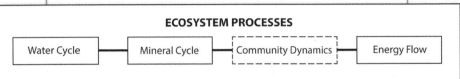

## ECOSYSTEM PROCESSES

| Water Cycle | — | Mineral Cycle | — | Community Dynamics | — | Energy Flow |

### ECOSYSTEM MANAGEMENT TOOLS

| Human Creativity | Technology   Fire   Rest | Living Organisms<br>• Animal Impact<br>• Grazing | Money & Labor |

### ACTIONS & DECISION MAKING

Objectives, Goals, Tactics, Strategies, Policies
Customary Selection Criteria (past experience, expert advice, research, etc.)

### CONTEXT CHECKS

| Cause & Effect | Weak Link<br>• Social<br>• Biological<br>• Financial | Marginal Reaction | Gross Profit Analysis | Energy/ Money Source & Use | Sustainability | Gut Feel |

### MANAGEMENT GUIDELINES

| Time | Stock Density & Herd Effect | Cropping | Burning | Population Management |

### PROCEDURES & PROCESSES

| Holistic Financial Planning | Holistic Land Planning | Holistic Planned Grazing | Holistic Policy Development | Research Orientation |

### FEEDBACK LOOP

Plan
*(Assume Wrong)*
Replan          Monitor
Control

# 10

## Introduction
### *The Four Fundamental Processes That Drive Our Ecosystem*

*Many scientists today speak of different ecosystems*—riparian, grassland, wet tropical forest, aquatic ecosystems, and so on. I've personally resisted doing so because, in managing them, too many people fail to remember that the boundaries used to define an ecosystem are an artificial distinction. A riparian ecosystem, for instance, cannot be managed separately from the grassland or forest ecosystem surrounding it, but time and again I've found that in practice it is. Each of these ecosystems only exists in dynamic relationship to the other and as members of a greater ecosystem.

While most ecologists appreciate this fact and so many managers don't, I find it more helpful to refer to one ecosystem, which encompasses everything on our planet and in its surrounding atmosphere, and probably more than that as well. Rather than distinguish lesser ecosystems within it, I've found it more practical to speak in terms of different environments within one ecosystem, each of which functions through the same fundamental processes: water cycle, mineral cycle, community dynamics, and energy flow.

The word *environment* doesn't, in my experience, seem to promote the idea of boundaries to the same extent that the word *ecosystem* does. When people talk of managing a riparian environment, I mostly find that the boundary between the riparian area and its surrounding catchment becomes seamless in their minds, and this is reflected in their actions. The idea is reinforced, of course, by their focusing on the four fundamental processes that are common to all environments, and through which the greater ecosystem—our ecosystem—functions. Ultimately these four processes are the foundation that undergirds all human endeavor, all economies, all civilizations, and all life. That is why they appear near the top of the Holistic Management framework, serving as the foundation on which the holistic context rests.

Deliberately modify any one of these four processes and you automatically change all of them in some way because in reality they are only different aspects of the same thing. It helps if you think of them as four different windows through which you can observe the contents of the same room, our ecosystem. You cannot have an effective water or mineral cycle or adequate energy flow without communities of living organisms, because you would then have nothing to convert sunlight to a form of energy usable by life. If you were managing a piece of land and wanted to improve the water cycle to increase forage production, you would plan which tools to use and how to use them. But before going further, you would also consider how those tools would affect the mineral cycle, community dynamics, and energy flow.

---

Deliberately modify any one of these four processes and you automatically change all of them in some way because in reality they are only different aspects of the same thing.

---

To perceive the unity of our ecosystem requires no scientific training or specialized education. The spread of acid rain across wide areas, the buildup of carbon dioxide and the breakdown of ozone in the atmosphere, the worldwide implications of a nuclear power plant disaster all demonstrate that isolated ecosystems don't exist. However, it is less obvious that if we are to head off the environmental disasters looming ahead, all of us—not just scientists, or farmers, foresters, and others managing natural resources—must begin to acquire a basic understanding of the fundamental processes through which our ecosystem functions. Hopefully it will soon be unacceptable for any economist, politician, or corporate CEO to remain environmentally illiterate, and thus ignorant of these processes and our connection to them.

You can read these words only because the sun shone on the leaves of a plant somewhere, and the leaves converted that sunlight energy to food and oxygen. You ate the food and inhaled the oxygen, both of which enabled you to read these words and understand them. The living organisms in our ecosystem are responsible for keeping atmospheric gases in balance so that the air we breathe remains conducive to life as we know it. Oxygen, for instance, makes up twenty-one percent of the atmosphere's gases. If it were to increase by four percent, the world would be engulfed in flames; were it to decrease by four percent, nothing would ever burn. Anytime oxygen is exposed to sunlight it reacts chemically with other gases and binds to them; thus free oxygen is constantly being depleted. To ensure that atmospheric oxygen remains at twenty-one percent, living organisms—the whole complex of plants, animals, insects, and microorganisms—must keep supplying it. Bear in mind that there was no free oxygen to support humans and other higher life until living organisms, of one form or another, created it.

Tragically, we are now less aware of our dependence on a well-functioning ecosystem than we were in earlier, less sophisticated, eras. Economists now have more leverage in the U.S. government than the farmers who formed it ever did. Accountants and lawyers serve as the chief advisers to the business world in which some corporations now wield larger budgets and more influence than many national governments. To be the specialists they are, most economists, accountants, and lawyers have considerable training in the narrow confines of their professions but less of an education in the broader sense, with some exceptions—ecological economists being one. As a consequence, most of these specialists exhibit little knowledge of the natural wealth that ultimately sustains nations, the quantity and quality of which is determined by how well our ecosystem functions.

All of us, however, have played a part in creating the environmental problems we now face simply because all of us at one time or another have made decisions that contributed to them. The decision to burn a field, poison the cockroaches, or buy a bar of soap often appears to be correct because the objective is accomplished. The field is cleared, the cockroaches are dead, and you are clean. But in the longer term, such actions can prove wrong, particularly in terms of how they affect the four processes through which our ecosystem functions.

If the whole you have defined includes land under your management, become intimately acquainted with water and mineral cycles, community dynamics, and energy flow; otherwise you will not be able to monitor the results of decisions you take that affect our ecosystem. Managing a piece of land in terms of these four processes may seem a daunting task, but it is simpler than first appears. An analogy provided by Sam Bingham, my coauthor on a previous book, helps make the point. It is one he borrowed from a German writer named Heinrich von Kleist who, sometime in the early 1800s, interviewed a famous puppeteer.

How, von Kleist wanted to know, can a normal person possibly manage the body and each individual limb of a marionette so it moves harmoniously like a real person instead of like a robot? How does the puppeteer learn that when he moves the puppet's leg forward, he also has to tilt its head slightly, bend the torso, and shift both arms in opposite directions?

The puppeteer answered that von Kleist had not understood the actual challenge, which was both simpler and more elegant. Of course no human could produce natural gestures by pulling any number of individual strings. No matter how skilled the puppeteer, the result would still look mechanical. On the other hand, a skillfully designed marionette had a well-positioned center of gravity, and simply moving that center of gravity would bring about all the other gestures automatically, just as when a human takes a step and all the other parts of the body automatically move in order to stay in balance.

In Holistic Management, the four fundamental processes are the puppeteer's center of gravity. The land manager who formulates a holistic context that describes a future landscape in terms of these four processes will find that in moving them in the direction of that vision, the land will come right.

# 11

## Water Cycle
### *The Circulation of Civilization's Lifeblood*

As FAR AS WE KNOW THERE IS A FIXED AMOUNT of water on the planet that constantly cycles from the atmosphere to the surface and back to the atmosphere. Much of this water is seawater, too salty for most uses until it evaporates and returns as rain or snow. Some of it becomes locked underground or in polar icecaps for vast periods of time before rejoining the cycle. But most water remains constantly on the move, becoming now liquid, now ice, now vapor.

Because water is fast becoming a limiting factor to the growth of cities, agriculture, and industry and because people are already killing one another over diminishing supplies, there is an urgent need to understand how it cycles and thus becomes available for use. I have drawn the basic pattern of the water cycle in figure 11-1. It shows the various paths taken by water falling on the land as rain, hail, and snow. Some evaporates straightaway off soil and plant surfaces back into the atmosphere. Some runs off into streams, rivers, dams, lakes, and eventually the sea while also evaporating. Some penetrates the soil, and of that a portion sticks to soil particles. The rest flows on down to underground reservoirs. There it may remain for millennia or find its way back to the surface in riverbank seepage, springs, and bogs, or possibly through deep-rooted plants that pick it up and transpire it back into the air or in some cases leak it out through surface roots, where grasses take it up and transpire it. Of water held by soil particles, a small portion remains tightly held, but

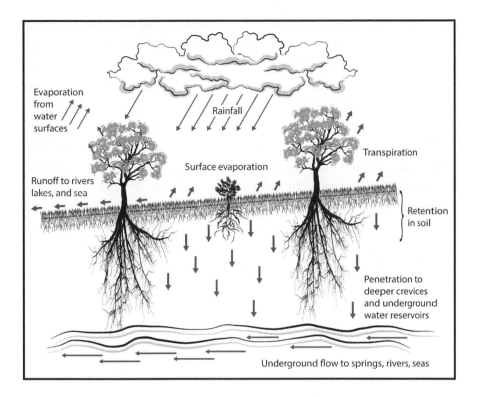

**Figure 11-1.** The water cycle.

the bulk is either attracted to drier particles or drawn away by plant roots and transpired. Thus, one way or another, all the water eventually cycles between earth and air.

Because the time water spends in the soil is critical to the growth and reproduction of plant life, which in turn is essential to most animal life, including humans, it is important to understand this stage of the water cycle in more detail.

## The Water Cycle at Soil Level

Any water that penetrates the soil will be strongly attracted to drier soil particles. This is why, after a while, no sharp edge between wet and dry soil remains, but rather a gradient from wetter to drier particles. The water will keep moving down until all of it has adhered to soil particles or passed on to underground reservoirs of free water. Short of drying soil in the sun or an oven, it is hard to remove the final film of water from soil particles. As water is drawn away from a particle by any means, that particle tightens its hold on whatever remains. However, particles hold each added increment of water more feebly than the last, just as you, if mugged on the street with an armload of parcels, will defend the last one you hold better than the first ones taken.

Plants absorb water, and essential nutrients dissolved in it, through root hairs. They can do this as long as their ability to draw water can overcome the grip on the water exercised by soil particles. As drying particles yield less and less water, the plant slows its growth rate. Eventually it begins to wilt in the heat of the day or curl its leaves to conserve moisture as its ability to obtain water from the soil is reduced. Much can be done, however, to retain more moisture in the soil and thus extend the time during which plants can grow vigorously before wilting point is reached.

To sustain humans and the maximum amount of life, in all but wetlands and true deserts, we need to maintain an *effective water cycle*. In an effective water cycle little water is lost to runoff or evaporation from the soil. Most water penetrates the soil, only subsequently leaving it through transpiring plants, by flowing down to underground reservoirs or out through rivers and streams. A good air-to-water balance exists in the soil, enabling plant roots to absorb water readily, as most plants require oxygen as well as water around their roots to grow.

In a *noneffective water cycle*, plants get minimal opportunity to use the full amount of precipitation received. Much is lost to soil surface runoff, and of that which does soak into the soil much is subsequently lost through evaporation from bare soil surfaces. What remains is often not readily available to plants because air and water are not in balance. Too much air and too little water in the soil does not suit many plants. Only slow-growing plants adapted to passing very little water through their pores, such as narrow-leaved grasses or cacti, can grow in these conditions. Too much water and too little air, and the soil becomes waterlogged. This can occur on grasslands, croplands, and of course in all wetlands, where an impervious layer of subsoil prevents downward movement of excess water. The water actually displaces air in the soil, and thus only plants adapted to a lack of oxygen around their roots can grow.

---

**Aeration and Effective Water Cycles**

*As a sugarcane farmer in Africa,* I experienced firsthand just how critical proper aeration can be. I had my crops under overhead irrigation and was told by the local extension service to plant my cane in hollows between high ridges made with a ridging implement. I questioned this advice and sowed a test plot of cabbages claimed by the seed company to produce two-pound heads.

I got the answer I sought quickly but let the crop mature in order to sell it. Heads planted in the furrows averaged one pound. Those on flat ground averaged two pounds. Those planted on ridge tops averaged *eight pounds*. Clearly several factors operate in such a situation, but most obvious was the fact that the soil in the ridges was never waterlogged and always well aerated; in the ridges the water cycle was always effective.

---

Some, such as mangroves, which grow around coastal wetlands, surround themselves with roots thrust up 30 centimeters (1 foot) or more above the water in order to breathe.

When creating your holistic context, describe the water cycle on the land you manage *as it must be functioning in the future* to sustain the quality of life you have described. If you are not actually managing land, but merely describing the land surrounding the community in which you live, this is not imperative, but it is desirable. In most cases you will want to have an effective water cycle, but perhaps not everywhere. You may be dealing with a wetland, or perhaps rice paddies, or with a desert plant and animal species that might require a landscape, or habitat, resulting from a less effective water cycle. If you are managing such a piece of land, the careful application of available tools (chaps. 16–23) and constant monitoring should produce this.

Whether or not you are actually managing land, it is helpful to know more about water cycles and what makes them more or less effective in different environments.

## Effective Water Cycles

Most land managers know the average rainfall their land receives and manage accordingly. Unfortunately, averages often mean little, particularly in the more brittle environments. In areas of erratic precipitation, as these environments often are, the average seldom occurs. Nearly every year the rain will be higher or lower than average. Even when it is average the distribution can be very different from the last year of average rainfall. Fortunately, *an effective water cycle tends to even out the erratic nature of the rainfall in any environment* by making the rain that does fall more effective.

To make precipitation as effective as possible means producing a cycle that directs most water either out to the atmosphere *through plants* or down through the soil to perennial river flows or underground reservoirs. In all the more brittle areas of the world where I have worked, rarely have I seen an effective water cycle. Typically of, say, 350 millimeters (14 inches) of rain received, only 125 or 150 millimeters (5 or 6 inches) is actually effective. In very rough figures it takes approximately six hundred tons of water to produce one ton of vegetation, so one can't afford to waste any of the rain that falls.

In less brittle environments with humidity well distributed throughout the year, effective water cycles tend to be more common simply because it is so much more difficult to create and maintain vast areas of bare soil. However, where soils have lost a large  proportion of their organic matter, and thus crumb structure,* they are unable to absorb much water, which runs off, and a less effective water cycle results. This problem is compounded on croplands, where much of the soil surface is deliberately exposed for long periods, or where subsoil compaction inhibits the downward movement of water, leading to waterlogging and a noneffective water cycle.

## Capping

From these remarks you can see that *the nature of the soil surface is vital to the effectiveness of the water cycle*. On bare and exposed ground, the direct impact of raindrops breaks down surface crumb structure, freeing the organic and

---

* *Crumb structure* refers to the presence of aggregated soil particles held together with "glue" provided by decomposing organic matter. The space around each crumb provides room for water and air, and this in turn promotes plant growth.

lightweight material to wash away while fine particles settle and seal, or cap, the soil surface.

A capped surface not only reduces water penetration but also prevents oxygen getting into the soil and carbon dioxide getting out. This in turn leads to a number of problems, one of which is nutrient deficiencies that show up in plants—and the animals that feed on them—even though the nutrients may be abundant in the soil. The air imbalance appears to affect the activity of the millions of soil organisms responsible for releasing nutrients in a form plants can use.

The initial capping is subsequently enhanced by a vast array of microorganisms and fungi that develop in it, providing more strength. You can see this if you lift a bit of the capped layer and inspect it closely. Even very sandy soils will develop a cap with this fibrous nature, though to a lesser degree. Some soils, particularly in the tropics where there is no freezing and thawing action to loosen the soil, develop a cap so hard it is difficult to break without a knife or some other hard object, as color plate 4 shows. If you tap such a severely capped soil with your fingers you will hear a hollow drumlike sound created by the air space beneath it.

Soil cover generally comes in two forms: low-growing plants that intercept rainfall so that drops hit the ground with less energy; and dead, prone plant material, or litter. Litter not only stops rain hitting the surface but also effectively slows the flow of water across the land, allowing more water to penetrate. (Water can flow quite fast between plants where no litter impedes it.) Snow also provides soil cover and can help lay dead standing plant material on the ground too.

In less brittle environments, maintaining soil cover is seldom a problem since plant spacing is naturally close, which tends to hold litter in place, and also because plant life establishes quickly on exposed surfaces. It is almost impossible to create thousands of hectares of ground with a high a percentage of bare soil between plants unless you use machinery and herbicides constantly, as many crop farmers do.

The more brittle the environment, the more the opposite is true. Bare soil develops easily between plants over millions of hectares in the absence of large herds of herbivores behaving as they once did in the presence of pack-hunting predators. Plant spacing tends to be wider apart, allowing wind and water to carry litter away.

## Creating an Effective Water Cycle

To enter the soil, water must first penetrate the soil surface, and this depends on the rate at which it is applied and the porosity of the surface in particular. Management tools that break up a capped surface or increase the soil's organic content and crumb structure speed up penetration. Management tools can create a surface that slows the flow of water across that surface, in effect slowing the rate of application of the water to that surface and thus increasing penetration. A loosened, rough surface or one covered by old, prone plant material achieves this.

### *The Importance of Soil Cover*

More than any other single factor an effective water cycle requires management that maintains soil cover, followed by soil life, organic matter, aeration, and drainage. As soil cover is the key to an effective water cycle, let's first look at the management tools available to us that can either destroy or promote it.

In the less brittle environments there is no tool whose application causes soil exposure over large areas other than technology—repeated use of machinery or herbicides on croplands, as mentioned. And on grasslands there is no tool whose application leads to vast areas of bare soil between closely growing plants. In the more brittle environments, however, which cover most of Earth's landmass, and particularly if the rainfall is low, a one-time use of machinery or a dose of herbicide can cause the majority of the surface to remain exposed for years. But when it comes to exposing soil between close-growing plants, especially soil-stabilizing perennial grass plants, millions of hectares of soil are routinely exposed through the application of three tools: rest, fire (periodically), and, to a lesser degree, overgrazing (the tool of living organisms misapplied). No other tools available can expose soil on such an extensive scale. Because these three tools are applied nearly universally in the more brittle environments, the staggering amount of bare ground and the resulting desertification come as no surprise. Even the earliest nomadic herders used the same tools. They tended to overgraze fewer plants, due to constant movement, but still partially rested the soil and used fire much too frequently, leading to the vast manmade deserts of antiquity.

In environments leaning toward the nonbrittle end of the scale, rest (partial or total) is the main tool available that can *produce* soil cover. To increase soil cover, land managers disturb the vegetation and soil as little as possible, if at all.

In more brittle environments, the only tool that can provide adequate soil cover over large areas is the impact of large herbivores behaving as they did in the presence of pack-hunting predators and before domestication. Today we refer to this tool as *animal impact*, which appears under the heading of *living organisms* in the Holistic Management framework. On both grasslands and croplands animals can be used to trample down old standing vegetation or crop residues to provide litter. Their hooves can be used to break up bare, capped soil surfaces, preparing a seedbed in which new plants can germinate.

## Fewer Droughts and Floods

When you have an effective water cycle, there are fewer floods and droughts and they are less severe, even where rainfall is very erratic. Those floods that do occur, as they will in very high rainfall years or years of rapid snowmelt, tend to rise more gradually and subside more slowly. The floodwaters tend to be clear, as they carry far less soil and debris.

The effects of droughts that do occur, as they will in a year when there is little or no rainfall in the growing season, are far less severe because moisture received in the previous year has likely been stored in the soil, and any received during the drought penetrates the soil surface more readily. In general, an effective water cycle will ensure that far more water is available over a longer time for plant growth. Plants will start to grow earlier in the growing season and more profusely, and will continue growing longer, even into the fairly long dry periods that can be present in the growing months. And there is of course far more water available to be released gradually to stream flow, bogs, springs, and underground aquifers.

What would it be worth to you as a rancher, farmer, or city planner concerned about urban water supplies to be able to double your rainfall? No doubt quite a lot. Fortunately, with a little understanding, you can greatly improve the *effectiveness* of your rainfall. And this, as I illustrated with the cabbage plants I grew, has an even greater effect than doubling the rainfall.

## Recognizing a Noneffective Water Cycle

When you have a noneffective water cycle, humanmade droughts occur more frequently, and years of very low or high rainfall tend to result in more serious droughts or floods because so much water is lost to soil surface evaporation or runoff.

Because so little attention is paid to the amount of bare ground between plants, which adds up to many billions of hectares in the world's more brittle environments, and because the amount of evaporation this leads to is often ignored, I like to use a simple example to illustrate the impact this has on the water cycle.

Imagine an area of grassland in a more brittle environment that receives a rainfall of 25 millimeters (1 inch) per month over the next three months. The land is level, and light regular showers produce no runoff. In figure 11-2 we have an effective water cycle on the left. The soil is covered and plants have healthy roots and are plentiful. On the right we have a noneffective water cycle where there are fewer, less healthy plants *and a great deal of bare soil between them.*

Assume that the first 25 millimeters (1 inch) of rain has fallen and all the water in both cases has soaked into the soil down to level A. We have 25 millimeters (1 inch) of water retained in the soil. Over the next month we receive no further rain, and the sun shines, temperatures are good, and plants grow. On the left they grow well, drawing out 12.5 millimeters (0.5 inch) of water in the process. No further water losses take place, and by the end of the month 12.5 millimeters (0.5 inch) of water still remains in the soil.

The plants on the right have their roots in poorly aerated soil due to the hard cap on the soil surface and have not been as productive. They have grown as well as they could, using 6.25 millimeters (0.25 inch) of water.

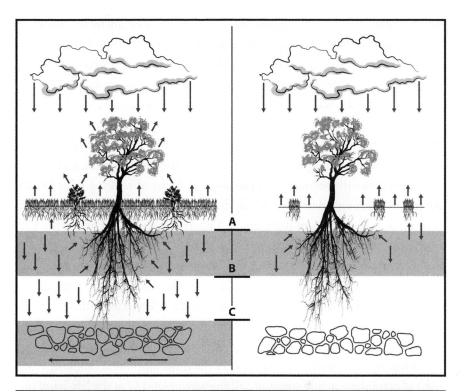

| Effective Water Cycle | | Noneffective Water Cycle |
|:---:|:---|:---:|
| Low | Soil surface runoff | High |
| Low | Soil surface evaporation | High |
| Low | Drought incidence | High |
| Low | Flood incidence | High |
| High | Transpiration by plants | Low |
| High | Seepage to underground reservoirs | Low |
| High | Effectiveness of precipitation | Low |

**Figure 11-2.** Effective and noneffective water cycles. Over three months, three rainfalls of 25 milli-meters (1 inch) each will wet soil layers down to level C on the left, where the water cycle is effective. The excess will trickle through decomposing rock fragments to join underground reservoirs. In the noneffective water cycle (right), most moisture will evaporate after each rainfall and never wet soil layers deeper than level A. Groundwater will receive no recharge this season.

Theoretically, then, 18.75 millimeters (three-quarters of an inch) should remain at month's end. However, *most of the soil surface between the plants is exposed*, and the sun shines and the wind blows directly on it, thoroughly drying the surface soil particles. The surface particles, directly in contact with particles below, can take water from them, and they do. As sun and wind continue to dry them, they draw yet more water from the particles below, which in turn draw moisture from the next layer down. At month's end, most of the water that the plants on the right didn't use has evaporated through the soil surface.

Now comes the second 25 millimeters (1 inch) of soaking rain. On the left it flows through particles already holding water and penetrates to level B, so we now have 37.5 millimeters (1.5 inches) in the soil. During the next month the plants grow well, again taking out 12.5 mm (0.5 inch), but by month's end 25 millimeters (1 inch) still remain.

The water on the right has soaked in, but the dry particles near the surface retained most of it, so it again only reaches level A. In the following month the plants again use 6.5 millimeters (0.25 inch), and sun and wind dry up the rest as before.

When the same processes repeat after the third rain, the 25 millimeters (1 inch) of precipitation will penetrate all the way to level C in the left-hand picture, and there the excess trickles through larger decomposing rock fragments to join underground reservoirs or eventually stream flow.

The water on the right has still not pushed beyond level A. *Groundwater will receive no recharge at all this season.* The soil has only 25 millimeters (1 inch) of water to carry plants through the long dry season to follow, and all of that will be lost soon through plant use and surface evaporation. Growth could well end before reduced temperatures limit it. In the following season when temperatures again rise, growth will have to await rainfall. People attempting to manage land such as this commonly say "the rains aren't what they used to be" or blame the lack of growth on "drought."

The soil on the left has almost 50 millimeters (2 inches) of water in it at the start of the nongrowing season. Plants will continue to grow until temperatures fall. The following year they will still have moisture enough for an early start when the weather warms, even though rain may not fall for another month.

## Dams Fill With Silt

I have deliberately devoted considerable space in this chapter to water cycles in the more brittle environments simply because these environments cover the majority of Earth's landmass. Today they are characterized by largely noneffective water cycles, and violence over diminishing water resources is mounting, as are environmental refugees who can no longer subsist on land that is drying up and turning to desert. Some of the world's largest cities also lie in these environments and are rapidly running out of water because noneffective water cycles have increased the severity and frequency of droughts and floods, and greatly reduced the amount of water once stored underground.

This problem of supplying cities with water is an ancient one that humans have tried to solve by building dams. The first large dam ever built was located on the Arabian Peninsula near the city of Marib, reputedly mentioned in the Koran as the original Garden of Eden. Built in 400 BC, the dam filled with silt and burst. It was rebuilt in 200 BC but burst again soon after, as the dam bed was still full of silt. The remains of the dam wall were still visible when I visited in the early 1980s and advised the World Bank team I was assisting not to rebuild the dam until they rectified the noneffective water cycle in its catchment. A new dam *was* built, however, 3 kilometers (roughly 2 miles) upstream of it in 1986, but I suspect it, too, will burst because the bare ground in the catchment has only increased and soil erosion accelerated.

Dam sites can really only be used once before filling with silt, and when they burst, as the Marib dam did twice, the bulk of the basin remains filled with silt, which is why the new dam has been moved upstream. Before using valuable dam sites we should first ensure that the water cycle in their catchment area is effective. Ironically, in the more brittle environments noneffective water cycles so reduce the amount of water available to cities that citizens demand dams be built. And we build them, rendering the few available sites useless for future generations.

Many countries have undertaken extensive engineering feats to overcome problems caused by noneffective water cycles. Some people, alternatively, advocate the digging of swales (shallow ditches along contours) to catch surface water flow, directing it to small areas where crops and trees can be grown. But this is simply harvesting noneffective rain from large areas to produce food on

---

### Water Cycles in Cities

*Look down on most cities* as though you were a falling raindrop
and you will find that about all you can land on are impervious
roofs, pavements, and roads. Because there are so few places rain or
snowmelt can soak into, the runoff from cities is extremely high.
A midsized city spread over 130 square kilometers and receiving
750 millimeters of rain each year will have to deal with 98 billion
liters of water (i.e., a 50-square-mile city receiving 30 inches of rain
will have to deal with 26 billion gallons of water). Very little of this
water is used where it falls but rather is channeled into storm sewers
and emptied into lakes, streams, or wetlands.

Imagine a city where building, road, and pavement codes met
standards that more closely imitated nature's water cycle: roofs that
catch water that runs into cisterns for use in homes and offices and
saves billions of gallons from being pumped into the city at great
cost; paving and road materials that absorb the water falling on
them and which are periodically treated with oil-eating bacteria so
that less rubber and oil is carried into underground reservoirs and
rivers. Such solutions are within our reach, with some already in the
development and testing stages in cities throughout the world.

---

small areas of land, and it is not a sustainable practice, as the Nabataean civilization learned two thousand years ago. No technology will ever achieve what simply putting the water cycle right does at a fraction of the cost. When we intervene with our earthmoving, we are dealing with water that has already started to flow. Dams only hold a fraction of it. Likewise, contour ridges only serve to spill the water on a more gradual gradient into the drainage pattern. The "keyline" contouring system developed in Australia by P. A. Yeomans at least spills that water back toward the ridges to allow more time for it to soak in instead of leading it into the drainage and thus off the land.[1] *Far better than any of these interventions is to prevent the loss of water from the land at the outset.*

## Conclusion

In terms of your holistic context, you can see why it is important to consider how the water cycle functions on the land surrounding your community, even when you do not manage that land yourself. If your community, town, or city, is prone to flooding or water shortages, the water cycle is probably not functioning well on the surrounding land. Having read this chapter, you now know what land generally looks like when the water cycle *is* functioning well. Bare ground is covered in vegetation, for instance, and rivers and streams run clear—even when in flood—because soil, rather than being carried into them, stays where it forms.

If you *are* managing land, you will need to describe how the water cycle must function to sustain the quality of life expressed in your holistic context. In most cases, you will want to describe what the land looks like when the water cycle is effective. If you are managing larger tracts of land, you will need to include separate descriptions for the different land types—croplands, grasslands, riparian areas, woodlands, and so on. If you want to maintain a wetland, you will in effect be describing a noneffective water cycle where soils remain waterlogged, but an effective water cycle in the wetland's catchment area.

# 12

## Mineral Cycle
### *The Circulation of Life-Sustaining Nutrients*

L*IKE WATER, MINERALS AND OTHER NUTRIENTS* follow a cyclical pattern as they are used and reused by living organisms. Nevertheless, because we don't see these nutrients so conspicuously in motion, we tend to ignore the extent to which our management can drastically alter the speed, efficiency, and complexity of their circular journey within our ecosystem.

A good mineral cycle implies a biologically active *living* soil with adequate aeration and energy underground to sustain an abundance of organisms that are in continuous contact with nitrogen, oxygen, and carbon from the atmosphere. Because soil organisms require energy derived from sunlight, but many do not come to the surface to obtain it firsthand, they rely on a continuous supply from plant roots and small animal life that transport plant litter underground. A good mineral cycle—one that provides a wide range of nutrients constantly cycling—cannot function effectively in a dying or dead soil.

On farms, agricultural chemicals may temporarily help produce higher crop yields, but all too often they destroy many soil organisms, and they inhibit others, such as those that fix nitrogen from the atmosphere or make nutrients available to plants. The net result is the destruction of living soil, something humans cannot afford to do. Turning over deeper soil layers, as we do when plowing, generally leads to the breakdown of organic material and destroys millions of soil organisms. The planting of crops as monocultures results in a less diverse root system and an environment that discourages diversity in microorganism species. All the same problems and possibilities for damaging soils and mineral cycles also exist in grasslands and in forests, but we tend to focus less attention on them.

To produce anything from the land at low cost on a sustained basis, soil and air should provide almost all the nutrients required by plants and animals, including humans. Some nutrients come in the form of minerals from newly decomposing rocks. Some nutrients come from the atmosphere via falling raindrops or organisms that convert gaseous substances, such as nitrogen, to usable forms. When the whole under management includes crop, livestock, wildlife, or timber production, the holistic context will usually describe a fully functioning mineral cycle in which we *strive to keep nutrients from escaping the cycle* and to steadily *increase the volume and speed of those cycling in the soil layers that sustain plants.*

This concept of high and increasing volume of nutrients cycling and available for use near the soil surface is easy to visualize (see fig. 12-1). To achieve it, however, requires a grasp of the natural processes that produced healthy growing conditions for eons before anyone thought of plows, chemical fertilizers or pesticides, or burning grasslands once maintained largely by herding herbivores and their pack-hunting predators. These processes are incredibly complex, but here we will look at the basic principles we need to know about to manage any mineral cycle well. Remember as we do that the mineral cycle does not function independently of the other three ecosystem processes. It is totally dependent on living organisms and the dynamics of the communities they inhabit and is inextricably linked to the water cycle and energy flow. If one process is malfunctioning all are.

To benefit humans, wildlife, and livestock, mineral nutrients have to be brought aboveground in living plants. To obtain maximum nutrient supplies in the active soil layers, minerals must continually be pumped up to the surface from deeper soil layers. Then, after use aboveground by plants and animals, they must be returned underground. There they will be held in the active root zones until cycled again, or lost downward to greater depths.

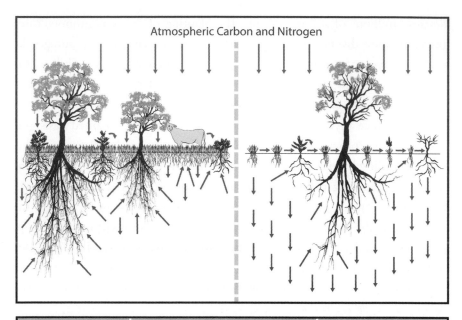

| Good Mineral Cycle | | Poor Mineral Cycle |
|---|---|---|
| High | Percent surface litter cover | Low |
| Mature, decaying | Nature of litter | Immature or oxidizing |
| Many | Surface insects/microorganisms | Few |
| Close | Plant spacing | Wide |
| Porous | Soil surface | Capped and sealed |
| Abundant | Plant roots | Reduced |
| Many levels | Root penetration | Mainly shallow |
| Porous with good crumb structure | Soil underground | Poor or no crumb structure, compacted |
| Abundant | Underground life | Reduced |
| Low | Surface mineral loss | High |
| Low | Mineral loss to leaching | High |
| High | Mineral turnover rate | Low |

**Figure 12-1.** Good and poor mineral cycles.

## Minerals to the Surface

Plant roots are the main agents for lifting mineral nutrients to surface soil layers and enabling plants to take a proportion of them aboveground. However, plants cannot do this without the mass of soil life working symbiotically with plant roots to provide access to nutrients. For a good mineral cycle, then, we need healthy soil life as well as root systems, with many of those roots probing as far as possible into the lower layers of soil and decomposing rock. In addition, we need a wide range of plant species to have many root structures. Just as you recognize plants aboveground by their appearance, so underground you could know them by their wide variety of rooting patterns. Some have abundant surface roots, whereas others probe deep, sometimes reaching below the soil itself into rock crevices and cracks to seek water and nutrients, which then move upward through the plant.

Where fibrous-rooted plants such as grasses, which include all grains, are the main line of production, some deep-rooted plants are generally essential to the health of the whole community. Incredibly small amounts of many trace minerals are critical to plants and animals, including humans, and they may lie beyond the reach of shallow roots. If you have ever noticed the many colors and textures in soil layers revealed by a highway or railway cutting, you have an idea of the variety of essential nutrients that might be found at widely varying depths.

Although plant roots are the main agents in mineral uplift, many small animals play an important role too. Earthworms are the obvious example, but in drier areas termites and other insects often help perform this function.

## Aboveground to Surface

Plant material aboveground, having obtained carbon and nutrients from the air and soil, finally returns to the soil surface in the form of dead leaves, stems, bark, branches, seeds, flowers, and crop residues to decay biologically on the surface. This may happen quickly or over a period of many years in the case of some plant parts, though the feeding, trampling, and other activity of large animals, birds, and small organisms generally speed the process.

Returning plant material to the surface, however, does not yet make it available for reuse. To be reused, nutrients have to move underground, and this does not happen until the dead material is broken down into finer particles by mechanical forces, such as rain, wind, hail or trampling, or through consumption or decay by surface-feeding insects and other soil organisms.

Rapid biological decay, rather than slow chemical or physical breakdown, ideally should play the lead role in the decomposition of annually dying plant material in environments right across the brittleness scale, with one major difference. In the less brittle, perennially moist environments, the soil surface typically supports extremely active communities of small organisms throughout most of the year that can break down old plant material without any contribution from larger animals. In the more brittle environments large herbivores hosting vast microorganism populations in their intestinal tracts become critical because, over the period of the year when fifty to ninety-five percent of the aboveground plant material dies, the microorganism and insect populations also die down. In such environments, large animals are needed either to trample the material down to the soil surface, where it will break down more quickly, or to reduce its bulk by grazing and digesting it. The gut of the grazing animal is one place that microorganisms do remain active year-round in brittle environments.

---

In brittle environments, large animals are needed either to trample the material down to the soil surface, where it will break down more quickly, or to reduce its bulk by grazing and digesting it. The gut of the grazing animal is one place that microorganisms do remain active year-round in brittle environments.

---

As you will see in later chapters, without adequate animal impact, the spacing between plants in the more brittle grassland environments enlarges and soil becomes exposed. This decreases biological activity, even in the humid months. Some mobile organisms, like termites, will build their earth structures out over the bare soil to reach animal droppings or leaf fall, but this activity alone cannot sustain a good mineral cycle. In an environment where plant spacings are wide and soil is easily exposed, it becomes difficult to hold plant litter in place against the forces of wind and water.

Fire drastically alters plant material, of course, and this may not always be bad, though we need to be aware of the pollution it creates and its tendency to ruthlessly expose soil. As chapter 19 explains, fire converts many nutrients that are vital assets when contained in the soil, to gases that become harmful when released into the atmosphere—hence its polluting effect. Soil exposure becomes increasingly important in very brittle, low-rainfall environments, where the soil cover so essential to healthy mineral cycling takes so long to rebuild. Even the minerals in the ash left on the exposed soil are generally lost to blowing winds or running water.

Physical weathering also functions differently across the brittleness scale. In the less brittle environments, where humidity tends to be higher and more consistent, weathering may play only a small part because biological decay proceeds so rapidly. Most plants, even the largest of trees, tend to rot at their bases first, fall over, and then continue to decay on the ground. By contrast, in more brittle environments, even those with high rainfall, most dead plant material breaks down slowly through oxidization and weathering in the absence of large numbers of grazing animals. Because weathering occurs from the top down, dead grass, brush, and trees do not readily fall to the soil surface, where microorganisms could help speed their breakdown. Dead trees in such environments can stand for a century or more while they gradually weather from the top down. The dead leaves and stems on perennial grass plants can stand for many decades. This can create a bottleneck in the cycle as nutrients remain tied up in dead plant material aboveground.

---

In brittle environments, in the absence of large animal populations, dead trees can stand for a century or more, and dead leaves and stems on perennial grass plants can stand for many decades, creating a bottleneck in the cycle as nutrients remain tied up in dead plant material aboveground.

---

Large accumulations of unrecycled, oxidizing plant parts also suppress plant growth, especially in perennial grasses, rendering them less able to absorb those nutrients that do eventually get below the soil surface.

In environments that are brittle to any degree, therefore, animal activity in various forms speeds the decomposition and cycling of the plant material essential to building mineral supplies in the top layers of soil.

## Surface to Underground

Once biological decay, fire, oxidation, or weather has broken down plant material, how do the critical nutrients move underground? Two agents, water and animal life, bring this about naturally. That explains why, when managing to enhance the mineral cycle, you will tend most often to apply tools that encourage water penetration and animal activity.

One further danger remains, however. The same water that carries nutrients underground can carry them on down below the root zone of plant types you hope to encourage. This is called leaching, and I believe that, more than any other factor, leaching caused the demise of civilizations in high-rainfall nonbrittle environments that were far from navigable waters, lacked the wheel or draft animals, and were dependent on shallow-rooted crops they could only plant nearby. The main factor that impedes leaching is organic matter in the soil. (Many tropical forests actually grow on poor underlying soils and are only lush and productive because of the mass of organic matter retained in upper soil layers by living and dead plant material.) The chemistry by which organic molecules bind mineral elements is extremely complex but derives from the same principles that allow organic matter to create the beneficial crumb structure referred to in chapter 11. The less organic material provided by dead plants and animals, and the less biological activity, the greater the tendency for leaching to occur.

Therein lies one of the great dangers of synthetic herbicides, pesticides, and fertilizers to our soils. The more we apply, the more we destroy organic material and living organisms in the soil, and the more we decrease the soil's water-retaining capabilities, the more we increase the loss of nutrients to both leaching and surface runoff. The leached minerals not only become unusable on that particular piece of ground, they may become highly dangerous pollutants as groundwater flow carries them to places they were never intended to be, such as in the water we drink.

## The Importance of the Soil Surface

The key to the health of the mineral cycle, like the water cycle, ultimately lies in the condition of the soil surface. An exposed surface, capped by the effects of rainfall, is a harsh microenvironment in which biological breakdown occurs slowly at best. This is seldom a problem in the less brittle environments, where new communities readily reestablish on bare soil surfaces, but it is ever present in the more brittle ones.

## Conclusion

In terms of your holistic context, if you are attempting to describe the land surrounding your community as it must be in the future, you would describe it much as you did to indicate an effective water cycle. But if that land is now characterized by miles and miles of monoculture cropland or grassland, you might want to describe it as also including more trees, brush, and forbs—all of which tend to have deeper roots than annual crops or many grasses, and thus enhance mineral cycling.

The same applies if you are managing land, though your description would be more detailed and could vary according to land type. If cropland, grassland, or forest soils are sterile now, you will want to describe them as biologically active. If you are in a brittle environment, you will want to be sure that minerals don't remain trapped in dead, oxidizing material and may want to describe what the land looks like when minerals are cycling rapidly (very little gray or black oxidizing material is visible by the end of the nongrowing season).

Let us now look into our ecosystem functioning through a different window.

# 13

# Community Dynamics
## *The Ever-Changing Patterns in the Development of Living Communities*

FROM THE MOMENT LIVING ORGANISMS ESTABLISH RESIDENCE on bare or recently disturbed soil, a rock, or in a newly formed pool of water, things are never the same again. Change begets change as the organisms interact with one another and with their microenvironment—the environment immediately surrounding them. Eventually a complex community* made up of a great many life forms develops and functions as a whole in an apparently stable manner.

Once any community has reached the highest level of development achievable in any environment, be that environment grassland, river, lake, coral reef, or forest, it can remain in that stable state for many years. However, closer inspection reveals a kaleidoscope of changing patterns within the mature community. Species composition, numbers and age structure, as well as numerous other factors are in a constant state of flux. Individual plants and animals are continually dying and being replaced, and varying weather conditions promote the well-being of some species and diminish that of others within the community. Because communities remain dynamic at every stage, from initial establishment to maturity, we refer to the process of their never-ending development as *community dynamics*.

Precisely what is taking place in any community at any one time is currently beyond human understanding and may always be. It is only relatively

---

*The collection of organisms that exists in any locality.

recently with the invention of high-powered electron microscopes that up to a billion or more organisms were found to be present in a cubic centimeter of soil or a spoonful of water. Most of these organisms have not even been named. And their relationships to one another and how they function within a community of organisms, which is far more important, is barely understood and difficult to imagine.[1]

In *The Redesigned Forest*, ecologist Chris Maser offers a glimpse of the complexity inherent in a northern temperate forest when he describes a relationship that exists among squirrels, fungi, and trees. The squirrels feed on the fungi and then assist in their reproduction by dropping fecal pellets containing viable fungal spores onto the forest floor. There, new fungal colonies establish. Tree feeder roots search out the fungi and form an association with them that enables the tree roots to increase their nutrient uptake. The fungi in turn derive sustenance from the roots.[2] Researchers are learning much more about such relationships, particularly the role played by soil organisms, which new technologies have helped make visible.

Of the four ecosystem processes, community dynamics is the most vital. Water and minerals cannot cycle effectively and solar energy cannot flow through life unless plants of some form—from algae to trees—first convert sunlight to usable energy for life and cover the soil. For this reason it is imperative that we learn to maintain healthy biological communities, whether they be associated with grasslands, forests, rivers, lakes, coral reefs, or oceans.

Although we still have much to learn about the dynamics of living communities, a few general principles have emerged based on the work of a great many ecologists and on the four key insights. Those mentioned here do not make up a comprehensive list, but they are the ones that bear most directly on day-to-day management. (My apologies to any ecologists offended by my simplification of some complex concepts.)

## There Are No Hardy Species

If we take *hardy* to mean that an organism is able to withstand very adverse conditions, then there are no hardy species. All living things are adapted to specific environments in which they can establish and thrive. Even though we may label a certain plant or animal "hardy" there is no evidence that it is.

You might think that a teenager from one of the toughest neighborhoods in Chicago is hardy because he thrives in an environment of crime, gangs, guns, and violence. But place him with a hunter-gatherer bushman family in the Kalahari Desert and he probably wouldn't survive a season. Take a teenage bushman out of the Kalahari and place him in a four-star hotel with hot showers, television, air-conditioning, and three meals a day and he would fail to thrive. The same applies to those so-called hardy plants that invade bare, baked, and cracked ground—the environment that suits them. If the soil was covered, damp, and cool, instead of being bare, baked, and cracked, they would have difficulty establishing themselves because they are not hardy enough to thrive outside of their environment.

When an organism establishes in a community, it will inevitably alter the microenvironment surrounding it. Even though the environment may be hot and dry, for instance, an incoming plant will create a little more shade and hold a little more dew and that might make the microenvironment hospitable for a new insect or microorganism, both of which in turn subtly alter the microenvironment as well. The germinating and establishment conditions may eventually become less favorable for that original plant and more so for other plant species, which begin to establish themselves. As these new species change the microenvironment, this change influences the types of animals, birds, insects, and soil organisms that find this habitat ideal, and their populations change too. Thus some species increase and others decrease as the composition of the community and the microenvironment above- and belowground make it more or less suitable for them over time.

## Nonnative Species Have Their Place

Biological communities develop as wholes over time as numerous species join them, interact with and change them, and depart them. But sometimes communities can be altered catastrophically within a short time by the accidental or deliberate introduction of a new species. We see this most noticeably on islands, particularly when the introduced species are predators or grazers. Cats, rats, goats, rabbits, mongooses, snakes, pigs, and humans have all wreaked havoc on islands where they were let loose, and on a number of continents as well. Many species in the community, having no defenses to ward off the introduced predator or grazer, are quickly killed out, and many species that depend on those species die out with them, leaving an impoverished community for years to come.

This is not to say that all introductions have been catastrophic. On the contrary, species *appear* to have been introduced successfully on many continents. One of the better known was the honeybee introduced from Europe to North America nearly four hundred years ago. That we regard this bee as a benign introduction might of course be a reflection of our ignorance.

Although I have focused on animal species, introduced plant species have followed a similar course, though those that have created havoc in the communities to which they were introduced don't seem to be of a certain type (like predators). Humans tend to refer to them, however, as weeds or noxious plants. Successful plant introductions have in many cases been so successful that we no longer think of them as nonnative: corn from South America has become the staple of Africa and potatoes the staple of northern Europe, and coffee from Ethiopia and tea from China have spread to every continent on which they can be grown.

The number of new species that humans have introduced to long-established communities pales in comparison to the species dispersed by nature. Plants, birds, insects, and microorganisms in particular have been spreading around the world unaided by humans for millions of years. Once any species establishes in a community it becomes a part of it, although the community always changes, and over time the species itself can change. If a new species causes major disruption, then it may take a very long time for the community to rebuild its former complexity and to stabilize. Fortunately, most introduced species, including the many *not* spread by humans, are either absorbed into the community without major catastrophe or die out altogether.

I mention these examples because there is today an unhealthy fixation on nonnative species in the United States and increasingly in other countries. I believe the term *nonnative* in most instances is purely a bureaucratic one, because those who use it assume that species that arrived after a specified date, often in the last century, do not function naturally in their new communities, and should thus be destroyed *at any cost*. If a species arrived even shortly before the specified date, it is considered legitimate; after that date it is nothing less than an illegal immigrant. Some animals, such as the elk of North America, pose problems for the classifiers. Because elk arrived on this continent at roughly the same time as the first humans, they are accorded legal status, as are "Native" Americans. Yet horses, whose ancestors were present in North America prior to the coming of humans and were probably killed out by them, are now considered nonnative, or illegal immigrants, when they flee domesticity and repopulate the vast rangelands of the West.

Once an illegal immigrant arrives it will, like the legal immigrant, fill any vacuum nature or, more commonly, mismanagement, provides. In the United States snakeweed, mesquite, sagebrush, cedar, a number of rodents, or grasshoppers (all legal immigrants) may fill the vacuum faulty management creates just as successfully as knapweed, leafy spurge, and fire ants (all illegal immigrants). No one has managed to annihilate either the legal or illegal immigrants, despite the expenditure of billions of dollars, nor has this been necessary.

Laws have been passed in various U.S. states that require landowners to poison, or otherwise "control" some of the nonnative plant immigrants. If the landowners refuse, the state will do it for them and send them the bill, even when the landowners can point out that the species in question is not causing any harm, is useful as forage for wildlife and livestock, provides ground cover, cycles nutrients, and generally plays nothing but a beneficial role. Unfortunately, once given the tag "nonnative" the species has to go, no matter what the cost to society, the economy, or the environment.

With greater enlightenment those managing holistically are finding a way to live productively with both legal and illegal immigrant plant and animal species, all of which at one time or another were "nonnative." The landscape you describe in your future resource base, when you reach it, may still include species you once considered pests, but in reduced numbers, and likely as not in a beneficial role as they contribute to the complexity, health, and diversity of the communities they inhabit, and ultimately to your prosperity.

None of what I've said here is meant to encourage the introduction of new species, including genetically engineered ones. We are right, I believe, to do our best to prevent the accidental introduction of species to new areas. However, once a species has established itself in a community, we are better off managing for the health of the whole community. I believe it is futile to spend the vast sums we generally do to try to eradicate it. This is particularly so when an introduced species invades a monoculture community, where it tends to become virulent. Our main option then is to drench the invader with poisons, which damage other soil life and further simplify the community, making repeated treatments inevitable, as will become clear shortly when we look at the link between simplicity and instability in communities.

---

We are right to do our best to prevent the accidental introduction of species to new areas. However, once a species has established itself in a community, we are better off managing for the health of the whole community.

---

## Collaboration Is More Apparent Than Competition

There is far more collaboration in nature than competition. What I am referring to as *collaboration* most scientists call *symbiosis*—the mutually beneficial relationships that occur among species in a community. For instance, there are apparently some nine hundred species of figs in the world, and each is dependent on a different species of wasp to aid in its reproduction. Lichens, a marriage of algae and fungi, are so bound together that for all practical purposes we think of them as one plant. Likewise, each human being is a colony of ten trillion cells, ninety percent of which are bacterial rather than human cells. Over the millennia in Africa a partnership has developed between two honey-eating species—humans and honeyguide birds. One has the ability to locate beehives and the other to break them open. The bird signals the human, who has learned to recognize its call and follow it to the beehive.

Many educators, based on an oversimplification of Darwin's ideas, continue to push competition as the driving force in nature, despite increasing evidence to the contrary. Studies of island communities, dating from the earliest observations of Darwin and his contemporary, Alfred Wallace, have revealed that over time new species will develop from existing species in an attempt to *avoid* competing for the same ecological niche. Personally, after years of working on several continents, I've been unable to find any *clear* evidence of competition in nature. Where problem plants are said to be outcompeting other plants and taking over, closer observation usually turns up evidence that shows mismanagement has created ideal conditions for the establishment of the offending plants, and less than ideal conditions for more favored plants. In many cases management had resulted in bare ground, and the problem plants were merely establishing on ground unoccupied by *any* plants. Being the only plants growing on the bare ground, however, they are automatically assumed to have caused the ground to become bare.

When you view competition as the driving force in nature you are compelled to take actions that may have ramifications you don't expect. The rancher who views the coyote, dingo, or jackal as a competitor (for calves and lambs) and shoots them out, may later find the predator helped keep small animal populations in check. With the predator gone, their numbers explode. In some parts of Africa where we believed crocodiles were competing with people for fish, we killed out the crocodiles but found to our surprise that fish populations then plummeted. We now suspect that while young crocodiles fed on insects, amphibians and then fish, older crocodiles had all along been limiting the numbers of fish predators.

*If your paradigm is competition, then you tend to see competition everywhere and manage accordingly.* But if you begin to think in terms of functioning wholes, collaboration, and synergy, you interpret and manage events differently.

## Stability Tends to Increase with Increasing Complexity

When biological communities are in the early stages of development or when they have lost biodiversity* because of a natural catastrophe or human actions, they are prone to major fluctuations, both in the composition of species and in their numbers. Disease outbreaks in plants, animals, and humans occur more frequently, as do outbreaks of weeds, insects, birds, or rodents.

This instability often correlates with weather patterns. In a high-rainfall year a mass of annual plants may germinate; in a low-rainfall year few plants may germinate at all. If the rains come early in the growing season one year, a particular weed may dominate; if they arrive later in the season the next

---

* Biodiversity is the diversity of plant and animal species—and of their genetic material and the age structure of their populations—within a given community.

year, another weed may dominate. Several seasons may go by when rodents or grasshoppers aren't a problem; in the next they become a terrible plague. Above-average rainfall years tend to become serious flood years; below-average rainfall years tend to become serious drought years. Inevitably we blame these conditions on the weather, rather than on a loss of biodiversity that has probably occurred as a result of our management.

The more complex and diverse communities become, the fewer the fluctuations in numbers within populations of species, and the more stable communities tend to be. As the number of species increases, so does the web of interdependencies among them, as illustrated in figure 13-1. In both higher and lower rainfall years, there are fewer outbreaks of any one species and lesser fluctuations in the mass of life, or *biomass*, present. An exception occurs in the true deserts where communities may be simple, but because the weather is consistently dry they appear remarkably stable. Even so, when the desert does get rain a mass of flowering annuals may suddenly appear, or swarms of locusts may develop and take flight.

When grasslands in the more brittle environments begin to die out—usually through overrest—some argue that this is a case where increasing complexity does *not* lead to stability. They are ignoring of course a major component of that complexity—the grazing animal species and their predators that have either become extinct or remain in far fewer numbers. In North America, for instance, only eleven large mammal species remain of the forty

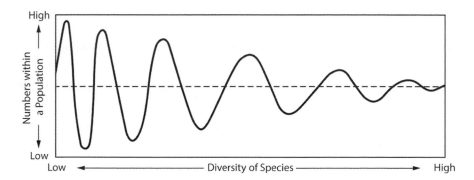

**Figure 13-1.** The more complex and diverse communities become, the more stable populations within them tend to be.

that were present when the first Americans arrived some ten thousand years ago.

Scientists have long believed that complexity in a biological community leads to greater stability, though they have had difficulty proving it. In 1996, however, a group of researchers in the American Midwest managed to provide convincing evidence, based on a study of 147 grassland plots, of a significant connection between diversity of species, productivity, and stability. And again in 2015, a much larger effort, conducted on 46 different grasslands in the United States, Europe, and Japan by more than three dozen researchers, found that the higher the plant diversity within those grasslands, the more stable productivity remained during extremely wet and dry years.[3]

## Most of Nature's Wholes Function at the Community Level

If we consider the world to consist of, and function as, wholes within wholes, it helps us to understand certain important relationships. Each individual plant or animal is a whole composed of billions of cells, each of which is itself a whole. Individual plants or animals, in turn, belong not to a whole population, but to a whole community composed of many species. I make this distinction because a *population* of any one species would not constitute an ecologically functioning whole, though in attempting to manage for a particular species as we often do we attempt to manage some populations as if they were. The population of any one species, humans included, cannot exist outside its relationship with millions of other organisms of different species.

The fundamental importance of the whole community, including the functioning of the four ecosystem processes within and surrounding it, can too easily be overlooked when we focus on rare, endangered, or preferred species. Some grasslands are still classified as being in good condition if the "right" (useful to humans) species are present. Many of those areas, however, belong at the other end of the scale because so many other species and so much biomass have been lost. In many cases, the microenvironment at the soil surface has deteriorated to such an extent that perennial grass species, though still present, can only reproduce asexually through surface runner stems (stolons), or underground running stems (rhizomes). Plants establishing through runners, rather than seeds, lack genetic diversity as they are merely clones of the mother plant.

Our inability to think in terms of whole communities is reflected in our efforts to save rare and endangered species through the use of increasingly draconian laws that pay little heed to the biological (and human) communities that support these species. In a few cases where slow-breeding, easily hunted species, such as rhinos, are involved, these laws can serve as temporary Band-Aids, but no more. Our ignorance is understandable, however, because it is based on some long-standing beliefs and widespread misunderstandings.

We now know that biological communities include all living organisms—from the most simple virus or unicellular organisms to elephants, humans, trees, whales, and corals. This of course includes the microscopic world within our soils where a complex web of life dwells among decomposing particles of rock, sand, clay, and "dead" organic material. And they include the near invisible world of our atmosphere as well. Many complex and mutually dependent relationships exist among all levels—belowground, aboveground, and into the atmosphere.

Some scientists suggest with good evidence that our whole planet is a living organism that modifies the atmosphere surrounding it through the activity of biological communities on land and sea. The composition of Earth's atmosphere, on which all life depends, can change gradually in conjunction with Earth's life forms as the two, life and environment, influence each other, but it cannot change rapidly without catastrophic consequences. Billions of years ago the very earliest communities would have been suited to a totally different environment from that of today, as there was little free oxygen until living organisms formed it. We have remnants of these earliest communities in the anaerobic life forms (those that do not use oxygen) such as some species of bacteria, which still exist in a variety of environments, including the digestive tracts of many animals.

Even though the emphasis here is on whole communities (the collection of organisms that exist in any locality), to be practical you can see that I often describe the environments they occupy in terms of the most obvious plants, such as grassland or forest, for the obvious reason that vegetation is most visible. This unfortunately fosters the misconception that plants, and sometimes only certain plants at that, are more important than the whole. In reality the insects that pollinate the plants are just as crucial to the survival of a community. So are many unseen species, particularly those underground, as well as species that may visit only once every few years.

## Most Biological Activity Occurs Underground

Any changes brought about aboveground are likely to cause even greater changes underground, simply because there is generally more life underground than aboveground. Figures vary widely with different soils, but on average, upper soil layers (the top 15 centimeters/6 inches) of grassland soils contain 17.5 metric tons of microorganisms, such as bacteria, fungi, earthworms, mites, nematodes, and protozoa *per hectare* (7.75 U.S. tons per acre). The richest soils can contain up to 33 metric tons of microorganisms per hectare (15 U.S. tons per acre).[4] Healthy European pastures carrying large numbers of cattle have been calculated to contain earthworm populations alone that are double the weight of the cattle.[5]

Plant roots also contribute to the biomass underground. The roots from a single tallgrass plant in the American prairie continue downward for anywhere from four to six meters (twelve to twenty feet), spreading through an area the size and shape of a tepee. If placed end to end, they would run for kilometers, with the roots of one tallgrass plant developing as many as fourteen billion fine root hairs. Scientists estimate that a full seventy-five to eighty-five percent of the prairie's biomass is underground.[6] Even when as many as sixty million bison roamed the prairies, and millions more elk, deer, and pronghorn, the underground organisms would likely have outweighed them by a considerable amount.

## Change Generally Occurs in Successional Stages

The process of change in biological communities from bare rock or new pool to mature grassland, forest, or lake is a gradual, often staggered, buildup of species diversity and biomass along with changes in the microenvironment. I like to compare this movement from simplicity to ever-increasing complexity to a coiled spring, which, whenever pressed down by human intervention or natural catastrophe, will, by its nature, rebound as soon as the pressure is taken away. Thus grass reclaims old battlefields. Jungle climbs the slopes of dormant volcanoes. And weeds occupy fallow ground.

This relatively orderly process of change has been given the name *succession*. The word entered the vocabulary of science through the work of botanists who observed that disturbed areas revegetated in successional stages— from bare ground to simple algae/lichen/moss, to grasslands, brushlands, and forest. Later insight took account of the fact that plants cannot exist in isolation, and thus we now think of succession in terms of entire communities made up of more than simply plants. A simple understanding of the basic idea of succession is easy to grasp if you visualize it in process on a tropical island lava flow.

After the lava has cooled and hardened, its surface remains a very harsh microenvironment. In rain it is very wet, minutes later it is extremely dry. At dawn it may be quite cold, but by midday too hot to touch. Only a few species find an environment of such extremes ideal, and thus initially establish. Other species will try and fail.

Without soil, only algae, lichens, and minute organisms dependent on them, will establish. The moment they do, however, the microenvironment becomes different. The meager collection of life will hold moisture a bit longer and reduce the daily temperature range ever so slightly, and moisture retained at the surface will now have time to begin dissolving the rock. When a few fine particles of dust catch on the algae and lichens, moss and other organisms are able to establish and the creation of simple soil has begun.

Gradually other organisms join the community as the microenvironment begins to favor them and their offspring. They further change the microenvironment. Succession accelerates. Moisture is retained longer. That breaks down the parent rock faster to join with living organisms in forming yet more soil. Anywhere physical weathering cracks the surface, the process speeds up in the microenvironment of the crack, which in turn affects the immediate neighborhood.

Complexity, productivity, and stability increase, and the microenvironment changes until something limits the successional process, typically climate or some obstruction to further soil formation. A cold-air drainage leading to heavy frosting or a subsurface rock layer, for example, might cause a patch of ground to remain at the grassland level. Otherwise, the lava of the tropical volcano will eventually advance to a rain forest community complete with its soil and the millions of organisms forming the whole forest complex. Elsewhere dry seasons, hard winters, limited sunlight, and the pattern and volume of precipitation will define the kind of landscape unfettered succession can produce. But whether the outcome be jungle, desert, savanna, healthy productive lake, or coral reef, the community is always dynamic as deaths, decay, and rebirth foster ongoing change within it.

The full implications of succession become clearer through an understanding of population dynamics at various successional levels—when do certain species thrive, in what numbers, and why? Typically a particular species will begin to appear and its population to build up as its requirements for establishment within the community are met. The community will be made of populations of many other species, each with specific requirements for their survival and each with specific contributions to the ever-evolving community. These populations will tend to build in numbers as their requirements become optimized through the changes in the microenvironment brought on by the growth of the whole community. But as the community advances, a population may find its requirements for successful reproduction are no longer ideal. It will decline in numbers and may even disappear as the successional process advances beyond it.

The range of conditions that are favorable to any one species varies greatly. The eland antelope of Africa, for instance, can thrive from the snowline to the deserts, while many other African antelope species are restricted to areas with specific vegetation patterns. Elephants or cattle can feed across a very wide range of plants, but koalas apparently only derive nourishment from the leaves of one species of eucalyptus tree.

### Succession and the Brittleness Scale

In very brittle environments, the microenvironment on exposed soil surfaces is subject to such daily and seasonal extremes that the successional process starts with the greatest of difficulty. On smooth surfaces that are steeply sloped or vertical, the process might never get beyond frail algal communities that are easily lost as a result of the physical action of rain, hail, wind, and animal life, or the movement of eroding soil. This is why the walls of the Grand Canyon cannot really stabilize with higher communities, although some plants and animals establish in odd niches.

In very brittle environments, the process starts more easily on soil covered by old material and on ground cracked by weathering or broken by the physical impact of animals or machinery that "chips" the surface. In both cases a better microenvironment results, with two notable exceptions.

If fallen material all lies in one direction, as in the case of lodged wheat, grass, or pine needles, it suppresses plant growth. The reason is not yet well understood, though farmers have long known that a straw mulch has to be scattered to be effective. On grasslands snow and wind will lay old, moribund bunchgrass in one direction, suppressing growth. Hail and animal impact tend to scatter it, encouraging growth. In forests a carpet of undisturbed pine needles may suppress plant growth. But new plants establish when animal impact (or some other action) disturbs the pine needles. Earthworms and a variety of insects help disturb deep litter surfaces. Turkeys, guinea fowl, or baboons often do so extensively.

The other exception concerns areas outside the tropics where certain soils become puffy and soft from alternate freezing and thawing. As color plate 5 shows, they may have very broken and rough surfaces, and yet succession does not progress easily due to lack of soil compaction. However, it will advance to grassland and greater biodiversity given adequate animal impact.

Such examples highlight once more the need for viewing the community as a whole. Managing plant or animal populations while paying little heed to the functioning of the whole community, including its soil life, is meaningless and likely to be damaging. Any brittle environment community that lacks large grazing animals is unlikely to be able to develop to its optimum level of complexity and stability. The presence of grazing animals in vastly increased numbers will be necessary if we are to facilitate the healing of the world's human-caused deserts.

By contrast, in nonbrittle environments succession starts with ease from any bare surface, often beginning with algae, lichen, and mosses, and plants will establish even on vertical slopes. The nature and distribution of temperature and humidity allow the rapid advance of succession nearly anywhere

---

### Advancing Beyond Capped, Encrusted Soils

*In a deteriorating brittle environment,* algae, lichen, and moss
are often the last plant life left. Once they encrust soil surfaces
succession stagnates unless the crust is broken so that other
species can establish and succession can advance. In less brittle
environments breaking the crust is not necessary. Succession will
advance of its own accord in reasonable time.

Failure to make this distinction has led to misleading public
information in several national parks in the western United
States where signs inform the public of the value of these "early
successional" algal crusts. They explain that the crust protects
the soil, provides nitrogen, and creates rough surfaces on which
grasses and other plants can establish, *as long as the crust remains
undisturbed.* Not surprisingly, because of the protection these areas
have received (close to a century in some parks), it's the *lack* of grass
plants and more complex communities that is most obvious (see
color plate 5).

That the park officials subconsciously question their own
wisdom is illustrated by their attempts to assist grasses to establish
by placing a woodwool mulch encased in nylon netting over
demonstration areas, as shown in color plate 6. Surely, if these
crusts were early successional, plants should be able to grow on
them unaided by humans. However, a year later all you can see
are the oxidizing remains of the mulch. The hoped-for grass
communities have not even begun to develop, nor will they until
the soil surface is disturbed to allow succession to advance to higher
plant and animal communities.

---

without the aid of some physical disturbance. It will start on the back of a
shower curtain, the abandoned bicycle seat in the garage, or the clapboards
of the house.

Figure 13-2 shows a once smooth concrete mantelpiece where a community rapidly developed to the point of supporting perennial grass, obviously

**Figure 13-2.** Community establishing on a concrete mantelpiece in a less brittle environment. Oregon.

without benefit of disturbance from fire, machinery, or herding animals. I have even seen succession start within a few weeks from algae on the sloping glass of a greenhouse window in the English countryside. In a more brittle environment the glass would remain bare for years, perhaps forever.

In environments leaning toward the nonbrittle end of the scale it is hard to stop the rise of succession. The coiled spring is powerful and hard to hold down. Clear a jungle or rest a pasture and watch how fast the community regains complexity. If a jungle has been cleared for farming and the soil structure and fertility destroyed or badly damaged, as past civilizations did in such areas, the return to full complexity as jungle may take decades, but the successional advance from bare to covered soil will proceed very fast indeed.

The successional process in nonbrittle environments is highly resilient to periodic drastic disturbances, be they weather or human induced. The coiled spring expands with such force. Without major disturbance, the community retains great stability while minor fluctuations in populations within it continue to occur.

## Community Dynamics and Management

Obviously an understanding of community dynamics opens all kinds of possibilities for better management of land, water, and all life. In the Holistic Management framework a dotted line surrounds both the ecosystem process Community Dynamics, and the tool Living Organisms to indicate that they are in fact the same thing. All life is successional and dynamic; therefore the future resource base described in a holistic context revolves around community dynamics. Our food comes from living organisms and so do most of our diseases. Our landscapes include living organisms. But to date we have too often managed living organisms in ignorance of the community as a whole. By continuing to do so we are endangering all higher life forms.

If you seek profit from livestock or wildlife, you may want a landscape that includes productive grassland. In one case that may mean advancing succession from desert scrub. In another it may mean preventing your pastures from returning to forest. Either way, certain plants, insects, predators, and other forms of life may become either allies or foes, depending on how you understand their place in succession.

If you wish to favor a species—game animal, plant, reptile, insect, or bird—then you must direct the successional movement of the community toward the optimum environment for that species, not by automatically intervening with some technological tool, but by applying whatever tools produce an environment in which that species thrives. Simply protecting the species, desirable as that might be, will not save it, although protection may be a necessary interim step.

If you start with a landscape that contains problem numbers of an undesirable species, the future landscape in your holistic context will specify a community that is less than ideal for that species and more suited to what you want to produce. To achieve this in practical terms you will need to know something of the basic biology of the species. What stage in the life cycle of the organism is its weakest point? What precise conditions does it require to survive at that weakest point? A little effort toward providing the appropriate conditions at that point will greatly influence whether the population of that species increases or decreases.

## *The Coiled Spring*

A successful approach to management should rest on the concept of the coiled spring. By *nature* succession moves upward, as does the coiled spring, toward greater stability and complexity. All prolonged downward shifts, or compressions in the spring, that I have experienced—and they are many on six continents—could be traced to human intervention in the process by the purposeful or accidental application of one or another of the tools listed in the tools row of the Holistic Management framework. The moment we reduce or cease that pressure on the spring it rebounds, and the community gradually regains its complexity and stability.

This is a *very* important principle as we currently spend billions of dollars annually worldwide on actions that compress the spring while chasing objectives that small advances in succession could produce. Attempts to eradicate a so-called pest plant or animal species with traps, guns, or poison generally symbolize our tendency to ignore the force of succession and deal only with its effects.

Some fluctuation of species is natural within a community, especially among short-lived organisms with high reproductive rates, which often characterize lower successional communities. Prolonged downward movement to lower successional levels of a whole community is unnatural, however, and, excepting the occasional natural catastrophe, virtually always betrays human intervention.

## Conclusion

If, in creating your holistic context you are attempting to describe the future land base surrounding your community in a healthy, productive state, you now have a better idea of how biological communities function and can begin to describe what is needed: mainly, communities that are rich in plant and animal species, or *biodiversity*. If you are a birdwatcher or like to fish, it may be important to you to describe, in general, the habitats that support a variety of birds and fish.

If you are attempting to describe a future landscape for land you manage, then, again, you will need to be more specific, but would likely describe all the environments you manage as complex biological communities in which many species thrive, both above- and belowground. Occasionally, you will need to describe communities that are less diverse in some areas to maintain greater diversity overall. Examples might include small areas you deliberately overgraze to create "lawns" for species that need open areas to thrive; or deliberately keeping a few areas bare by resting them (in a brittle environment) because they are important to birds for dust baths or breeding grounds, or to other animals for "socializing" or as "licks" when the soil is highly mineralized.

We have explored simply the relationships between water, soil structure, mineral availability, and communities of living organisms. The next chapter will explore the flow of energy that animates all these relationships.

# 14

## Energy Flow
### *The Flow of Fuel That Animates All Life*

*LIFE REQUIRES ENERGY.* And all living organisms, apart from a few dwelling near thermal springs deep in the ocean, depend on the ability of green plants to capture that energy from the sun and convert it to a form they can use. This chapter addresses both how that energy is employed as it moves through our ecosystem, and what can be done to increase its availability.

Because all life depends on the plant's ability, through photosynthesis, to convert sunlight energy into edible forms, so does every economy, every nation, and every civilization. The importance of this statement warrants a little reflection in today's world of electronic marvels, corporate takeovers, and online banking. Most economists, not to mention the rest of us, have all but lost sight of our dependence on the plant's ability to harness sunlight.

---

Because all life depends on the plant's ability, through photosynthesis, to convert sunlight energy into edible forms, so does every economy, every nation, and every civilization.

---

Photovoltaic, hydroelectric, wind, and tidal power sources also convert energy for practical use, but not directly into forms usable as food for life. Nor do geothermal and nuclear power plants produce food for living organisms. Fossil fuels, which represent solar energy converted by green plants long ago, can be used to produce a variety of products, but the fuels themselves are nonrenewable and inedible to all but the simplest bacterial life forms.

## The Energy Pyramid

Traditionally, the flow of sunlight to food for life is represented conceptually as an energy pyramid, as shown in figure 14-1. Of the sunlight striking land and water, some is reflected back immediately, some is absorbed as heat to be radiated back later. A very small portion is converted by green plant life into food for the plants' own growth and that of other organisms in the food chain. Thus green plants form the base, or Level 1, of the energy pyramid and support almost all other forms of life, including, of course, humans.

On land, all of Level 1 energy conversion is at or above the soil surface, where algae and the green parts of plants convert the energy. In aquatic environments it is slightly different. Around the shallow edges where observable plants can grow and protrude above the water's surface, energy conversion still takes place as it does on land. But over the rest of the area covered by a

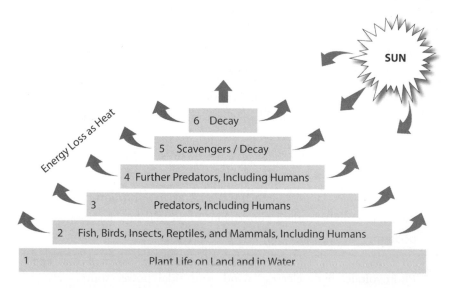

**Figure 14-1.** Basic energy pyramid. Sunlight energy must first be converted by the plants at the base of the pyramid (Level 1) before it can be used by other life forms. Therefore, expand the base of the pyramid to increase energy flow through the ecosystem.

body of water, energy is also converted by simpler plant life below the surface at depths sunlight can reach. Other than this difference, the concept of the energy pyramid is somewhat similar in all environments.

Level 2 represents the energy stored by animals that eat the plants of Level 1—fish, insects, birds, and mammals, including humans. It is smaller by the amount of energy expended as heat in the living processes of the feeders. This is no small amount. *Roughly ninety percent of the energy is lost as heat as you move from one level to the next.*

Level 3, the realm of predators, again including humans, that eat the eaters of Level 1, is smaller still for the same reason.

At Level 4 we again find humans and some other predators dining on fish and the other predators that fed on Level 2. Once more the living processes of the feeders has diminished the bulk of energy remaining in usable form by a further ninety percent.

By Level 5 humans drop out of the pyramid. Scavengers and organisms of decay reduce the bulk of stored energy yet further, and beyond that perhaps another level or two of decay organisms will use and convert to heat the last remaining useful energy. The last of the complicated organic molecules assembled by the original green plants will finally have been broken down, having made energy from sunlight available to many organisms.

At all levels, of course, a portion of the energy passes straight on to decay levels through feces and urine in the animals and through microorganisms that feed on the plants. Thus, in real life, the energy pyramid is not exact or tidy. And its form, due to the high loss of energy between levels, is much flatter than is shown in figure 14-1, which is vertically exaggerated so it fits on one page. However, the concept of ever-decreasing volume in usable energy holds throughout. *None of this energy is actually destroyed or used up; its form merely changes to heat that is no longer usable as food for life.*

Humanity's position in the pyramid covers three possible levels. One person can actually dine on all three in a good fish chowder, in which the potatoes represent Level 1, the piece of grain-fed salt pork Level 2, and the boiled cod Level 3.

Where high human populations exist on restricted land, people tend to feed directly off Level 1 rather than sacrifice the energy lost by first passing the food through the animals in Level 2. The animal protein they do consume probably comes from animals that do not compete with them for food. Fish or other animals may feed at the same level but off plants that humans cannot eat, or at higher levels from animal, including human, wastes. In some of the so-called developed countries the pyramid becomes messy when animals that would normally only feed at Level 1, the herbivores, are fed offal (waste meat, feathers, and blood) to supplement the protein in their rations, and are thus forced to dine on animals at their own level or higher.

The energy pyramid also extends belowground where the energy flow greatly affects the health of the other three ecosystem processes—water cycle, mineral cycle, and community dynamics. All three require a biologically active soil community, which in turn requires solar energy to be conveyed underground, mainly by plant roots or surface-feeding worms, termites, dung beetles, and others (fig. 14-2).

## The Energy Tetrahedron

The four key insights have enabled us to see that the old two-dimensional pyramid diagram does not reveal the possibility of much sophistication in the management of energy flow. Clearly, the broader the base of the triangle, which the face of the energy pyramid represents, the larger the whole structure, and the more energy available for use at every level. This two-dimensional view, however, suggests very few ways of broadening the base. On cropland, we have done it by increasing acreage, producing better-yielding crop strains, irrigating, planting two or more crops on the same land, and so on. On rangeland, we have attempted to do it through such technologies as brush clearing, grass reseeding, and so on. In forests, we have attempted to do it with fire and machinery to regulate timber stands. In aquatic environments, we have all but ignored the energy base by creating an artificial one in fish hatcheries and shrimp farms. In each case, especially in industrialized countries, we have accomplished this through heavy use of resources in a nonrenewable manner to fuel machinery and manufacture fertilizers and chemicals to kill unwanted life.

Even discounting the fact that many of these methods tend to damage natural water cycles, mineral cycles, and biological communities to the

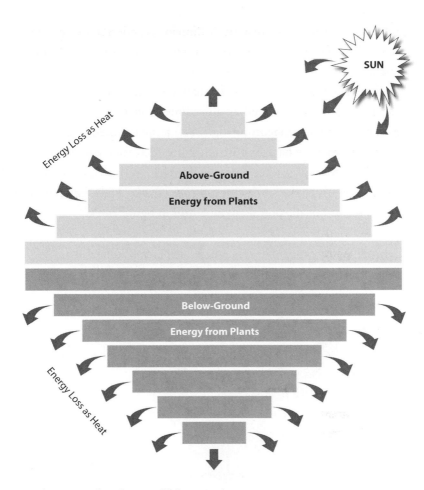

**Figure 14-2.** Energy flow above- and belowground.

extent that only increasing outside energy input can compensate, most present technology quickly reaches the point of energy debt: broadening the base requires more energy than it returns in captured sunlight. In countries where fossil fuel remains abundant and cheap, it is far too easy to ignore this fact. But in countries where inputs are costly it is already a question of life and death. It underlies much of the continuing American farm crisis manifested in ever fewer family farms and an increasing reliance on heavily subsidized corporate agriculture. In the case of the vast but minimally productive rangelands, it already prices technical solutions (if there were any) far out of reach.

The problems can only get worse until humanity understands and starts to manage energy flow as an integral part of community dynamics and water and mineral cycles.

Based on what we now know through the four key insights, we have come to view the energy pyramid as multidimensional, above and below the surface, that is, as two tetrahedrons (fig. 14-3) joined at their bases. In applying this concept we now have opportunities for increasing energy flow at the vital first level—the soil surface—greatly. Level 1, in this three-dimensional diagram, now has three sides, which I call *time*, *density*, and *area*, as indicated on the cross section of the double tetrahedron in figure 14-3.

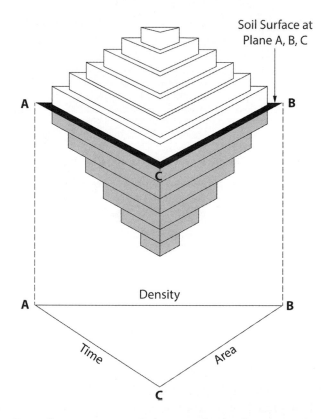

**Figure 14-3.** Energy flow seen as two tetrahedrons joined at their bases (A, B, C). We can increase energy stored in the above- and below-ground bases (shown in cross section) in three ways, by (1) increasing the time during which vegetation can grow and the rate at which it grows, (2) increasing the density of plants on a unit of ground, and (3) increasing the leaf area of individual plants.

---

**What Attention to Energy Flow Can Mean in Management**

*Once, while consulting on a ranch* struck by a serious drought, I found that my attempt to explain the importance of increasing energy flow appeared to bore the rancher to distraction. He wanted to discuss less theoretical questions like the hay he would have to buy and the animals he would have to sell. The problem really was a matter of energy flow, and so I had to demonstrate rather than explain why.

We went out onto his land, where I broke the capped soil with my fingers and pushed down some of the dead and oxidizing grass to cover the surface, much as any gardener would do to prepare a seedbed. I asked him if he thought an extra ounce (twenty-eight grams) of grass could grow on this small bit of "prepared" ground if he received only one inch (twenty-five millimeters) of rain. He said he could. Then I had him calculate the amount of hay he would have to buy to replace the one extra ounce of grass per square yard (or meter) that he could grow if he had his animals prepare the ground for him.

All thought of buying hay was dropped in favor of bunching four herds and replanning the grazing. The increase in energy flow that resulted saved him thousands of dollars in hay purchases and herd reduction.

---

On land, the right management can increase the volume of energy stored at Level 1 not only by increasing the *density* of standing vegetation on a unit of ground, but also by lengthening the *time* during which that vegetation can grow and increasing the *rate* at which it can grow, and by expanding the leaf *area* of individual plants to capture more energy. In aquatic environments we have still to learn how we might apply some of this thinking.

Clearly, the more we can extend any of the three sides of the base, the greater the volume of energy humans can harvest at levels 2, 3, and 4. On the other hand, shortening any single side decreases the volume of energy we can harvest all the way up, and the same effect ripples underground.

Perceiving the full ramifications of energy flow requires a thorough understanding by the land manager, so let's look at the three sides of the energy tetrahedron in more detail.

### Time (Duration and Rate of Growth)

The energy converted by plants while they are green and growing must support all life both above and below the surface throughout the year. The longer plants are growing, the more productive the community as a whole. We can increase the growing time—the *time* side of the base—by lengthening the growing season or by increasing the growth rate within a given time.

In practice, producing a better mineral cycle and water cycle and greater complexity in the biological community will extend growing time in both these ways. In the management of grasses, the growing time can also be used more efficiently if most of the grazed or cut plants are not taken down too far. The less taken from a plant during its active growth, the faster it regrows, as it has more leaf area with which to convert sunlight immediately.

In chapter 6, I described my frustrations in learning about time as a dimension in grazing on brittle grasslands and how we had to plan its manipulation to neither overgraze plants nor depress the performance of the animals. Later, under the management guideline of time (chap. 33) we will see how to actually prevent severe grazers from reducing the energy flow on grasslands, although they graze some plants severely within the first hour or so on the land as they select their diet.

Taking grassland as an example, naturally anything that creates better growing conditions through improvements in water and mineral cycles will allow plants to grow more rapidly. The role of the water cycle, however, deserves special attention. Under good management of the four ecosystem processes, moisture will remain available in the soil after falling temperature and diminishing daylight hours end the growing season. As chapter 11 explains, that enables plants to start growing the instant the new year restores those conditions. Given a noneffective water cycle, growth won't start until the first rain. Given a more effective water cycle, plants can also continue to grow for a much longer time during dry spells in the growing season because less moisture is lost through soil surface evaporation.

The opposite problem—too much water due to overirrigation or poor drainage and aeration—also occurs, cutting time out of the growing season. Every hour with adequate temperatures in which plants cannot grow at their best potential rate because of poor aeration means lost energy conversion on croplands or pastures under rainfall or irrigation, as I illustrated with the example in chapter 11 where, given the same daylight hours, I grew eight-pound cabbages in the well-aerated ridges and one-pound cabbages in the waterlogged furrows.

The degree of species complexity in a community also has a fundamental relationship to growing time. The most obvious examples are grasslands that support both cool-season and warm-season grasses. A complex, healthy grassland will include enough of both to ensure that some part of the plant community will be growing as long as any growth is possible. The annual grasslands produced by poor management, such as we find over most of California and in many parts of Africa and in southern Australia, illustrate this clearly. On these grasslands, where the dominant grasses are now annuals, there are prolonged good growth periods in every year in which there is no growth at all because the perennial grasses that could and should be converting energy have largely disappeared as a result of such practices as burning, overresting, and overgrazing of plants.

Color plate 7 shows some of the few remnant patches of perennial grassland that still survive on one California ranch. Even as they continue to grow actively and convert energy to fuel the rest of the life in the community, the annual grasses that have come to dominate the surrounding land have long since finished their cycle, dried off, and died. A change in management, using the knowledge now available, could bring back the perennials that once covered the land and restore months of productivity to the area.

The same principles for extending time apply equally well on croplands and offer even more management opportunities. Figure 14-4 shows the poor growth of plants in an overirrigated field in New Mexico alongside excellent growth in the same plants on the edge of the field where aeration is accidentally good. The plants on the edge reached cutting stage a month before the main crop.

**Figure 14-4.** Sparse growth of grass in an overirrigated hay field but with excellent growth in the foreground on the accidentally well-drained edge of the field. New Mexico.

Many farmers are aware of the difference in time efficiency rendered by applying nitrogen fertilizer in periodic light dressings as opposed to a single large dose. Good selection of heat- and cold-tolerant crops and planting dates can also extend effective growing seasons and thus the time base of the energy tetrahedron. Some of the same principles could well apply in aquatic environments.

## Density (of Plants)

The density side of the base refers to the number of plants growing on each square meter (or yard) of land. Ten plants growing on the average square meter of ground can probably convert more solar energy than three. Farmers have long recognized that plant spacing, or density, can greatly affect energy flow in their fields and have planted accordingly—striving for the density that produces the highest yields.

In environments that lie closer to the nonbrittle end of the scale, the spacing between plants growing in the wild, which is close, is a reflection of climate. Management can affect it, but plant density is naturally high. By contrast, the more brittle the environment, the more plant density is affected by the disturbance, or lack of disturbance, from large animals.

Figures 14-5, 14-6, and 14-7, photos taken on a ranch in west Texas, show the difference disturbance by animals can make in a brittle environment. Figure 14-5, taken in 1982, shows the bare spaces on a piece of land within the ranch boundaries that was fenced off and thus undisturbed for some years. A series of fixed-point photos, including figure 14-6 (taken in 1985), reveal that these plant spacings have widened continually as old plants stagnated from overrest and new seedlings failed to establish on the bare, capped soil. Figure 14-7, taken in 1984, shows nearby land where the same rancher has planned the grazings and periodically subjected the land to disturbance in the form of

**Figure 14-5.** A long-rested site fenced off from grazing animals showing sparse grass and some bare ground between plants, 1982. Texas.

**Figure 14-6.** Same view as shown in figure 14-5 but taken three years later in 1985. Grass has become more sparse, and bare ground has increased under continued rest. Texas.

**Figure 14-7.** View on the same ranch, close to the site shown in figures 14-5 and 14-6. Here, where very heavy animal impact has been applied periodically close to the waterpoint (on left), the grass has thickened up and the bare patches have disappeared, 1984. Texas.

high animal impact. Its proximity to a livestock watering point intensifies the animal impact. As you can see, plant densities show a very positive correlation to this periodic disturbance.

Our measurements confirmed that where animal impact was highest, near the water, plant spacing was closest. Plant spacing widened as the distance from the water increased and the same periodic impact was lighter. Generally, more energy can be converted where more plants grow.

Traditionally people have believed that such plant spacings were a function of climate, which is correct toward the nonbrittle end of the scale, and thus lay beyond their control. However, in more brittle environments, management can be crucial, as these photos show. Chapter 22 describes animal impact and its management in more detail. Suffice it to say here that animal impact does provide a means of increasing the density of plants per unit area of land, particularly in environments that are brittle to some degree, as is the case for roughly two-thirds of Earth's landmass.

In aquatic environments one needs to be wary of promoting too high a plant population when adding nutrients to water, as we do every day when we dispose of detergents and other wastes. When these nutrients arrive in large quantities, algae populations can explode, consuming so much oxygen in the process that little or none is left to sustain other life forms.

## Area (of Leaf)

Leaf area matters because a very dense stand of narrow-leafed plants may trap less energy than a moderately dense stand of broader-leafed plants. So to expand the area side of the base, you would have to increase the number of broad-leafed plants.

Plants adapt themselves in three major ways to suit different growing conditions. *Hydrophytic* (wet-environment) plants thrive in soggy, poorly aerated ground. *Mesophytic* (middle-environment) plants grow best when air and water are balanced in the soil. *Xerophytic* (dry-environment) plants survive where water is scarce, though aeration may be good.

In some ways the wet- and dry-type plants resemble each other more than they do the middle-type plants. Both often have cuticles or fairly impervious skins over their leaves (like cactus plants and water lilies), and adaptations surrounding their breathing pores that result in them passing little water through their systems. Both may have narrow leaves or leaf-stems (green stems that convert energy), which reduce the area exposed to sunlight and thus the area transpiring water. Some dry-type plants may have broader leaves, but they are often tightly rolled for the same reason.

The wet-type plants, such as water lilies, bulrushes, cattails, many sedges, and some grasses are generally found in sites that remain wet due to high rainfall or to poor drainage. They can also be found in sites that, in terms of rainfall and soil depth, should be occupied by the middle-type plants, but where severe soil capping has caused poor aeration. Likewise, dry-type plants, such as cacti, euphorbias, and some grasses, are generally found in sites where moisture is minimal. However, they can also be found in sites where rainfall is high but quickly shed or evaporated from a capped surface. Among the dry-type plants, many of the perennial grasses often stand out clearly in the dormant season, when they dry off to a white or very pale color. Both wet- and dry-type plants tend to grow slowly and thus store a limited amount of solar energy in a given time.

The middle-type plants are very different. They produce generally open, flat, and broad leaves that curl only when wilting under moisture stress. They do not have thick protective skins or well-developed mechanisms to shut off breathing pores. They also tend to grow rapidly when moisture and temperature are favorable. In contrast to their cousins in the dry category, many of the middle perennial grasses cure to red or gold in the dormant season, which happens also to reflect more nutritious dry forage for animals.

Oldtimers in South Africa made the distinction between *witveld* and *rooiveld* (white range and red range), the latter being far more productive. In North America this difference can also be seen along many western highways. Protected from overgrazing and kept alive by severe defoliation as a result of periodic mowing, these roadside grasses when dormant often have a definite reddish or deep gold tinge. Yet, just over the fence, where plants are overgrazed and soils insufficiently disturbed, the grasses are pale or dead white in color. The sharp change from red or gold to white along the fence line shows up for miles.

Obviously, to increase the area side of the energy tetrahedron on land in more brittle environments you need to shift the community to the middle

---

### Using Technology to Increase Energy Flow

*We can also increase energy flow* through direct use of technology in many forms—machinery, drainage, irrigation, chemicals, and genetic engineering, to name a few examples. However, such direct intervention in one of the fundamental processes can be dangerous because we are dealing with complex interrelationships, of which we understand little. We should intervene with technology only in ways that allow for simultaneous development, and never damage, of the water cycle, mineral cycle, and biological communities.

No one ecosystem process can safely be bolstered at the expense of the others. Enhancing energy flow through heavy inputs of fossil fuel products, which damage biological communities and water and mineral cycles, has been the cornerstone of North American agriculture, but we are paying a heavy price for our ignorance: increasingly severe floods; food and water riddled with life-threatening chemicals; rising rates of reproductive disorders, cancer, and other diseases; accelerated erosion destroying millions of years of biological capital; millions in public funds spent yearly to kill insects and other increasingly resistant pests. Ultimately thousands of farmers and ranchers leave the land, followed by once-healthy small businesses and rural communities.

---

plant that spreads broad leaves to the sun and grows fast. As in the case of plant spacings, most people have always felt that plant type depended on soil type and lay beyond their control. Occasionally something like an impervious layer of clay or rock below the surface can indeed kill any chance for the middle plant, but more often a poor water-to-air balance resulting from sealed or capped soil and poor water cycles is the problem, and management can change that.

In less brittle environments where soil capping is far less a problem, the maintenance of more broad-leafed plants can often be improved by drainage. In Zimbabwe I have seen, in the course of twelve years, a patch of eighty-four percent *Loudecia* grasses, known for their fibrousness, poor forage quality, and association with badly drained ground, change to an eighty percent mix of productive middle grasses associated with good drainage. We had used animal impact to break up the hard soil capping and create closer plant spacings, and planned the grazings to prevent overgrazing. The change in species was an unexpected by-product and led to more observations of the same kind.

In addition to causing grass plants in more brittle environments to grow closer together, animal impact and severe grazing (without overgrazing) cause many species to produce more leaves and fewer stems, which in turn increases the flow of available energy to animals and humans. The same holds true for grasslands and pastures in less brittle environments.

## Conclusion

In terms of your holistic context, if you are attempting to describe the land surrounding your community as it must be in the future, describe what it would look like if energy flow were high: soil would be covered in vegetation, plants would stay green and continue to grow much longer than they do now, and there would be a variety of them. Wildlife would reap the bounty and be more plentiful as a result. In essence, if water cycles are effective, minerals are cycling rapidly, and biodiversity is high, then energy flow would tend to be maximized.

The future landscape that land managers describe will generally require the highest energy flow possible, whether grassland, cropland, or forest be their concern. In most cropland situations, we should strive to manage for an effective water cycle, a good mineral cycle, a highly complex biological community (above- and belowground), and thus a high and sustainable energy flow. We will plan actions that maximize the time side of the energy tetrahedron's base by ensuring good daily growth rates and lengthening the season through polyculture cropping or at least two or more crops per year whenever possible. We will maximize density by planting with close spacings. We will maximize the area of leaf that is open and exposed to sunlight by creating good drainage, good crumb structure, and abundant organic matter in the soil and providing adequate soil cover. In the future, we will be better able to manage our cropland soils and increase the energy flow to the microorganisms that populate them by keeping soils covered year-round and by incorporating animals into cropping strategies.

In most forest situations, we should strive to maximize energy flow by improving water and mineral cycles, and by increasing the diversity of plant and animal species, particularly in forests that have been simplified through industrial-style forestry practices. In aquatic environments, we will maximize energy flow by reducing pollution and sustaining highly complex biological communities, and by ensuring that, on the land that surrounds them and that catches much of the water that feeds them, water and mineral cycles and biological communities are healthy.

In most grassland situations, we will increase energy flow by manipulating the tools of grazing and animal impact, with both livestock and wildlife, to produce and maintain maximum growing time, plant density, and leaf area. Innovative farmers are also drill-planting a grain crop into grazed grasslands, boosting energy flow even more through what they call pasture cropping (see chap. 35). The amount of energy we might have to buy from other producers on other land to supplement what our own land does not provide would be the measure of success or failure.

The chapters in the next section describe in detail the tools we can use to alter any one of the four fundamental processes. An understanding of how each tool affects their functioning is essential to the land manager and will influence many of the decisions he or she makes. The non–land manager also utilizes some of these tools and, although not doing so to influence the ecosystem processes directly, does so indirectly. Thus we need to learn more about these tools so we can plan to use them wisely.

# PART 5

# THE TOOLS WE USE TO MANAGE OUR ECOSYSTEM

# Holistic Management Framework

**WHOLE UNDER MANAGEMENT**

Decision Makers — Resource Base — Money

**HOLISTIC CONTEXT**

(Statement of Purpose) — Quality of Life — Future Resource Base

**ECOSYSTEM PROCESSES**

Water Cycle — Mineral Cycle — Community Dynamics — Energy Flow

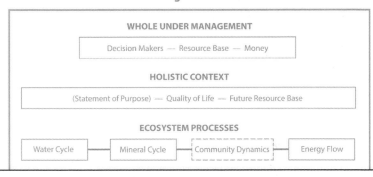

## ECOSYSTEM MANAGEMENT TOOLS

Human Creativity

Technology  Fire  Rest  Living Organisms
· Animal Impact
· Grazing

Money & Labor

**ACTIONS & DECISION MAKING**

Objectives, Goals, Tactics, Strategies, Policies
Customary Selection Criteria (past experience, expert advice, research, etc.)

**CONTEXT CHECKS**

| Cause & Effect | Weak Link · Social · Biological · Financial | Marginal Reaction | Gross Profit Analysis | Energy/ Money Source & Use | Sustainability | Gut Feel |

**MANAGEMENT GUIDELINES**

| Time | Stock Density & Herd Effect | Cropping | Burning | Population Management |

**PROCEDURES & PROCESSES**

| Holistic Financial Planning | Holistic Land Planning | Holistic Planned Grazing | Holistic Policy Development | Research Orientation |

**FEEDBACK LOOP**

Plan
(Assume Wrong)
Replan              Monitor
Control

# 15

## Introduction
### *From Stone-Age Spears to Genetic Engineering*

*BELOW THE ROW OF ECOSYSTEM PROCESSES* in the Holistic Management framework stands the row of tools, but the word here gets a broad definition. It includes everything that gives humans the ability, which most organisms lack, to significantly alter our ecosystem in order to enhance or sustain our lives. In fact, every action you take will involve the use of some tool, either directly or indirectly, even if the whole you are managing is limited to your own personal life.

All tools available to humans, from Stone Age spears to computers and genetic engineering, fall under one or another of the headings in the tools row. Be you politician, economist, engineer, farmer, herder, housekeeper, gardener, widget maker, or whatever, you will not find a tool currently known to humanity that is not included within these six general headings.

Human creativity as well as money and labor bracket the other four tool headings in the framework because they come into play in the use of technology, fire, rest, and living organisms, and nothing can be done with them other than through one of the tools inside the brackets. We list money and labor together because the once simple combination of labor, creativity, and resources frequently operates through the agency of money. The capitalist's investments, the labor of a commune, or the unpaid children on a family farm all function according to similar principles to be covered in later chapters.

Of the four tools listed between the brackets, technology alone is the prime tool employed in urban or industrial businesses and professions and by most households, few of which intend to modify our ecosystem through the use of technology, either directly or deliberately, but often do nonetheless. Chapter 18 elaborates on this theme.

If you are not managing land, then do you need to concern yourself with the chapters on fire, rest, and living organisms? Absolutely. Read these chapters (19 through 23), no matter what your walk in life, because the information in them will be critical to sustaining your business and your community. If you support nonprofit organizations and their programs to enhance community vitality, mitigate climate change, save threatened wildlife populations, alleviate hunger, or assist environmental refugees, you should know more about how these remaining tools can be used to eliminate or exacerbate such problems. Armed with this knowledge you can begin to demand change where change is needed. Organizations, including governments, cannot and will not change their policies and practices until public pressure compels them to.

In any land management situation, *fire* and *rest* are included with *technology* as the standard tools for modifying our ecosystem. However, none of these tools can begin to reverse the desertification occurring in environments that lean toward the more brittle end of the scale—the majority of Earth's landmass.

A typical example is the civil engineer commissioned to stabilize an eroding catchment in order to save an important dam or irrigation project from silting up. The average engineer's tool kit contains only technology. Thus the average engineer may contour all the slopes, build silt traps in all the valleys, or try to channel rivers to no avail because, in the more brittle environments, where most dams and irrigation projects are developed, that can never constitute more than a Band-Aid on a dying patient. The catchment in question has, in all likelihood, been subjected to the influence of three tools whose damaging influences the engineer was unaware of—rest, periodic fire, and living organisms (as overgrazing by livestock)—all of which tend to expose soil and increase the amount of water that runs off it. Inevitably, the dam will silt up, as I, the son of a civil engineer, have seen repeatedly and as past civilizations have illustrated abundantly.

Were engineers to expand the number of tools in their kits, they would greatly increase the possibility of success. The more brittle the environment, the more the need for some form of periodic disturbance over millions of hectares of the catchment, and for assistance in decomposing billions of tons of plant material every year without the use of fire. There is no technology, and likely never will be, that can do this in a more environmentally friendly way than living organisms, in the form of animal impact and grazing provided by herding animals. And as they perform these tasks they remove the cause of the erosion on the catchment that leads to the silting of the dam. Fire and rest cannot do this and generally only make matters worse. If, on the other hand, the catchment of a rapidly silting dam lies in a nonbrittle environment, herding animals are not required. The land only needs to be rested to cover over with vegetation that keeps soil in place.

Sociologists, economists, environmentalists, and politicians can similarly move beyond the tools traditionally available to them within their professions. But this only becomes possible when they begin to work together with others outside their professions—something more and more people are doing as they find that very few problems can be solved within the confines of individual disciplines. These integrated teams of experts can overcome major hurdles by sharing knowledge of the various tools available within their professions, but success will still elude them, as I emphasized in chapter 3, if they fail to see the whole first and manage within a holistic context.

The tools of the future might well break new ground. For instance, a young cousin of mine, after watching a television program where Israel's Uri Geller bent iron with his "mind" and a light stroke of his fingers, picked up his grandmother's steel nail file and did the same thing himself. He was of course too young to know that "you can't do that." For most of us, such mind-over-matter phenomena fall completely outside the tool chest, and just reading about it may make us wince. Nevertheless, that is just the attitude we must avoid at all costs, for who knows what tools we may harness for use in the future?

When managing holistically, all tools are equal and no tool is good or bad. Make no judgments on any tool or action outside the context of the whole under management. Only when the holistic context and the position on the brittleness scale are known, together with the many other factors that bear on the situation, is any tool finally judged suitable or unsuitable in that particular situation at that time. Fire, for instance, is good when it keeps my hands warm on a cold morning, but it is bad when it is burning down my house.

In Chapters 19 through 23 we'll examine how fire, rest, and living organisms (both in general and specifically as animal impact and grazing), *tend* to affect each of the four ecosystem processes relative to the brittleness scale. Analyzing the impact of a single tool when many other processes and other tools may be at work at the same time would appear to be an impossible task. Where a cow places her hoof today, for instance, begins a chain of reactions that ensures that spot will never be exactly the same again. A solution to this dilemma rests on the hypothesis that the tendencies of the tools, when chosen and applied in a certain way, function in the ecosystem like the ripple patterns of pebbles thrown into a still pool.

It is a fact of physics that, even though multiple ripples appear to create disorganized chaos on the pool's surface, the orderly ripples produced by pebbles thrown individually still exist. Each pebble does in fact impart a predictable *tendency*. If we throw in two pebbles of very different sizes, we can, in fact, see what each pebble's ripples tend to do. A larger pebble may overcome a smaller one's ripples but the smaller one's ripples will still have a visible effect.

In considering which tools to apply, either singly or in combination, we think of them like those pebbles and ask ourselves, Will it start a ripple that pushes the community toward more complexity? How will its ripples tend to change water and mineral cycles and energy flow? Though there may be countervailing ripples that diminish and partially obscure the force of the ones we start, it is not likely that the power of two ripples moving in the same general direction will combine into an entirely new and opposite force.

Once these tendencies are acknowledged, a careful consideration of the context checks and management guidelines covered in later chapters helps us then judge which tools are best to apply now. Even then we always assume we could be wrong when managing the environment; thus we monitor to ensure that the tools selected achieve what we intend to achieve.

# 16

## Money and Labor
### *At Least One of These Tools Is Always Required*

ONCE UPON A TIME people supported themselves by applying creativity and labor, or brains and brawn, directly to the raw resources of our ecosystem. Many societies still do this, as do farm and ranch families who don't pay for the labor of family members. Both then and now we used our creativity to obtain the maximum effect with as little labor as possible.

Because money and labor are often linked (e.g., cash can be exchanged for labor) and because neither can be used other than through another tool, we group them together in the Holistic Management framework. Ideally, our natural tendency to economize on labor should apply equally to money. But money is a more complicated matter, and this isn't often the case.

Some years ago, I spoke to a group of agricultural economists and asked them to define wealth. They grappled over that question and in the end only defined wealth as money. Well, once upon a time money probably did perfectly represent wealth, but that was a long time ago, and the fact that many experts still believe it is a disturbing aspect of modern times.

Money has been the oil that has kept the wheels of society turning and allowed the complexity of our present civilization to develop, but credit, the centralized creation of money, interest, and particularly compound interest, have seriously destabilized the relationship between money and the goods and services, or wealth, it originally represented.

## Wealth versus Money

In my lifetime alone, the distinction between wealth and money has probably become more blurred than at any time in history. High interest was usurious when I was a child; now that's seen as quaintly old-fashioned. Major banks move headquarters to states with more lenient usury laws and still retain customer confidence. Where it was once unacceptable for lenders to advertise or engage in aggressive promotion, it is now commonplace. Money itself has become a commodity (like grain or oil) that earns money and can be traded internationally. The use of credit cards and the electronic speedup of monetary transactions has blurred the distinction further.

Advancing technology has provided ever more creative ways to make money rapidly available, from online banking to cell phone banking that enables pastoralists living remotely to conduct transactions in the capital city. But some bankers have also found clever ways to create wealth that is entirely artificial, with destructive consequences for those taken in by the ruse, such as the one million Americans who lost their homes in the subprime mortgage crisis of 2008.

Today fortunes can be made overnight on the international trade in money, or currency speculation, where real goods or services play no part at all. The amounts of money involved in this trade in any one day—5.3 trillion dollars per day in 2013, or 200 billion dollars per hour—exceed that of most nations' annual budgets.[1]

It becomes easy as one stands in a plush, air-conditioned bank, humming with electronic activity, to lose sight of the underlying reality of wealth in the financial resources we manage. Whatever the source of the money—real goods and services, including information, or corporate takeover—it all looks the same as we stare at the dollar bills or the computerized spreadsheet.

The urban life most people lead seldom reflects any distinction between money and wealth. Those managing global financial markets are unaware of agriculture's role in sustaining economies, and of the environmental destruction created through unsound practices, although there is a small movement of former managers attempting to change this. In the meantime, many farmers and ranchers are learning to distinguish between the types of money, or wealth, available and are pursuing strategies to create an agriculture our planet can sustain.

Most of the holistic contexts we create, whether they apply to us as individuals or to a nation as a whole—should involve a sustainable source of wealth, because the reality (wealth) is more vital than the symbol (money) in the long haul. However, to manage wealth as it has become today, where a dollar (pound, euro, or rupee), regardless of source, can purchase the same things as any other dollar, we must first understand the three most basic sources of wealth the dollar represents: mineral dollars, paper dollars, and solar dollars.

## Mineral Dollars

Money can be derived from a combination of human creativity, labor, and natural resources that are mined, used once, or sometimes recycled. I call these mineral dollars. Coal, oil, gas, gold, and other minerals fall into this category, hence the name.

Depending on how they are used, other natural resources can fall into this category, although they should not. Soils in modern industrialized agriculture are being mined until exhausted or eroded away; thus dollars generated from them are mineral dollars. Ocean life, too, is being mined rather than harvested sustainably, and money gained from such practices would be mineral dollars. If, after use, water is too polluted for reuse, any production from it would yield mineral dollars.

## Paper Dollars

Many of us acquire money through human creativity and labor alone. I refer to this source as paper dollars. The beauty of such income is that it consumes no other resources. All we have to do is apply our creativity in thousands of different ways to the many avenues open for investment: speculation in futures markets, stocks, bonds, corporate takeovers, and so on.

On the other hand, various services—many of them essential—also fall into this category. Lawyers, consultants, educators, accountants, civil servants, armies, and so on, do not actually make anything or produce the kind of elemental wealth that supports life. But they do enhance and protect that wealth, and life without them would be inconceivable today. Professional speakers, entertainers, athletes, and many others also reap paper dollars for the services they provide, and although they produce no tangible goods, they make life genuinely more pleasurable.

In some cases, the money generated in this category has the fascinating characteristic of apparently instant and unlimited accessibility. We can make fortunes in a day with nothing but our creativity and minimal effort in the stock market or overnight currency trading. On the other hand, this money can vanish as quickly as it appeared. Paper dollars are backed by confidence in the government and the banking system, and when that confidence is lost, paper dollars can lose their value overnight. I have lived through such a calamity and still possess several 100-trillion-dollar bank notes barely worth the paper they were printed on.

## Solar Dollars

Third, we can generate income from human creativity, labor, and such constant sources of energy as geothermal heat, wind, tides, wave action, falling water, and, most of all, the sun. I call this last class of money solar dollars. Such energy as a source of wealth is noncyclical, but it is apparently inexhaustible. A characteristic of wealth derived from this combination is that it tends not to damage our life support system or to endanger humankind as far as we know.

A further characteristic is that wealth in this category is the only kind that can actually feed people. Unfortunately, this requires the conversion of solar energy through plants that themselves depend on water and biologically active soils. (Since more than ninety-nine percent of our food comes from the land and less than one percent from marine and aquatic ecosystems,[2] maintaining and augmenting the world food supply depends on the productivity and quality of our soils.) *Only when plants grow on regenerating soils would the money earned from timber, crop, or forage qualify as solar dollars.*

Keeping the three categories of money in mind enables us to see the extent to which failure to do so governs our society now. Economists daily engage in juggling paper dollars, sublimely unaware of what those dollars actually represent in terms of real wealth. On the advice of these same economists, via business advisers, agricultural extension officers, or salespeople, farmers diligently pursue mineral dollars while consumptively mining their soils to do so. Some seventy-five billion tons of soil erode from the world's agricultural lands each year.[3] In other words, for each half ton of food we produce (the amount required to feed one person for a year) we are producing more than ten tons of eroding soil.

## Conclusion

Sooner or later the underlying basis of a nation's, or an individual's, quality of life asserts its nature. A country rolling in oil revenue today must ask itself to what end the cash flows in. The nation that thrives by burning the oil must ask what that does to the greater ecosystem that sustains us all. What will happen to the nation's long-term quality of life and productive base? If the wealth from oil goes to accumulating paper dollars and to support unproductive legions of bureaucrats, accountants, soldiers, and others who consume and keep transactions going but do not enhance the nation's ability to increase or maintain its biological capital, is that sound? In other words, is it sound if the nation's resource base deteriorates while it wallows in paper dollars, or produces yet further mineral dollars from its agriculture? Shouldn't some of the dollars from nonrenewable mineral wealth go to develop ways to reap solar and mineral dollars on a sustainable basis?

One of the seven checks used to ensure actions are in context (see part 6) asks you to consider the source of the money and how it is to be used. If you are a land manager this is a reminder to consider seriously the source of dollars invested or reinvested in the business, and how that relates to what you have expressed in your holistic context. All forms of money will figure in your plans, but the wealth that ultimately will sustain your business, community, and nation is that derived from solar dollars produced from plants growing on regenerating soils. There is nothing inherently wrong with either mineral or paper dollars, but being more aware of their limitations is essential, particularly for those involved in directing global financial markets who appear to remain unconcerned by the staggering loss of biological capital it took to grow those markets. Only solar dollars can produce the biological capital that can sustain humanity in the long run.

Whatever forms of wealth you control, your success depends on how creatively you use them. Thus we look at *human creativity* next.

# 17

## Human Creativity
### *Key to Using All Tools Effectively*

WHEN HOLISTIC MANAGEMENT WAS IN ITS INFANCY we needed a name for the reasoning and judgment attendant on any use of labor and resources. I first used *brainpower* but soon realized that word didn't cover the ground. A person who can mentally add and subtract six-digit numbers may have great brainpower but lack creativity, common sense, and humanity.

Just as notable in Holistic Management is that an idea that enables one person to attain maximum effect from his or her labor and money may not work for another, or even for that person in the following year. Thus *every situation requires management that must be an original product of human imagination, and even that must evolve as the situation changes.* Creativity, not brainpower, is the crucial element, and it is needed constantly.

In *Meeting the Expectations of the Land,* farmer-essayist Wendell Berry noted that a whole generation of farmers has been brought up to use their heads to advertise others' products (on their caps) and to phone the extension service to be told what to do. The extension service in turn employs a generation of advisers whose university education trained them in how to *do* rather than how to *think.* This problem is not limited to farming; it has become increasingly common in all fields.

The context checks listed in the Holistic Management framework have been developed to help us assess the possible consequences, relative to our holistic context, of using any available tool. As many of the consequences are not quantifiable, this is not a task for a computer. Any human responsible for management will have certain feelings about it and will encounter the feelings of others. Love, fear, hopes, dreams, and interpersonal conflicts very much affect any management situation in ways no computer yet devised can understand.

The Holistic Management framework itself, however, is much like a bit of software to help organize your thinking and planning. Its successful application depends entirely on your ability to think and be creative. Fortunately, creativity is not simply a genetic endowment. It depends on your mental, emotional, and physical health; on your environment; and most of all on how deeply you desire the quality of life described in your holistic context.

## Creating an Environment That Nurtures Creativity

Although individuals can and should ideally manage their lives holistically, natural resources are rarely managed by one individual. More often a family, a company, a tribe, or a nation has this responsibility. The levels of emotional maturity and relationships between the people involved can be located at all points along a continuum from very stressful to very caring, and the state of these relationships has a bearing on the creativity of each individual and the group as a whole. However, it's the person at the top (the owner, manager, chief, etc.) who sets the tone. That person's beliefs and behavior have the greatest impact on the creativity of the group.

The most vital responsibility any manager ever has is creating an environment that nurtures creativity. People tend to protect and hide their creativity to avoid making mistakes or looking bad in front of leaders and peers. Managers can help overcome this by providing encouragement and recognition to their coworkers, and by demonstrating how to give and receive feedback in a way that moves the team and its efforts forward.[1,2] Creativity of the group, no matter what its size, tends to be greatest when the leader's everyday actions display trust and confidence in members of the group, when the work is meaningful, when all feel free to express ideas and to be creative, and when all feel valued.

Such a spirit cannot be faked. The person at the top really must value his or her coworkers as human beings and not mere tools for making profit. Corporations that saw profits rise when they treated employees as their prime resource saw them tumble when employees sensed they were merely being manipulated.

At present the majority of farmers and ranchers, in my experience, take human creativity for granted and do not see it as something that must develop through the family and social and work environment. We rarely see it as a tool that governs our success or failure. The male head of a small farming or ranching family leaving the land in America today, as thousands are doing, no doubt blames the banks, the interest rates, the prices, the government, the weather, and looks to society to help him and his family. It is difficult indeed for him to see that, although he labored long and hard with his hands, if he had also used his head and the creativity of family members and friends it might have helped him survive. (I've used a gender-biased example here because the male head of household dominated decision making in many of the financially strapped families I assisted over the years, discouraging the creativity of other family members.)

## Poor Time Management Stifles Creativity

Subconscious worries and stresses all too often completely sap our creative energies. Unfortunately, this is a common and subtle factor that affects many of us, but fortunately it is entirely within our control. Many of our stresses are allied to crisis management—where one crisis after another hits and drives your management, leading to ill-considered decisions that perpetuate the crises. Holistic Management can help circumvent any tendency toward crisis management and reduce the stress associated with it—as new practitioners commonly report.

Poor time management also leads to crisis management and a stifling of creativity. It has certainly proved to be a serious obstacle in many of the situations where I have consulted. Typically, a person would call on me to help stave off impending bankruptcy. Within a very short time I would realize that my talk of planning for the future guided by a holistic context was meaningless, because my client's mind was on the tractor that needed a new clutch, the pickup that needed new wheel bearings, the boundary fence that was down, or the dam that had burst the night before. To get him out of the crisis he needed all the creativity he could muster to carefully plan the next crucial months. But his worries prevented him from planning his time or anything else.

Over many years I had perfected a good system for managing my time that had served me well and that I shared with clients in these sorts of cases. I would assist them plotting ahead and allocating their time for all the concerns that had piled up and were believed to be of equal importance, so much so that nothing at all was being done because everything, they thought, had to be done today. Once we established priorities, including family time and holidays, and allocated more than enough time over the next few months to complete them, there would always be many days left with nothing to do! The clients would be immensely relieved. But this plan inevitably was not followed because they lacked the self-discipline, or so I thought. On reflection, I realized it was deeper than that. What was lacking was an established routine, or habit, and this could be learned.

Whatever time management system one uses, the keys to its success are *habit* and *trust*. A habitual procedure must be established whereby all the ideas that come to mind and all the commitments made are immediately recorded in one place, rather than on scattered scraps of paper, so they can later be retrieved (and understood) and acted on. Once this habit is formed you cease to worry about commitments or ideas you might forget. You begin to *trust* the process and let go of your subconscious, or conscious, worries. This then creates the energy that enables creative thought.

## A Change in Mindset

Although creative ideas can and do emerge at any time, they are especially needed when preparing the annual financial plan. This is when you will be brainstorming new enterprises and developing ways to cut costs and boost income, and it is where creativity is needed most. This is also where I have seen it save many a business.

Often I've worked with ranchers who see themselves as "cow-calf producers." While their knowledge and skills in running this type of breeding operation are impressive, they have struggled to stay afloat financially. When a family member, often from the next generation, suggests that the land they manage could be put to additional productive uses, it pays to listen. In one case, just such a family had young sons who realized the potential income they could generate from the trout stream they had fished all their lives, and were allowed to make the case for creating a business doing something they loved in addition to raising cattle. Their guided fly fishing and ranch stays grew into a successful enterprise that complemented the cattle operation and kept the family on the land. This is not an isolated example. The possibilities become endless with nothing but a change in mindset and the creative possibilities this opens up.

## Conclusion

Most of us complain that we just don't have enough time for creative thought, or in fact to do all that we want to do, and we marvel at those who seem to find it. Often we feel others have less to do while in fact they achieve far more than we do. In truth, every person in the world has exactly the same amount of time. How we manage it makes the difference in the quality of our lives and in what we achieve through our creativity. But there's more to it than that. When we want something badly enough we manage our time accordingly, and our creativity knows no bounds. If you truly want the quality of life described in your holistic context, the creativity will flow.

Think of creativity as the key to using money, labor, and the other tools of management successfully and as being the *only* tool that can produce the vision and goals you aspire to and that can be used to plan their achievement.

# 18

## Technology
### *The First- and Most-Used Tool*

*HOW MANY TIMES HAVE WE HEARD IT SAID THAT TECHNOLOGY,* the hallmark of modern humans, holds the key to the future? No doubt this belief has been reflected for many years, first around campfires and later in boardrooms and cabinet meetings. Technology will feed us better. It will provide lightning-fast transportation and communications. It will heal our wounds and cure our diseases. As a tool for modifying and controlling an environment, we have not seen the beginning of its potential. On the other hand, it is the tool we look to first to address the problems wrought by technology, including climate change.

The twentieth century provided a constant stream of wonders that strengthened our faith in technology, but only recently have we had to entertain the possibility that technology does more than simply produce better and better appliances, artificial organs, weapons, spacecraft, and entertainment. It now forces humanity to make choices not imagined since the beginning of time.

We now know that our pursuit of technological triumph can have dire consequences, particularly for the ecosystem that sustains us all. In fact, renowned astrophysicist Stephen Hawking, expressing concerns over population growth, resource depletion, and environmental damage, warned in 2010 that if the human race doesn't colonize space within the next two centuries humans will become extinct.[1]

Far too much of the technology available today is used to address the problem at hand without thought of larger implications. Many of the products we use daily—detergents, dyes, automobiles, pesticides, and so on—can affect the environment in ways we never anticipated. The effects can be delayed by days, months, or years, and may express themselves far from the site of application where an innocent public cannot connect them to their source.

Most of our most hazardous inventions have existed for less than a century. Humankind did remarkably well for many thousands of years using simpler technology for many of the same tasks, but such is the nature of human advancement that in many areas there is no going back. We cannot return to rudimentary living where a much smaller human population exists on subsistence agriculture, nor can we abandon our cities. Yet recognition that going forward will demand wisdom and humility is a breakthrough in thinking more significant than the notion of space travel.

## Technology and the Ecological Quick Fix

Much technology subconsciously stems from our desire to dominate nature, a desire that goes back a long way and that has generated its own philosophical justifications and patterns of thought. In resource management, agriculture, health care, and many other fields, all but a few professionals define their work entirely in terms of their technological tools. Their education and professional traditions do not even consider the broader principles that govern our ecosystem. Such people naturally devote their best energy to quick, unnatural answers, and often achieve immediate, dramatic, popular, and profitable results. Yet such quick fixes can prove very costly in the long term. Nowhere is this more apparent than when we use technology to bolster production on deteriorating land, or to drastically modify an environment to better suit our purposes. Consider the following examples:

- *Water Cycle.* When water cycles become less effective, we can use machines to scour out contour ridges, or swales, to slow, spread, or harvest runoff rainfall, and ditches to drain waterlogged soils, deep rippers to reverse the compaction caused by heavy wheels, and irrigation pumps to put water back where it came from. In doing so we show little understanding of how water cycles function, and our immediate successes often result in long-term failure.

- ***Mineral Cycle.*** When mineral cycling is poor, we might turn to the local agrichemical dealer who can supply any chemical treatment he or she thinks our land requires, and with the help of some diesel fuel, the old John Deere can plow it in. In implementing these measures we so damage soils, through loss of soil life and crumb structure, that the mineral cycle only becomes poorer. The treatments need repeating in ever-stronger doses.

- ***Community Dynamics.*** Suppose we want to change the successional level of a plant community. On unproductive rangeland, machines or herbicides will clear the brush and scrub, and we can drill in seed. Where a mixed forest makes logging inconvenient, we can remove the native trees and plant uniform stands of faster-growing species that permit mass processing. When our chosen plants falter, we can kill their enemies and fertilize their soil. Some even attempt total control in the form of plants genetically engineered to thrive in an artificially fertilized environment, chemically rendered lethal to *everything* else. All these actions conflict with how nature functions. Successful as they appear, in the end they generally fail, often generating new, more severe problems.

Such thinking overlooks two important attributes of nature. First, our ecosystem is not a machine but a living thing that energetically moves and reproduces itself according to its own principles. Second, the life that we artificially suppress or take to extinction may have contributed to our own survival.

At the end of the last millennium when the threat of climate change began to register in the minds of the public, the popular press asked scientists how they proposed to address it. Not one mentioned anything but technological fixes on a grand scale. And none referred to the impact these fixes could have on the diverse and complex biological communities inhabiting our planet and that also sustain us. Even now, mainstream thinking still flows in that channel.

## Agriculture's Addiction to Technology

To ignore these attributes triggers the same mechanisms of dependency familiar from cases of drug and alcohol abuse. The parallel is striking. The clinical stages of alcohol or heroin addiction—becoming hooked, denial, degradation, skid row, death—occur routinely now in industrial-model agriculture in every nation. Farmers get pulled in by sales pitches on the wonders of fertilizers and pesticides. The pushers themselves get hooked on profits or research grants and everybody feels great. However, rapidly breeding microorganisms and insects adapt far faster to new conditions than do humans or many of the predators that once provided natural control. Once the chemicals we use show signs of failure, we simply increase the strength and quantity of our attack or attempt to isolate new compounds.

In the United States alone, where pesticide production is a 12.5 billion dollar industry, over one billion tons of pesticides are used each year—that averages about three tons per person.[2] Half of these pesticides are in the form of herbicides, most of them are inadequately tested for human safety, and virtually none are tested for their impact on our ecosystem as a whole. Crop damage continues to increase, even as evidence mounts that we are now poisoning ourselves and other slow-breeding creatures while strengthening the pests we set out to kill. Statistics on condemned water sources and sales of bottled water alone are powerful indicators of the unacknowledged external costs of doing business, which are borne by society.

We have now reached the denial stage. Government and industry point fingers every which way but at the real problem. Sympathetic people and organizations offer stress counseling for farmers and ease relocation from countryside to city. Yet suicide rates in farmers remain the highest of any occupation.[3]

No one wants to talk about a debilitating dependency because we can no longer conceive of life without it. Perpetual monocultures, inadequate rotations of monocultures, genetic engineering to support chemical treatments, and heavy machinery have become standard practice, but have so simplified soil communities and structure that, like a junkie's worn-out body, the land demands even harsher stimulation to produce the same high. And not surprisingly, many farmers become desperate raising money for that fix. Their cash flow can't stand a cold turkey withdrawal either, because the dying soil won't grow enough to pay last year's debt.

Agriculture has for too long looked to technology, in the form of chemistry and fossil-fueled machinery, to provide consistent or increased yields of crops and forage. Not only has this fueled an addiction problem, it has caused us to turn away from biological solutions that enhance community dynamics, especially soil life. On the other hand, we can never hope to feed future generations without sophisticated technology, and we certainly can apply it in ways that don't become pathological. Research into biological pest control, machinery for handling intersown crops, cultivation techniques and grazing practices that neither expose soil nor destroy its life, and even better mousetraps in place of stronger poisons represent a healthier direction for development.

## Developing a Collective Conscience

Having been a farmer and householder myself, I understand the frustrations, pressures, and urge for quick fixes. "They are eating my crops! What else can I do?" "Last night I went to the kitchen, and you should have seen all the cockroaches when I turned on the light!" A change in attitude can lead to simple ways to supplant destructive practices. Householders, builders, and appliance manufacturers can significantly dent the stocking rate of kitchen vermin by sealing cracks behind refrigerators and stoves to deny them a sanctuary. Physical traps for cockroaches and mice cut populations without cumulative damage. The State of California alone spends millions of dollars annually poisoning Californians unlucky enough to share their environment with roaches. Since the bugs have survived from the age of the trilobites, smart money says this will give them an even greater long-term edge over humans than they enjoyed before.

Given a holistic context and a way to check decisions for their economic, social, and environmental soundness, both short and long term, we can expose nonsolutions among technological remedies. Sometimes we might choose to solve a short-term problem, out of urgent necessity, with a form of technology that has adverse long-term effects. But where the wrong thing has to be done today in order to survive until tomorrow, we now know we have to quickly work on ways to prevent a recurrence of the situation. In many other cases, a holistic context that truly reflects the quality of life people desire and the life-supporting environment and behaviors that help assure it will encourage them to forgo technologies that provide immediate gratification in favor of those that provide lasting gain.

At present humanity does not have a shared land ethic or collective sense of conscience and responsibility, either to our fellow humans or to other life, and our governments reflect this only because our governments reflect us—they make decisions the same way most of us do. The context checks covered in part 6 offer a way to assess technology and foresee where it is likely to lead to crisis. It is my hope that they will also contribute to a new political attitude toward technology that embraces everyone, from the householder who shampoos his hair and poisons roaches, to multinational cartels that dam rivers and chainsaw jungles. I would like to know for certain, for example, that the pages of this book do not reflect profiteering on bad forestry, destruction of land, air and water pollution, exploitation of people, or cruelty to animals. They could as easily represent solar wealth and a stable community in an ever-vital landscape supporting abundant life now and even more in the future. Technology can help achieve that, but only public will and holistic management, in some form or another, can assure it.

# 19

## Fire
### *An Ancient Tool Tied to Ancient Beliefs*

BE WARY OF THE ARGUMENT THAT SAYS FIRE, because it occurs naturally, can be used as a management tool without adverse consequences. *No fire lit by a human is natural* and the effects of fire will vary greatly depending on how frequently an area is burned, what other tools are associated with its use, and how brittle the environment is.

Although fire has existed ever since green plants amassed pure oxygen in our atmosphere millions of years ago, its use as a tool by humans for modifying our ecosystem is a relatively recent phenomenon. Nevertheless, having had no other tool for modifying whole landscapes for ninety-nine percent of human existence, we have used it with such abandon that whole continents have been transformed. The scale and frequency of fire have almost certainly undergone a geometric increase against the background of the millions of years it took many biological communities to evolve.

Natural fires started by lightning, spontaneous combustion, or volcanic activity occur infrequently in comparison to the number of humanmade fires. In addition, most lightning fires occur with rain and thus spread less than those lit by humans, who often burn well before rain is expected. Although humankind has had the ability to make and use fire for about a million years, booming populations and, more recently, matches and government agencies and environmental organizations that advocate using fire, have radically increased its use in modern times.

It is my firm belief that this increased frequency of fire, combined with a reduction in the disturbance to soil surfaces and vegetation caused by dwindling animal herds and their predators, is one of the prime factors leading to desertification in the world's brittle environments, which make up most of the world's land surface.

We know that in North America the earliest people used fire a great deal and that they significantly altered the landscape by doing so. This, together with the decimation of most of the large animal populations some nine thousand years ago, could only lead to profound change. Where large numbers of some species did survive the earlier human onslaught, as in the case of the bison on the prairies, their presence would have diluted the damaging effects of fire. In fact, it was that combination of factors—infrequent fire, grazing, and predator-induced animal impact by vastly more species and numbers— that produced the lush grasslands found by early Europeans on the American prairies, *not fire alone*, a point I'll return to.

---

A combination of factors—infrequent fire, grazing, and predator-induced animal impact by vastly more species and numbers— produced the lush grasslands found by early Europeans on the American prairies, not fire alone.

---

The same could be said for much of Australia as well. There, over eighty percent of the large mammal genera became extinct following the arrival of humans some fifty thousand years ago. Today enormous areas of Australia, like Africa and North America, are dominated by vegetation that is fire dependent, owing to frequent burning by humans that caused previously abundant fire-sensitive species to disappear.[1]

For millennia fire has played a vital role in human life, touching not only our hunting and agriculture, but also our religions and rituals. This makes it difficult to consider this tool objectively, especially its potentially adverse effects on our ecosystem. Some years ago, for instance, I attended a U.S.

Natural Resources Conservation Service (NRCS) training session on "prescribed," or controlled, burning for the purpose of eradicating woody vegetation. Discussion centered on such things as time of day to burn, appropriate wind velocities and temperatures, and safe widths for firebreaks. We learned how much warning to give neighbors in order to avoid litigation, argued whether legally the rancher or the civil servant should hold the match, and probed the legal fallout of fires that get out of hand.

Throughout the day not a word was uttered about the effects of fire on the four ecosystem processes, nor its contribution to atmospheric pollution, nor how burning would reduce the effectiveness of the available rainfall. No one brought up the troublesome fact that fire invigorates many woody shrubs in the adult form. Every supporting argument rested either on ancient beliefs, or short-term research that had focused on plant species at the expense of soil and biological communities as a whole.

The use of fire has so many ramifications that the majority opinion, whether for or against, can seldom be right. Fire, like any other tool, *can only be judged according to the holistic context created for the whole under management and the current state of the four ecosystem processes relative to what is desired in that context.* I find any discussion about the use of fire to manage vegetation, wildlife, or crop residues without this basic information to be academic and as likely as not to end up in pointless argument.

## Effects of Fire on Biological Communities in Brittle Environments

Much of society is generally aware of the damage we do to biological communities and our planetary atmosphere when we burn nonbrittle tropical forests, all too often to make way for monoculture crops or cattle production. But there is much less awareness of the damage fire can produce in brittle environments. Here, the burning of billions of hectares of dry tropical forests, savannas, and grasslands is supported by environmental organizations, or in some cases by government policy. For this reason my focus in this chapter is on the brittle environments.

Fire is used far less in nonbrittle and less brittle environments simply because the higher year-round humidity inhibits fire, and it is more difficult to get vegetation to catch and stay alight. When fire is used in these environments they recover more quickly from its adverse effects than do brittle environments.

Deciding whether or not to use fire in brittle environments in any year requires an objective understanding of what it does and does not do, and the effects it tends to produce in biological communities. Those effects will always vary depending on how brittle the environment is and how high or low the rainfall.

### Soil Surface

First, and of primary importance, fire tends to expose soil surfaces. As soil surface management is central to the healthy functioning of all four of the ecosystem processes, this tendency must be kept in mind before all others. Bare ground is conspicuous right after a fire and until new growth appears to hide it. More critical, however, is the time it takes to build up the litter between plants. That depends on such things as the brittleness of the environment, the amount and pattern of rainfall, the amount of grazing or overgrazing by livestock or wildlife concentrating on the burned area (as they tend to do), and the degree and timing of animal impact.

Fire appears to have the most lasting impact where soil cover takes longest to form, the lower rainfall, very brittle environments. The lower rainfall produces less vegetation that might restore cover, but the fact that bare soil makes rain *less effective* compounds that effect because the entire biological community is only operating on a portion of the rainfall received, the rest being lost to both surface evaporation and runoff. The effects of fire on low rainfall grasslands can persist for years, creating scarred patches of land that are clearly visible from the air. While flying across Botswana in a light plane over many years, I drew these scarred areas in on my otherwise featureless maps and navigated by them for nearly a decade. If total rest follows fire in these environments, as is commonly advised with prescribed burns, or low animal impact (partial rest), soil cover accumulates even more slowly. The guidelines for burning detailed in chapter 36 discuss using other disturbances with fire instead of the unnatural two years' total rest so commonly recommended.

---

Fire appears to have the most lasting impact where soil cover takes longest to form, the lower rainfall, very brittle environments.

---

## *Plants*

Fire affects plants in different ways. Some sensitive perennial grasses disappear if burned. In brittle environments, the majority, at least as mature grass plants, thrive when burned because burning removes the old, oxidizing material that when allowed to accumulate prematurely kills grasses. Some grasses thrive on periodic fire, having seeds adapted for establishing after a fire. A number of grass seeds have awns, or tails, that actually twist and drill the seed into exposed soil when they become moist, suggesting an association with fire, which exposes soil.

Woody plants, too, may respond in many ways. Some are extremely sensitive, others resilient. In many countries I have observed that most of the trees and shrubs considered problem species are resilient when burned. Though they may appear dead immediately afterward, they soon resprout more stems than before, as figure 19-1 shows. This plant in the Arizona chaparral once had about six stems but after burning has thickened up to a great many more.

Many tree species are damaged by fire, yet some can still survive in the shrub form where burning is prolific. Mopane trees, common in the southern

**Figure 19-1.** Burning killed off the main stems of this shrub, but a great many new ones have sprouted. Arizona.

African tropics, once carpeted land I wanted to irrigate for sugar cane. I easily cleared the twelve-meter (forty-foot) trees by building a small fire at the base of each tree and leaving it undisturbed for several days. The whole tree burned down, and as long as the fire was left undisturbed so that blow holes in the ash were not closed, the roots burned out far underground. Yet where many past fires had kept the mopanes down to short multistemmed shrubs, the same trick failed, as they were completely fire resistant, and a bulldozer had to pull out enormous root systems.

Fire that is not followed by some form of soil disturbance in brittle environments tends to cause major changes within a community by creating bare soil that favors establishment of the few plant or animal species adapted to it. In a community where some mature organisms survive the fire, the new species that establish on the bare soil may initially increase diversity. Frequent repetition of the burning, however, will provide a largely similar microenvironment over large areas for so long that complexity diminishes. Gradually, the original community of diverse populations is replaced by those adapted to the fire-maintained uniform microenvironment. Thus, where a periodic fire can create greater diversity, frequent fires tend to do the opposite. A uniform microenvironment leads to fewer species generally, and often a near monoculture of low stability. Test plots in both Zambia and Zimbabwe that were burned annually for many years were eventually dominated by one or two species of grass with self-drilling seeds adapted to charred, cracked, and bare ground.

---

Where a periodic fire can create greater diversity, frequent fires tend to do the opposite.

---

A noteworthy corollary to this effect stems from the tendency of boundary areas to support particularly complex communities. Thus a healthy diversity may thrive on the edges of burned areas, due to tongues of unburned patches extending into them that provide "edge effect"—where two or more types of habitat meet.

Managers often choose to use either a hot or a cool burn, depending on the effect they want to produce. Chapter 36 gives guidelines for using both. Hot fires imply a lot of dry material that burns fiercely with large flames. Limited dry material produces a slow, creeping "cool" fire with small flames. The immediate positive effect a hot or a cool burn has on certain plants sometimes obscures the long-term adverse effects of exposing soil.

For much of my life cool fires were used in the teak forests on the Kalahari sands of Zimbabwe, Botswana, Namibia, and Zambia to prevent hot fires later in the season. This policy appeared so successful in protecting large, mature trees that its impact on the forest floor went unnoticed. The greatly altered microenvironment at the soil surface was now inhospitable to teak seedlings, or to seedlings from the other hardwood species of commercial value, the protection of which was the aim of the forest managers.

To convince foresters in Zimbabwe that this was indeed the case, I suggested they see if they could find any teak seedlings over the thousands of hectares they managed. When they were unsuccessful, I took them to a small area in one of the forests alongside a railway line where for many years large numbers of cattle had been offloaded from the trains for watering. On the ground where the cattle had trampled and milled around, fire had not occurred for many years and hundreds of teak seedlings had established, as well as seedlings from other species that were dying out in the areas burned to protect those same species.

Experts have argued that the slow burning of "useless" dead wood cuts the risk of hot fires and allows nutrients to cycle faster as ash than they would through decay. Others note the importance of dead wood in creating habitat for wildlife species important to the forest's health.

## *Animals*

Animals, like plants, also vary greatly in their response to fire. And like plants, the mature members of a population may thrive in the short term, but the population as a whole may suffer in the long term. Many animals do not escape easily. Many others do. Some are attracted to fires for the easy pickings of food from fleeing insects. It is a myth to think the larger game animals of Africa always panic and flee from fire. Though people may drive them to panic with flames and noise, left alone and undisturbed they usually just get out of the way calmly. Once during a three-day battle against a grass fire in the Rukwa Valley of Tanzania, three companions and I barely escaped encirclement by plunging through a weak point in the line of fire. Only a few meters beyond the fire line we found a group of reedbuck that had just had the same experience and had lain down calmly on the warm ground to watch. Some animals will seek out burned areas very soon after the passage of the fire, especially when the first green regrowth appears.

Throughout history we have made the mistake of noting only the immediate impact of fire on the adult populations of plants and animals. The teak forests of the Kalahari sands were a case in point, but we tend to treat grass, trees, birds, reptiles, game, and other organisms in the same way. We only ask, Were they hurt by fire, invigorated by it, or attracted to the green regrowth? Did food supplies increase? We have not watched and formed opinions on what happened to the ecosystem processes in terms of what these things we value need in order to reproduce over prolonged time. A short-term benefit for adult populations can encourage further burnings that may destroy that population in the long run.

---

We have made the mistake of noting only the immediate impact of fire on the adult populations of plants and animals . . . not what they need in order to reproduce over prolonged time.

---

## Fire Alone Does Not Maintain Grasslands

In chapter 4, I mentioned my first questioning of the beneficial fire myth when we used it in a government-sponsored program to help clear big game from tsetse fly areas in Africa so people and livestock could be settled there. We had wiped out the large game herds, had not eliminated the tsetse fly, and had not introduced cattle. Yet the grasslands, which looked lush enough from the window of a Land Rover traveling forty kilometers (twenty-five miles) per hour, were deteriorating seriously, as a close consideration of the four ecosystem processes plainly showed. We were only using the tools of fire and rest, having shot out the game. The temporary grassland we made masked a serious long-term desertification process, as time proved dramatically.

By the time the once complex and healthy grasslands were ready for occupation by people with cattle, they were in a bad state of degradation. Active erosion between plants had set in, seedlings of many perennial grasses had become scarce, monocultures of mature plants abounded, and solar energy flow had dropped severely. While the damaging effects of periodic fire in the past had been slowed by high populations of game providing some soil disturbance, fire and no game speeded the damage. Ironically, it is now my belief that the fires we used in the tsetse fly operations were probably one of the main causes of the tsetse fly's spread, as the fire-induced, less effective water cycles led to a dramatic increase in the breeding sites for this slow-breeding insect.

The extent of the destruction caused by fire alone is symbolized for me by the oil sump of a Land Rover. In 1959 while in charge of the government burning and shooting operations in southeastern Zimbabwe, I drove out daily from my bush camp. After some months I was under the vehicle changing the oil and could not help noticing that the grass had polished the underside clean. In fact the front side of the brass sump plug had worn down so far and so smooth it would scarcely hold a wrench.

Six years later, because of the enduring fly, the area still lay largely unoccupied by large game, livestock, or people, though the burning had continued, so I made my old camp a base for training army trackers. After three months of continual driving over the same ground and remembering my past experience, I checked under all our vehicles and found the oil sumps caked with dusty grease. On level ground the grasses persisted weakly, but on slopes and where soil had been shallow before, naked earth and exposed pebbles characterized the scene. Such profound changes, when gradual, escape notice unless an observation like this forces us to think.

The Kruger National Park in South Africa is perhaps one of the best conventionally managed national parks in the world. Tragically, fire is used every few years on average as managers attempt to maintain grasslands that are turning to brush. Because park managers believe that the overgrazing and overbrowsing that are also occurring are due to too many animals, they remain with no solution but to keep burning and to cull animals believed to be too numerous. This of course only increases the amount of vegetation that is overrested and subsequently burned. I know of no soils anywhere that can withstand burning of this frequency.

Tragically it is not only Kruger National Park that is shooting animals and then, in effect, using fire to replace the role the once abundant animals played. In Zimbabwe, where elephants are destroying magnificent centuries-old baobab trees (fig, 19-2), park managers have for years been shooting elephants by the thousands. No one appears to be asking why even greater numbers of elephants in the past did not destroy the baobabs in these same areas. Why have elephant feeding patterns changed? I strongly suspect they have due to a combination of changed behavior in the parks (they linger in them), plus the increased use of fire, which has greatly increased the amount of bare ground, simplified the communities from which elephants derived their sustenance, and gradually transformed the vegetation they once favored into more fibrous, sharp-seeded, less nutritious grass species. Elephants are by nature largely a grazing animal, which most people today find hard to believe, so changed is their diet.

**Figure 19-2.** For centuries undamaged by vast elephant populations, magnificent baobab trees are now being tusked to death. Frequent burning has caused the grasses elephants once favored to be replaced by more fibrous, less nutritious varieties.

## Fire and Atmospheric Pollution

Studies on the atmospheric pollution created by forest and particularly grassland fires in Africa and North America add good reason to seriously look at alternatives to fire wherever possible. Until quite recently, most atmospheric pollution was attributed to excessive use of fossil fuels. But data since gathered by researchers via satellite and fieldwork indicate that biomass burning is a significant contributor as well. Scientists have calculated that the emissions *every second* from a vegetation fire covering 0.5 hectares (1.5 acres) are equivalent to the carbon monoxide emissions produced per second by 3,694 cars, and the nitrogen oxides produced per second by 1,260 cars.[2]

Today, 750 million hectares (1.85 billion acres) of the world's grasslands are deliberately set on fire each year, releasing about 3.7 million metric tons of carbon into the atmosphere. This is three times more carbon than is released through forest burning.[3] While the burning of grasslands is common wherever vegetation is dry enough to burn, most inevitably occurs in Africa, owing to the sheer size of the continent (the whole of the United States, China, India, Europe, Argentina, and New Zealand would fit inside the boundaries of Africa with land to spare). In Africa three-quarters of the grasslands go up in smoke every year. Pollution—ozone, carbon monoxide, and methane—from grass fires in southern Africa drifts thousands of miles within weeks to Australia and Antarctica, where atmospheric pollutants have created an ozone hole.

Some range scientists argue that the amount of carbon dioxide produced by these massive grassland fires is not an issue because once plants regrow they will absorb the same amount. But they fail to consider the adverse effects of fire on plant spacing, composition, soil cover, water cycles, and so on, which generally lead to the production of less biomass, as I found when the burned areas no longer grew enough grass to wipe our vehicle oil sumps clean. When less biomass is produced, less carbon dioxide is absorbed.

In the mid-1990s, scientists at Germany's Max Planck Institute found that fires in Siberian forests, Californian chaparral, and South African savannas were producing disturbingly high levels of methyl bromide. The bromine in methyl bromide is potentially fifty times more efficient than the chlorine in chlorofluorocarbons, or CFCs, in destroying upper-level atmospheric ozone and has a global warming potential five times greater than carbon dioxide.[4,5]

While there is still much to learn, and the sheer scale of biomass burning makes accurate estimates impossible, we shouldn't ignore the warnings provided by these studies, and those since, on biomass burning's contribution to climate change. Unfortunately, many scientists continue to promote "prescribed" burning, with little or no consideration given to its effects on the atmosphere.

Few indeed are the people interested or even willing to explore alternatives to fire. On a ranch I was advising in New Mexico, we initially struck all the problems one would expect to find on land that for years has been subjected to partial rest and overgrazing of plants. Over eighty percent of the ground was bare, and the few perennial grasses that had survived were mainly of three varieties that could withstand high levels of rest. Most of these plants were dense, dark-gray-to-black masses of old material surrounded by eroding soil. To begin putting the situation right, we needed to knock down the old dead material so the plants could again grow freely and the soil could be covered. Naturally we wanted to use animals rather than machinery or fire because their trampling would also churn up the capped surface so new plants could establish.

However, we immediately ran into problems because the majority of the land was leased from the government and regulations prohibited us from running animals at the density and with the numbers required. We could only use fire. Using fire would, of course, remove almost all of the plant material, expose even more soil, and do nothing to break the capped surface. At the same time it would release carbon into the air when we wanted to keep it cycling through plants and soils. We eventually worked with the government agency concerned to find a temporary way around their regulations, but it caused several years of delay.

## Conclusion

When considering whether to use the tool of fire, know what you are trying to achieve. If other tools could achieve what you want without exposing soil, reducing the effectiveness of the rainfall, or polluting the atmosphere, then consider them too. Remember that the viability of the whole population

### Effects of Fire in Brittle and Nonbrittle Environments

*Here is a summary of the effects of fire on the four ecosystem processes at the extreme ends of the brittleness scale. The summary assumes the fire is followed by a period of rest, the most common practice.*

*Nonbrittle Environments*

**Community Dynamics**

Fire is difficult to light and is generally used in forests, where it simplifies the community and pollutes the atmosphere. The higher humidity ensures recovery on undisturbed land.

**Water Cycle**

Fire tends to damage the water cycle by exposing soil.

**Mineral Cycle**

Fire appears to speed the cycling of nutrients, but this effect is often an illusion. The biological decay so necessary to maintaining carbon in soil organic matter gives way to rapid oxidation and the air and atmospheric pollution associated with it. For thousands of years fire was the main tool in slash-and-burn agriculture. Although the fire freed many nutrients in the community for human use by growing plants in the ash, such agricultural systems, sound as they appeared, broke down fast unless there was a twenty-year or more rest between fire-use periods.

**Energy Flow**

Fire disrupts energy flow in the short run, but recovery follows if the entire community is rested. In slash-and-burn agriculture, more energy is temporarily directed to immediate human use.

*Very Brittle Environments*

**Community Dynamics**

Fire destroys litter and exposes soil. In low-rainfall areas where plant spacings are wide, new cover develops slowly. Fire stimulates the growth of woody plants that have grown beyond seedling stage and kills only a few. Repeated burning leads to fire-dependent grasses in simplified communities.

### Water Cycle

Fire reduces the effectiveness of the water cycle, thus increasing both drought and flood tendencies. It does this by exposing soil and destroying the litter that slows water flow, and the soil life that helps maintain soil surface crumb structure and aeration. The lower the rainfall and the more frequent the fire, the greater this tendency.

### Mineral Cycle

Fire speeds the mineral cycle in the short term by converting dead material to ash and polluting the atmosphere. Because it exposes soil and changes the soil surface microenvironment that supports the organisms of decay, fire reduces mineral cycling. The drier the area and the more frequent the fire, the greater this tendency.

### Energy Flow

Fire may produce an immediate increase in energy flow by removing old material that hinders the growth of grasses and brush. However, the soil exposure leads to less effective mineral and water cycles, which reduce energy flow in the long term. The drier the area and the more frequent the fire, the greater this tendency.

structure, *not* merely adult plants and animals, is critical. If they fail to reproduce and their young fail to thrive due to changes in the ecosystem processes resulting from the fire, you may not achieve what you were ultimately seeking.

There will be times when fire is the best tool to use for the job, such as creating habitat variety to encourage species diversity, including the survival of fire-dependent species, or burning firebreaks to help isolate and extinguish any accidental fires. The guidelines for burning outlined in chapter 36 will help you create a plan. Once your plan is made, of course, you must monitor to make sure that the results you expect do materialize.

Because it is impossible to use fire as a tool in managing grasslands without also using one of the tools covered in the upcoming chapters—rest or living organisms, in the form of grazing and animal impact—read on.

# 20

## Rest
### *The Most Misunderstood Tool*

WHEN SPEAKING OF REST AS A TOOL that can be used to modify the four eco-system processes, I am referring to rest from major physical disturbance that applies mainly to plants and soils. Disturbance comes in many forms. Large animals, domestic or wild, but particularly those that exhibit herding behavior, impact both soils and vegetation. So can machinery. Fire disturbs vegetation greatly but disturbs the soil surface slightly, if at all, while on occasion, hailstorms and natural catastrophes disturb both but fall outside our management control. A policy or practice of withholding all of these forms of disturbance completely for considerable time amounts to applying the tool of *total* rest. *Partial* rest is applied in the presence of livestock or wild grazers, but with such calm behavior in the absence of pack-hunting predators that a large proportion of the plant life and soil surface remains undisturbed despite their presence and grazing.

Partial rest is a new concept that some find hard to grasp. How can the land be resting while high numbers of livestock or wildlife are grazing on it? Its effects, however, vary little from the effects of total rest and are evident anywhere livestock or wildlife seldom bunch; the more scattered they are, the greater the degree of partial rest. Figure 20-1 shows partial rest on one side of a fence, where Navajos have grazed their sheep for hundreds of years, and total rest on the other, where the U.S. National Park Service has excluded

**Figure 20-1.** Despite different management—partial rest under scattered flocks of sheep, and total rest (no livestock) for more than seventy years—the results, on both sides of the fence, are the same. Chaco Culture National Historical Park, New Mexico.

livestock for more than seventy years. We would expect to see a sharp change at the fence line, given that Navajos have long been accused of overgrazing and overstocking their land, and the Park Service has eliminated this possibility by excluding all livestock. But there is no contrast at all. The partial rest applied by the Navajos grazing scattered flocks on their side of the fence has produced the same result as the National Park Service applying total rest on the other. We'll return to this subject later.

That rest in either form (total or partial) might function as a tool of the same order as a fire or plow comes as a new concept. We considered rest natural and thus beneficial everywhere until we registered the fact that brittle and nonbrittle environments react to it in very different ways and that the major grazing areas of the world, particularly before humans developed the ability to use fire and spears, seldom, if ever, experienced rest to the extent of today.

When humans first settled in permanent villages and thus drove wild herds from the surrounding area, they unintentionally but decisively subjected those lands to rest. We now intentionally rest land in the hope that it will recover from the effects of fire, overgrazing, or overtrampling. Though we justify this as "leaving things to nature," we have changed natural relationships no less than those first settlers did. To understand why, we must look at the different effects produced by rest at either end of the brittleness scale.

## Effects of Rest in Nonbrittle Environments

In nonbrittle environments (1 to 3 on a scale of 1 to 10), old plant material *by definition* breaks down through biological decay. In the tropics this might happen in a few days. In temperate climates it could mean breaking down by the start of the new growing season. Such "rotting down" or "decomposition" starts close to the ground, where the microenvironment supports the highest populations of insects and microorganisms engaged in the decay process. Initial decomposition close to the ground particularly suits grasses, because dead leaves and stems weaken at the base and quickly fall aside, allowing light to reach new, ground-level growing points. Figure 20-2 shows an example of

**Figure 20-2.** Less brittle grassland showing last year's old growth of leaves and stems (pale color) already fallen and decaying and not choking new growth. Alaska.

this. Last season's leaves and stems have toppled, and new growth is occurring unimpeded.

This rapid, bottom-up decay process on dead woody vegetation as well as grasses allows nonbrittle biological communities under total or partial rest to maintain a high degree of stability and complexity of species in grassland or forest. Water, if it runs off soil surfaces, carries little silt or other debris. Even very prolonged rest from the rare fire or physical impact by machinery or large animals has little or no adverse effect on the water cycle, mineral cycle, community dynamics, or energy flow.

Nonbrittle environments reduced to bare ground by some natural or humanmade catastrophe respond rapidly to rest, and biological communities return relatively quickly to their former complexity and stability, whether jungle, forest, or grassland. Desertification is seldom a long-term danger, although deforestation can cause enormous damage in the short term.

That all this works so straightforwardly in nonbrittle environments has obscured the fact that it does not in very brittle ones. Some of the greatest environmental tragedies in human history have ensued from the false assumption that all environments respond the same way to rest. In nonbrittle environments it is virtually impossible to expose vast areas of bare soil between plants and keep them bare. You can burn vegetation, poison it, bulldoze it, or overgraze all of it, and the soil surface will still cover over again quickly if rested, which is why these environments do not desertify. In a nonbrittle environment rest is a powerful tool for restoring and sustaining biodiversity and healthy land.

## Effects of Rest in Very Brittle Environments

In very brittle environments (8 to 10 on the 1 to 10 scale), *by definition* old plant material, in the absence of large naturally functioning grazing animals, breaks down slowly and the successional process advances slowly, at best, from bare soil. Most organisms of decay, especially in communities that have lost biodiversity, are present in high numbers only intermittently, when moisture is adequate. In some arid areas, termites help break down plant material and thus play a vital role. However, when fire and partial rest combine to produce the bare soil that favors many termite species, termite numbers can increase to such an extent that they consume all the soil-covering litter.

Under conditions of either partial or total rest if animal numbers are also low, much of the old plant material, particularly on perennial grasses, does not decay biologically but breaks down through oxidation and physical weathering. Being most exposed to the elements—wind, raindrops, and sun—the tips break down first. While this top-down breakdown has little adverse effect on woody plants, it severely hinders perennial grasses with ground-level growth points that often remain shaded and obstructed by several years of oxidizing material. Old plant material that lingers even through the next growing season weakens most perennial grass plants, and several years' accumulation can kill them. It also adversely affects the feeding of grazing animals that try to avoid oxidizing material as they balance their diets, making old oxidizing material difficult to remove by grazing.

Some perennial grass species can withstand high levels of rest. Commonly these will have growth points aboveground, reflected in branching stems, and this allows enough growth to keep the plant alive even when a mass of undecayed material chokes most of the plant. Tobosa grass, common in the western United States, is a good example. Other rest-tolerant species can be very short in stature, such as the grama grasses, or can have sparse thin leaves that, despite an accumulation of old growth, allow some light to reach ground-level growth points. Many of the species from the *Aristida* genus have this characteristic. In the sparsely populated brittle environments of the western United States, where there are so few animals on the land, a few perennial grass species with a high tolerance for rest dominate vast areas.

When the soil surface remains undisturbed, as much of it is throughout Africa, India, Pakistan, Australia, China, and the western United States, new plants do not easily replace those that are dead or dying. The capping that usually forms on rested soil offers poor opportunities to the germinating seed. Outside the tropics, prolonged rest allows freezing and thawing of the upper soil layers of exposed soil to create a puffiness that also inhibits the establishment of young, fibrous-rooted grasses.

Very brittle, low-rainfall environments subjected to extended periods of rest (partial or total) characteristically have wide bare spaces between vestigial perennial grass plants. The remaining plants survive because light can reach the ground-level growing points around the edges of each plant, but the centers may already be dead. Color plate 8 illustrates these symptoms on a once healthy, but now totally rested perennial grassland on the Sevilleta Wildlife Refuge, which lies in a very brittle 225-millimeter (9-inch) rainfall environment of New Mexico. Most remaining grass plants are barely alive, having only a few green leaves around the edges of the dead stems. Despite many years of seed production no grass seedlings have established on the bare, undisturbed soil surface. What seedlings are present are all forbs.

After adequate disturbance of the soil surface, closely spaced perennial grass plants often do establish. If such land is then partially or totally rested, however, the closely spaced plants kill one another off prematurely, as old growth that has accumulated on them shades even the edges of the neighboring clump. Figure 20-3 shows such a grassland in a 375-millimeter (15-inch) rainfall brittle environment in northern Mexico. Whole handfuls of dead grass can be pulled out by the roots with ease. Color plate 9 shows the same thing in a 750-millimeter (30-inch) rainfall brittle area of Zimbabwe.

**Figure 20-3.** Brittle environment grassland, that had developed close spacing under animal impact, now showing mass deaths under rest. Most plants are gray and oxidizing and can be pulled up easily. Coahuila, Mexico.

Most of the brush and tree encroachment many consider a problem today owes its existence to heavy doses of the tool of rest, mainly partial.

If the landscape described in your future resource base requires dominance by woody plant species and the animal organisms associated with them, then you would consider continued rest for these lands if they received enough rainfall to support good, soil-covering stands of woody plants. The dead clumps of grass provide a good microenvironment for seed germination and the dead roots lacing the soil make an excellent medium for penetration and establishment of the woody plant seedling's taproot. Color plate 10, taken on the same ranch as figure 20-3 in Mexico, illustrates the different microenvironment that exists between areas where overrest is causing most grass plant deaths and where overgrazing is doing so. Overrested sites tend to favor the establishment of woody forbs (weeds), brush, and trees. Overgrazed sites tend to favor the establishment of herbaceous forbs (weeds). If your holistic context does not involve this kind of change, don't apply either partial or total rest. Most of the brush and tree encroachment many consider a problem today owes its existence to heavy doses of the tool of rest, mainly partial.

Figure 20-4 shows a piece of land within the very brittle Chaco Culture National Historical Park in New Mexico, which had been rested more than fifty years when the photo was taken. Most grasses are dead. The shrubs, too, are now dying, and surrounded by bare, eroding soil. In a few places where the land is flat enough, algae and lichens persist as the main plant life left.

Towns and cities in very brittle environments are also subject to the adverse affects of rest. When the catchment areas surrounding a town are rested, as inevitably happens when parcels of land ranging in size from two to eight hectares (five to twenty acres) are zoned for residential sites, the effect is entirely good as long as those towns are sited in environments low on the brittleness scale. On the land surrounding London, Paris, Sydney, or Washington, DC, rest generally leads to an increase in soil cover and biodiversity. But towns and cities in the more brittle environments cannot afford to rest their catchment areas, particularly when rainfall is also low. Cities such as Albuquerque, Los

**Figure 20-4.** Fifty years of total rest in a very brittle low-rainfall environment. Most grass is already dead, bare ground and soil erosion are extensive, and many shrubs are dying. The main plants left are algae and lichens. Chaco Culture National Historical Park, New Mexico.

Angeles, and Perth are surrounded by minimally developed parcels of land that are being rested to death. The resulting amount of bare ground produces an enormous amount of runoff that commonly inundates these cities with floods anytime rainfall is heavy. Their dams and rivers are dangerously full of silt as well. The bed of the Rio Grande, which flows through Albuquerque, is now significantly higher than the original town center.

Few city or county planners in more brittle environments consider these consequences when they allow catchment areas surrounding towns to be divided into small ranchettes or summer homes, so popular with well-off retirees. Even if the owners keep horses and other animals, the tools they will be applying daily will be partial rest and overgrazing. Unless they apply different tools, urban residents will have to live with the destructive consequences and bear the costs for years to come.

## Effects of Rest in Less Brittle Environments

The effects of rest are more difficult to decipher in environments that lie in the middle range of the brittleness scale (around 4 to 7 on the 1 to 10 scale). You cannot be sure at which point the effects of rest change over from enhancing soil cover, energy flow, and the health of perennial grass plants, to damaging them—just as you cannot be sure, when standing atop a watershed, which way a drop of rain will flow. In these less brittle environments, the effects of either form of rest take longer to show up. If rest does have adverse tendencies, it may become apparent only after a good many years.

### *Grasslands*

The evidence is easier to show photographically in grassland than in forest. Figure 20-5, from the Crescent Lake Wildlife Refuge in Nebraska, shows land that appears to lie in the middle range of the brittleness scale and has been

**Figure 20-5.** Land lying in the middle range of the brittleness scale that has been totally rested for twelve years. Although damage appears minimal, a high proportion of old material is beginning to weaken these grass plants. Crescent Lake Wildlife Refuge, Nebraska.

totally rested for twelve years. Although there appears to be no great damage, a high proportion of old material is weakening the plants. All four ecosystem processes show early signs of adverse change.

Figure 20-6 shows nearby land on the same refuge totally rested for fifty years. By this stage the water cycle, mineral cycle, community dynamics, and energy flow have all visibly declined. Grasses that earlier provided ground-cover are clearly dying, and many are dead. All new plants are small herbaceous ones that many would consider weeds but are serving a purpose in that they are covering bare ground. Still, a large proportion of the ground has become bare, and once the dead grass litter breaks down this area will expand.

This environment obviously lies to the right of the midpoint in the scale. It is too brittle to sustain grassland without repeated disturbance, and it receives too little rainfall for the successional process to move it to wood-

**Figure 20-6.** Land lying in the middle range of the brittleness scale that has been totally rested for fifty years. By this stage many plants have died and bare ground has increased significantly. The few new plants are forbs, not grasses. Water and mineral cycles are seriously impaired and energy flow is greatly reduced. Crescent Lake Wildlife Refuge, Nebraska.

land with adequate leaf fall to provide soil cover. Thus, when totally rested, bare soil is the eventual result. If this land had been partially rested under low numbers of game or livestock it would only have slowed the rate of deterioration.

### Savanna-Woodlands

Some savanna-woodlands and grasslands in high but seasonal rainfall areas show clear signs of brittleness in that old grass parts oxidize rather than decay, dead grass and trees break down by weathering from tips rather than rotting near the base first. Yet, if rested long enough these communities commonly pass on to generally open forest with adequate leaf fall to provide soil cover and stable soil.

### Riparian Areas

I have also seen places where the rainfall was much lower, but tap-rooted woody plants provided soil cover and achieved the same stability because subterranean moisture allowed a dense enough stand of brush or trees. This situation is common along riparian (streamside) strips. If rested for prolonged periods, such environments tend to pass on to stable woodlands. But do not be fooled into believing all riparian areas remain healthy when rested. Some riparian areas lack sufficient subterranean moisture everywhere to maintain dense, woody cover and will deteriorate seriously when rested, as shown in color plate 11.

Even when the area adjacent to the stream or river is less brittle, it may be surrounded by very brittle catchments. Because the health of the stream depends entirely on its catchments, or watersheds, management has to cater to this larger whole.

## The Effects of Rest in Brittle Environments Take Time to Show

Most governments dealing with desertification have long been advised that the deterioration was largely due to overgrazing, which in turn was caused by the presence of too many animals. The recommended solution was to drastically reduce animal numbers so the land could heal, though most politicians lacked the willpower to actually force such a measure. To demonstrate the advantages of total destocking (rest) the U.S. government in the 1930s fenced off demonstration plots of land throughout the western states.

Once protected from overgrazing and unimpeded by old growth, the plots indeed grew lush and became the justification for often draconian campaigns to reduce stock. Figure 20-7 shows an example in New Mexico taken

**Figure 20-7.** Government demonstration plot established in the 1930s to prove rested land would recover. Grass on the left was fenced off from grazing animals, and within three years was more lush than on the right where livestock were present and overgrazing many plants. Note Cabezon Peak, which appears in the background. Rio Puerco Valley, New Mexico.

from the Council on Environmental Quality's report on *Desertification of the United States*, published in 1981.[1] Although the photograph lacks detail, within the plot we can see a good stand of vigorous grass, freed from the overgrazing of plants evident on the other side of the fence. Outside the fence we see overgrazed grasses typical of land under continuous grazing. The caption that accompanied this photo in the report reads: "Range improvement in the Rio Puerco Valley, Sandoval County, New Mexico. Grass on the left is protected from overgrazing (Soil Conservation Service)."

This is a typical situation where the officials were misled by two things, one being the power of paradigms, and the other being the time delay before the effects of rest became apparent. Although the report was written in 1981, a 1930s photograph was used—one that illustrated the immediate benefits of plant recovery from overgrazing, much as your lawn would look if left unmowed. In fact, the remaining demonstration plots in the Rio Puerco in

**Figure 20-8.** View of one of the remaining demonstration plots in the Rio Puerco Valley. Fifty years later (1987), the continued resting of this land had resulted in serious deterioration. Cabezon Peak again appears in the background. New Mexico.

1981 all looked like what you see in figure 20-8. The report writers *knew* that land recovers when destocked and rested, and probably felt no need to check on the current status of the plots; they merely used the 1930s photograph already on file. Had a 1980s photograph of any of the demonstration plots been used to illustrate the report instead of a 1930s one, a very different conclusion might have been reached. Because of total rest these plots are de-sertifying as badly as any land I have seen in Africa, Australia, or the Middle East. The authors were not being dishonest, just human. Why question what we already *know*?

### Rest versus Recovery

The critical distinction between rest as a long-term tool and rest as the time it takes a damaged plant to rebuild a root system and recover has caused confusion for many years. We correctly observed that animals in certain cir-cumstances overgraze and damage plants. If we run high animal numbers on an area for prolonged time and overgraze most plants, as many did in the western United States at the turn of the century, and we then remove the animals, the land appears to recover quickly and dramatically, as those early research plots showed. In fact we had stopped two animal-produced effects, one positive and one negative. The result was that the positive and immedi-ate effect of allowing overgrazed plants to *recover* colored our ability to see the eventual damage we created by eliminating beneficial animal impact and *resting* the land.

With fewer animals on the land, partial rest increases, but overgrazing of plants, being a function of time rather than numbers, still continues. Grasslands become dominated by bare ground between widely spaced rest-tolerant peren-nial grasses of the types mentioned earlier, brush, and so-called weeds. Seeing this decline, governments have for generations advocated cutting animal num-bers even further, and in effect, resting the land even more, while in some cases wasting millions of dollars in futile spraying of the resulting noxious plants, and in others causing social upheaval through the forced removal of pastoral people.

### Partial and Total Rest Have Nearly the Same Effect

By fencing off so many plots of land in the 1930s, the U.S. government un-intentionally provided excellent evidence for another important point—that partial rest can be nearly as destructive as total rest in the more brittle environments. The government research plots excluded all livestock, while on the land outside them livestock numbers were low and declining, wild grazers were few in number, and pack-hunting wolves and Indians had long since vanished.

It is not surprising that the government at the time wanted to reduce livestock numbers, given that the nation's attention was riveted on the Dust Bowl, which in the 1930s covered a large part of the country. Over the next decades, livestock numbers would continually be reduced in an effort to restore the land to its former productivity. The reductions have continued through the present but without any real land recovery. Now when you examine the totally rested land inside the old exclosure plots and the land outside them where partial rest and overgrazing have continued, there is very little difference.

Color plate 12 shows the boundaries of three experimental plots, each of which is in an environment successively less brittle than the first. In each case, the four ecosystem processes have seriously malfunctioned inside the plots after forty to sixty years of rest. Outside the plots the land is no better. Roughly the same amount of ground is bare. The fact that some plants outside the plots were overgrazed has had little impact on the total picture. The level of rest and the position on the brittleness scale were a much greater influence. Inside the first plot, no living perennial grasses remain; outside it at least some perennial grasses have been kept alive by grazing, even though they are overgrazed. In the other two plots, two or three species of rest-tolerant perennial grasses have managed to survive both inside and outside the plots.

Unfortunately, researchers who take it for granted that rest is natural have compared the surrounding land to the rested sites and, because there is so little difference between the two, have pronounced management outside the plots successful. The land, they say, has reached the highest level of development of which it is capable because it matches the totally rested land inside the plots. *They were unaware of or ignored the awkward fact that the plots at first got better and only subsequently declined.*

### Rest and Crisis Management

*Misunderstanding of the damage* partial rest does in brittle environments often leads to crisis management on land people intend to preserve in pristine condition. I have seen more than one environmental organization take over management of damaged brittle grassland slowly change their management of it from a dogmatic hands-off, leave-it-to-Nature approach to application of the most drastic techniques available. It happens in a predictable sequence.

Following years of overgrazing by livestock, the plants respond vigorously to rest and all looks good. The increase in volume and cover benefits many creatures and complexity builds up. Lists of small mammals, birds, and insects become impressive as more species reap the new bounty. Gradually, however, moribund grasses turn up in the log books, various weedy plants increase in number, and bare spots begin to open up.

When the problem does not go away, the managers conclude that fire should be used as "fire is natural, kills woody vegetation and it maintained grassland in the past." The first unnatural means of returning the land to Nature—rest—has led to another. Again the situation "improves" because fire does keep mature grasses alive, often reduces invasive forbs and shrubs, and the "right" species still appear on naturalists' checklists. But, given that rest also tends to expose soil, and that the cause of the old vegetation accumulating has not been removed, the situation predictably worsens as fire use becomes too frequent. Technology often comes next in the form of seedings, plowings, plantings, check dams, ditching, and the like. However, the problem remains insoluble until the managers understand the implications of rest in brittle environments.

## What Is *Natural?*

When confronted by the argument for reintroducing animal impact as a more natural influence than technological intervention or fire, environmentalists in such situations typically respond, "But no bison ever roamed here." Or "Cattle are not native to this environment." This ignores the fact that many animals maintained grasslands, not just bison, and that there were many more species of animals, including ancestors of cattle, as recently as nine thousand years ago in North America. Even in the last few centuries, species other than bison, such as deer, elk, pronghorn, and bighorn sheep, could break a period of rest, but so little is known about the actual number and distribution of these animals, even a century ago, that debates about them tend to be academic. In some parts of the American Southwest where people insist that large herds never occurred, we now have evidence that hunting peoples thrived even relatively recently (within the last two hundred years). In fact, aerial inspections have revealed a remarkable density of not so ancient permanent pronghorn traps very similar to those observed in the great humanmade deserts of the biblical lands. Although many, including myself, far prefer wildlife to livestock, today we have to employ the tools at hand or accept the dreadful consequences of desertification.

Invariably some people at this point close their minds because grassland does exist in parts of America, Australia, and Argentina, for instance, without any evidence of help from herding animals in the recent past. In my experience grassland will always reflect its level on the brittleness scale and the net result of the major influences on it. This will show in its biological complexity, plant spacings, effectiveness of water cycle, and overall energy flow. If plant spacings are close and age structure is good in the absence of disturbance by fire or animals, then that grassland lies lower on the brittleness scale. If plant spacings are wide, age structure is poor, and reproduction is predominantly asexual in the absence of disturbance, then that grassland lies higher on the brittleness scale. If we were in fact to discover a grassland that, based on the climate and annual distribution of humidity, should be high on the brittleness scale and which had developed great complexity, stability, and closely-spaced plants with no disturbance, we would have discovered a distinct new environment with rules of its own, and the concept of brittleness would have to be revised.

### Regaining Eden

Eons ago, before humankind controlled fire, livestock, and technology, there was an Eden to which we alas can no longer return. Logic tells us that the living soils and all plant and animal life developed together in complex, synergistic wholes including microenvironments, communities, and climate. The brittleness scale, although we did not recognize it, existed for millions of years. The extraordinary expansion of deserts in what were grasslands and savanna-woodlands a few thousand years ago can only be the work of humans.

We go to great lengths to avoid this conclusion. The archaeologists delving into the secrets of ancient ruins in New Mexico conjecture that the climate changed, that overpopulation led to a collapse of agriculture in fragile bottomland, that war destroyed the social fabric, that the civilization grew too large to administer without the power of the written word, that people really did not live in the large ruins but only came to pray there, and so on, endlessly. They point to ancient tree rings that indicate prolonged drought without acknowledging that this could possibly be a symptom of an ineffective water cycle. They do not consider that even a primitive population, by the very act of their settlement, would displace game, and that, plus their use of fire to drive game while hunting them in the catchments, might have upset the water cycle enough to affect tree growth.

To regain any part of Eden now means reproducing as closely as possible the conditions under which various biological communities, microenvironments, and climates developed. Wherever we manage to do that, the life that flourishes will be natural whether or not it represents what existed in that place at any given moment during the history of "unnatural" human influence.

Humans have so changed the environments they inhabit that today we are forced to make choices in managing them. To do nothing in the more brittle environments of today is choosing to use the tool of rest. Thus, for any piece of land, even a wilderness area or national park, we need to decide what it is we really want.

---

### Effects of Rest in Brittle and Nonbrittle Environments

*The effects of rest at either extreme of the brittleness scale are as follows:*

*Nonbrittle Environments*

- **Community dynamics**—biological communities develop to levels of great diversity and stability.

- **Water and mineral cycles**—build and maintain high levels of effectiveness.

- **Energy flow**—reaches a high level.

Rest is the most powerful tool we have to restore or maintain biodiversity and soil cover in nonbrittle environments.

*Very Brittle Environments*

- **Community dynamics**—biological communities decline and greater simplicity and instability ensue. The lower the rainfall, the greater the adverse effect.

- **Water and mineral cycles**—become less effective.

- **Energy flow**—declines significantly.

In very brittle environments, rest (partial or total) is extremely damaging to biodiversity and soil cover. At the midpoint on the 1 to 10 scale, partial or total rest shifts from being increasingly positive to increasingly negative in terms of maintaining soil cover, energy flow, and healthy perennial grasses. Because rest has such clearly different tendencies at the extremes, the condition of rested land generally indicates the underlying brittleness of any area.

---

Canyonlands National Park in Utah contains a site called Virginia Park surrounded on all sides by perpendicular cliffs that have excluded large animals for a long tick of geological time. The grass cover is unstable and very fragile. It needs the tool of rest because the American people quite rightly have chosen to preserve a landscape created eons ago. But Virginia Park does not represent what more accessible places once were or might become in such brittle environments.

The visitor's center at the nearby Arches National Monument passes out literature that describes the grassland that existed at the time the monument was set aside for preservation. The total rest that has been imposed since then has resulted in a dying grassland where widening bare patches are obvious. Even an unobservant visitor's attention is drawn to areas where officials have placed nylon mesh holding wood mulch to the ground (mentioned in chap. 13) in an attempt to get grasses growing again. Personally, although I, like so many environmentalists, once detested herds of cattle, I find them a more natural tool to use for restoring grassland health than nylon netting.

## Conclusion

All too often people assume that resting any environment will allow it to recover, but that assumption does not hold true in the brittle environments of the world. Nonetheless, rest remains a key strategy of conservation organizations attempting to protect the life in brittle environments. Because the adverse effects of rest are so hard to see when the belief that rest is natural is so deep, it qualifies as the most misunderstood tool in our tool kit. Partial rest, which has been even more difficult to decipher, is, I believe, the greatest single contributor to desertification in the brittle environments of the world.

The antidote to rest, especially in brittle environments, lies in the impact of the large grazing animals that coevolved with the pack-hunting predators, plants, soils, and other living organisms in these environments—the subject of the next three chapters.

# 21

## Living Organisms
### *Biological Tools for Solving Management Challenges*

THE PHRASE *LIVING ORGANISMS* may seem an ambitiously broad way to define a tool, and it is. *Plants and animals* sounds earthier, but it doesn't force us to consider the utility of bacilli and viruses in the same breath as trees, corn, or sheep, and we must when managing our environment.

In the Holistic Management framework, a broken line surrounds both the ecosystem process of community dynamics and the living organisms tool because they merely represent two aspects of the same thing. The dynamics of any biological community is manifested in living organisms. And living organisms cannot be isolated from the communities they nurture and that nurture them.

The relationship of this tool to the others in the tools row looks a bit clearer from the viewpoint of the earliest cave dwellers. Assuming they could have analyzed their situation and made decisions holistically, they would have seen that they required a certain landscape from which they could produce the food, cover, and water necessary to sustain the quality of life they desired. They would have recognized that all this depended on the same four ecosystem processes that sustain us. But their tools row was nearly empty. Their technology was made of sticks, bones, and stones, and they had no fire. They knew implicitly that the dynamics of the community they inhabited controlled them absolutely and defined what they could do in their environment.

When they harnessed fire and made their first spears they may have thought they had escaped this controlling influence, but the widespread use of these tools initiated many of today's deserts, proving that line of reasoning faulty. With increasingly sophisticated technology we continue to think we can escape the influence of the biological communities we inhabit, but we cannot any more than our earliest ancestors could. Yet, to the extent that we understand this, we can use living organisms to our advantage, as a tool, and the breadth of our options is remarkable.

The domestication of livestock about ten thousand years ago offered new possibilities for the use of living organisms in the service of management, but livestock were not recognized as such until very recently. Because their power to transform whole landscapes through their physical impact and grazing is so significant we break them into two tools (*animal impact* and *grazing*) within the living organisms category and discuss their use in detail in the next two chapters.

It could be argued that plants, in the form of cultivated crops, or tree belts, for example, also have this power to transform and should be listed separately too. But where plants remain static once established, animals move on, and this has implications for management. Unlike the diesel-powered bulldozer that can be parked once it has completed its job, the solar-powered herd used to help regenerate soils and reverse desertification needs to keep moving, both to feed itself and to escape ground it has fouled and trampled. These movements must be planned and there are specific guidelines for doing so.

The remainder of this chapter focuses on general principles and some of the other ways living organisms can be harnessed as tools for managing our ecosystem.

## Creating Biological Solutions

As a tool, living organisms offer potential biological solutions to a problem against technological ones: community complexity against pesticide and herbicide; crop polycultures and intercropping rotations against monocultures; composting against manufactured fertilizers; healthy water catchments against silt traps and other mechanical measures built to protect dams or cities from devastation. Also, it encourages us to treat the whole complex of life in our environment as a whole rather than as a menu of pesky or beneficial creatures that we may kill or husband at will. Most important, all living organisms both create and are created by the communities they inhabit and will not thrive outside them unaided.

## Regenerating Our Soils

The organic agriculture movement grew out of a desire to retain and improve on the knowledge developed by past cultures in their use of living organisms to enhance food production. However, in more recent times, so much food labeled "organic" is the result of soil-destroying practices that a new generation of *regenerative* farmers has emerged in response. Their aim is to produce food while enhancing soil life, which in turn produces healthier crops while ultimately decreasing production costs. To achieve this end they strive to keep soil covered year-round, do not turn it over (which destroys soil life), plant diverse crops, include livestock in their crop production, and more, as covered in detail in chapter 35. Above all, they view the soil itself as a living organism that produces abundantly when well cared for.

One excellent example of one small-scale farmer's success in harnessing living organisms to boost yields is Japanese philosopher Masanobu Fukuoka, who understood at a deep level the principles of biological succession. He used the organisms associated with particular successional levels to produce high yields of small grains without manufactured fertilizers, compost, pesticides, soil disturbance, or weeding. He succeeded because his understanding of community dynamics allowed him to enlist a great number of plants, insects, birds, small animals, and microorganisms as tools in creating an environment where his grain thrived in the protection of such complexity.[1]

Unlike Fukuoka, farmers operating in brittle environments generally have to enlist larger animals to aid in regenerating soils and soil life and to boost yields, as the next two chapters show. But these farmers, too, owe their success to a growing understanding of how biological communities function and the role of living organisms in keeping their land productive.

## Enlisting the Aid of Smaller Wildlife Forms

*Simple, inexpensive measures* can help increase the diversity of life. I once saw a dramatic example of this on a Mexican ranch where the owner had built a small concrete ramp up the outside of and down into a water tank, as shown in figure 21-1. One night, while we camped nearby, a terrific noise aroused us and we took flashlights to investigate and discovered a massive mating of toads in the tank. By dawn they had dispersed to resume their pursuit of bugs and flies around the ranch. The simple ramp to water in fact enabled a great variety of insects, birds, rodents, and other small animals to survive and contribute to the complexity and stability of the whole community, which also included larger wildlife—black bear, deer, turkey, and javelina—all of which were important in terms of that family's holistic context. By contrast, figure 21-2 shows a poorly constructed watering point provided to increase game populations in a national forest. While the intention was good, these young warthogs were finally able to water only because a pipe nearby leaked.

**Figure 21-1.** Concrete ramp into cattle watering trough that provides access to water for small animals and birds and prevents their drowning. Coahuila, Mexico.

**Figure 21-2.** Two young warthogs trying unsuccessfully to drink from a poorly constructed water trough provided for wildlife in a national forest. Zimbabwe.

Figure 21-3 shows a leaking water pipe on a Namibian ranch. Water pipes run along the fences to frustrate the local porcupines' appetite for plastic but many leaks are created by baboons, who find it easy to bite through the plastic when they want a drink. When the rancher articulated a holistic context involving complex living communities to sustain his family, he saw opportunity in the leaks. Half drums below them created additional watering points for thousands of birds, insects, and small mammals. Previously this rancher had thought such creatures had no connection to his family's wealth or to cattle ranching. But then he changed to jokingly talking about his enormous unpaid force of millions of "little people" all busily working for his family.

**Figure 21-3.** Water pipeline strung along a fence to avoid damage by porcupines. A leak has developed and been turned to advantage as a watering point for birds and small animal life. Namibia (courtesy Argo Rust).

## Biological Pest Control

The use of biological controls in lieu of chemicals represents a positive marriage of modern science and our knowledge of how communities function. Clear examples are breeding ladybugs to prey on aphids and nurturing certain bugs that eat problem plants. Parasites that attack fly larvae can decimate fly populations or people can simply run chickens in conjunction with pigs or cattle, not only increasing productivity but controlling flies and the diseases they spread. Screwworms have so far been controlled through the use of sterile males. When they mate, the female dies without reproducing.

When prickly pears were introduced to Australia in the early 1920s they thrived, so much so that vast acres were so heavily infested the land was considered useless. The cost of removing the cacti mechanically or poisoning them with chemicals was more than the land itself was worth, so entomologists ransacked their American homeland to find an insect that might help control the pest. They found it in the larvae of a small moth, which proved to be voracious eaters of the cacti. Within five years after being released in Australia the moth had done a spectacular job of destroying the vast majority of the cacti.[2]

Such measures are not without risk. The Asian carp introduced to the southern United States in the 1970s to control weed and parasite growth in aquatic farms escaped into the Mississippi River system, where they have devastated native fish stocks and have migrated upstream so swiftly that they now threaten the fishing and tourism industries in the Great Lakes region a thousand miles north.[3] Fortunately, there are many examples of biological pest controls that have proved less damaging to the environment.

## Genetically Engineering New Organisms

Modern breakthroughs in genetic engineering open the door to tremendous possibilities, but also equal temptations and dangers. Civilization might have spared itself some grief if it had gained more wisdom about the four ecosystem processes before acquiring this new power to intervene on an even bigger scale than previously. The new genes have escaped the test tube, however, so we must do the best we can to avoid embarking on a new Green Revolution more faulty than the last. A fundamental understanding of the reasons for our massive agricultural failures and the management that caused soil destruction and the bedeviling spread of deserts should temper the creation of new forms of life that agriculture does not need.

The current attempts to escape natural laws by genetically engineering crop plants that survive ever more powerful herbicides used to kill all other plant forms represent the wrong kind of thinking, in my opinion, and will not solve any of the world's problems. Such thinking considers soil as simply a medium for holding plants upright, while humans feed and nurture them in an artificial, hydroponics-type situation. Geneticists who have lost their connection with the land may dream that success lies in that direction, but reality dictates that living soil must do far more than physically support plants if civilization as we know it is to survive.

Up until now we have been enamored of the technology and ignored the web of relationships that define any biological community. The neglect and destruction of genetic material, even in the few principal plants sustaining civilization today, is in my mind criminal. Yet we continue to spend billions to genetically engineer ever more specialized plants to suit the artificial environments created by industrial agriculture.

Genetic engineering could also be placed under the heading of technology, but I prefer to position it in living organisms because it is the organism's use that is most relevant to management, not the technology used to create it. Genetic engineering may become a powerful tool for good in certain circumstances if handled with wisdom, and I hope that Holistic Management will play a role in assisting us to that wisdom.

## Planting Trees to Hold Back the Desert

In 1978 the Chinese, noting that the Gobi Desert annexed to itself over 1,500 square kilometers (600 square miles) annually, began planting greenbelts of trees, an overt case of enlisting living organisms as tools in their struggle against the advancing sand. By 2006, they had planted 25 million hectares (62 million acres). By 2050 the goal is for 100 billion trees to cover 4.1 million square kilometers (1.6 million square miles)—more than a tenth of the country. Yet, despite the number of trees planted to date, the Chinese government admits that roughly one million tons of desert dust and sand currently blow into Beijing each year.[4]

What so many of the practice's promoters overlook is that trees planted to prevent deserts spreading, effective as they may be as wind breaks, *cannot grow and reproduce independent of the level of development of the biological community as a whole*. In fact, scientists outside China have questioned the ability of the trees within the "great green wall" to reproduce and maintain themselves without human assistance.[5]

There are numerous examples, some dating back to antiquity, of capturing water runoff to plant trees in an effort to hold back or reclaim deserts. Israel's recent efforts to reclaim the Negev by using machinery to channel rainfall runoff into small pockets of land planted with trees is as likely to be self-sustaining as the similar, but failed, efforts of the ancient Nabataeans referred to in chapter 11. The Israelis have spent ten thousand euros per hectare planting trees outside their natural environment, one that receives 200 millimeters (8 inches) or less of rainfall annually and is more hospitable to grasses. In the same region, the government of Abu Dhabi (United Arab Emirates) has spent over thirty billion U.S. dollars planting fourteen million trees and using the latest in drip irrigation technology supplied by 5,651 wells and a network of more than 55,000 kilometers (34,000 miles) of irrigation pipes.[6] Despite spending millions of dollars each year to maintain these "forests," the sand is blowing right through them (fig. 21-4).

**Figure 21-4.** Planting trees to hold back the desert in low-rainfall areas more suited to grasses is futile. The sand is blowing right through these "forests." Abu Dhabi, UAE.

These areas were once grasslands, not forests, and it is grass plants, those great soil stabilizers, that will most quickly revegetate and reclaim these deserts. But to succeed and sustain that success, livestock, properly managed, will be essential to help re-create and maintain the grassland communities that gave rise to their wild ancestors.

---

These areas were once grasslands, not forests, and it is grass plants, those great soil stabilizers, that will most quickly revegetate and reclaim these deserts. Livestock, properly managed, will be essential to help re-create and maintain the grassland communities that gave rise to their wild ancestors.

---

## Conclusion

Looking to the future of science and resource management, it is clear that we must turn far more to studying ecological processes and how they function so we can better understand the relationships that exist among the living organisms, including ourselves, that populate any biological community. This will enable us to concentrate more on preventing the problems our ignorance has led to, and less on developing cures that, in damaging ecosystem processes, only create more problems.

Isolated people and organizations have thought more about our connection to the communities of living organisms that sustain us, but for several thousand years the mainstream of human interest has flowed in other channels, with the result that civilizations have come and gone, and now all civilization is endangered. Avoiding the inevitable outcome of such a threat will require more humility than past or present generations have shown and a greater acceptance that the unknowns in nature and science still far outweigh the knowns.

This brings me to the two critical tools within the living organisms category that are essential for reversing desertification and successfully addressing climate change—the physical impact and grazing of the hoofed animals, which today, for all practical purposes, means livestock.

**Plate 1.** In the 1950s, the banks of the Zambezi River were stable and well vegetated, despite the high numbers of game and the presence of hunting, gardening humans. Zimbabwe.

**Plate 2.** By the 1980s, the banks of the Zambezi River within the national park were nearly devoid of vegetation, even though game populations had been culled heavily and the hunting, gardening humans removed. Zimbabwe.

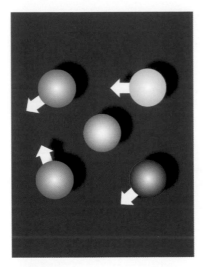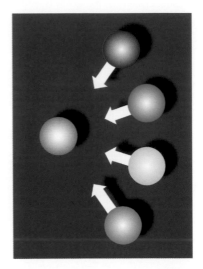

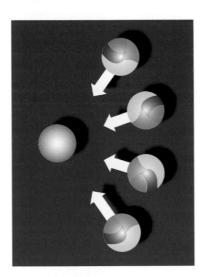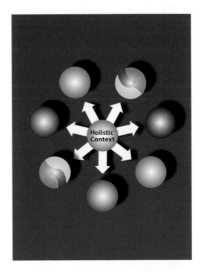

**Plate 3.** (Left to right): 1. People in isolated disciplines have considerable knowledge of green, yellow, red, and blue, but no knowledge of gray (the "whole"). Management is not even focused on gray, as arrows indicate. 2. People in a multidisciplinary team are focusing on gray from their perspective, as arrows indicate, but still with no knowledge of gray. 3. To overcome communication problems, people trained in several disciplines form interdisciplinary teams that then focus their attention on gray, as arrows indicate. Knowledge of gray is still lacking. 4. We now take the perspective of the whole (gray), by creating a holistic context that decision makers use to look outwardly at the knowledge available to determine which best serves their management needs.

**Plate 4.** Capping. Some soils develop a cap so hard it is difficult to break without a knife or some other hard object. Zimbabwe.

**Plate 5.** Typical rested soil surface dominated by algae and lichens and roughened by freezing and thawing. Natural Bridges National Park, Utah.

**Plate 6.** Unsuccessful attempt to get grass established with nylon netting and woodwool as litter. Dead sticks are holding down netting. Canyonlands National Park, Utah.

**Plate 7.** Remnant perennial grass patch in annual California grassland (courtesy Richard King).

**Plate 8.** U.S. Fish & Wildlife Service officials inspecting rested and dying grassland with large, bare ground spaces opening between plants. All seedlings within view are forbs, not grasses. Sevilleta Wildlife Refuge, New Mexico.

**Plate 9.** Perennial grassland in a 750-millimeter (30-inch) rainfall, very brittle environment rested one year. All plants are smothered by a mass of gray-to-black oxidizing material that is preventing light from reaching growth points. To the right, the same species are kept healthy by periodic mowing (performed by road crews). Zimbabwe.

**Plate 10.** Brittle environment community shifting from grassland to woody forbs (showing green) and trees in overrested patches (background), while overgrazed patches (foreground) show excessive erosion. Coahuila, Mexico.

**Plate 11.** Nature preserve alongside the Rio Grande. This very brittle riparian area, rested for more than thirty years when this photo was taken, has in many places deteriorated to bare, eroding ground. Almost all plants have died. (Metal structures were built to minimize flood damage.) Albuquerque, New Mexico.

[A] **Land at approximately 9 or 10 on the brittleness scale.** Following more than forty years of total rest inside the plot (right side of fence) and partial rest and overgrazing/overresting of plants outside it (left side of fence), no perennial grasses remain inside the plot and only a few have survived outside it. Central California.

[B] **Land at approximately 7 or 8 on the brittleness scale.** Following more than forty years of total rest inside the plot (right) and partial rest and overgrazing/overresting of plants outside it (left), only three species of rest-tolerant perennial grass remain on both sides. Western Arizona.

[C] **Land at approximately 4 to 6 on the brittleness scale**. Following sixty-five years of total rest inside the plot (right) and partial rest and overgrazing/overresting of plants outside it (left), plant spacing is much closer both inside and outside, but the few species of perennial grasses that remain on both sides are those that can withstand high levels of rest. Southeastern Arizona.

**Plate 12.** Research plots established by the U.S. government in the western states over the last fifty or so years demonstrate the effects of total rest (inside the plots) and partial rest (outside the plots) on land ranging from very brittle to less brittle.

[A] September 2004. Bare, hard capped, and eroding for several decades, regardless of how much rain fell, this site was thought to be incapable of growing grass. September is about six months into an eight-month nongrowing season.

[B] September 2005. One year after a herd of five hundred cattle were concentrated on the site over seven nights. Despite it being a dry year, grass has started to grow.

[C] June 2006. The rains in the 2005–2006 season were above average and well distributed and, without any further animal impact treatments, the grass has thickened up and is covering nearly all the bare ground.

**Plate 13.** An area impacted by a cattle herd late in 2004, and the restoration that occurred over a two-year period with no additional treatments. Zimbabwe.

**Plate 14.** By planning the grazing (and doubling livestock numbers) on the land in the background (beyond the fence), the grassland improved despite the brush, which was not the cause of the bare ground. In the foreground, the government continued to tackle symptoms by destocking and brush clearing, but the brush continued to thrive. Lebowa, South Africa.

**Plate 15.** One of a few severely grazed perennial grass plants among millions of plants after one horse had grazed for one hour in a paddock. New Mexico.

**Plate 16.** The only grass on the entire ranch appeared when this airstrip was graded. One perennial grass plant is visible under the wing of my plane. South Africa.

**Plate 17.** The same site as that shown in plate 16 three decades later. Grassland, in which the desert bushes still thrive, now covers the entire ranch. South Africa.

**Plate 18.** Herd effect periodically applied in the foreground; never applied in the background. Montana, USA.

**Plate 19.** The response to herd effect. Nearly every plant was grazed or trampled down (top photo) when four hundred cattle were crowded into this five-acre test enclosure for the better part of a day. One year later (bottom photo), the growth inside the enclosure was more lush than any outside it. Arizona (courtesy Dan Daggett).

**Plate 20.** Colin Seis harvesting a "pasture-cropped" oat crop, and revealing new perennial grass growth beneath the harvested crop (courtesy Colin Seis).

**Plate 21.** Cattle moving through a diverse cover crop mix that was seeded, using zero-till, into an old perennial hay land planting. The cattle were grazed at a stock density of 300,000 pounds live weight per acre and were moved to a fresh bite twice daily (courtesy Brown's Ranch).

*22*

# Animal Impact
## *A Tool for Regenerating Soils and Shaping Landscapes*

ANIMAL IMPACT REFERS TO ALL THE THINGS HOOFED ANIMALS (ungulates) do besides eat. Instinctively we have considered the dunging, urinating, salivating, rubbing, and trampling of these animals as generally inconvenient conditions of their presence. Fortunately, nature has not been so shortsighted, and we have recently discovered in the lumbering, smelly, but powerful behavior of these animals a tool of enormous significance for reversing desertification, and for better management of water catchments, croplands, forests, and wildlife.*

The Holistic Management framework grew from the discovery of the four key insights described in earlier chapters. Two of them—the fundamental differences between brittle and nonbrittle environments, and the role of herding animals and their predators in maintaining biological communities in the more brittle ones—led to a recognition of animal impact as a specific tool within the living organisms category, as described in the last chapter.

The following examples give an idea of the power and versatility of this tool. Though critical to the maintenance of the more brittle environments

---

* *The grazing marsupials of Australia, although they aren't hooved, have hard feet and exhibit these same behaviors.*

and most often called for in those conditions, it is useful in less brittle environments as well:

- In a brittle environment, overrest has allowed grass plants to accumulate several years of old material. Animal impact can be used to help restore biological decay, something neither fire nor any technology can do.

- In a less brittle environment a farmer is faced with trying to maintain soil structure and cover while having to dispose of crop residues in time to plant her next crop. She can do so using animal impact rather than machinery or fire. Now the crop residues are used to feed the animals, and nutrients are returned to the surface in the form of dung and urine.

- We need a firebreak. Herding or a fine spray of very dilute molasses or saline (salt) solution will attract and bunch a herd of cattle enough to make a firebreak through almost any kind of country at minimal expense without exposing soil or creating an erosion hazard.

- Leafy spurge, knapweed, thistles, or some other noxious plant is filling the vacuum created in a grassland as grass plants lose health and vigor, and thousands of dollars have already been spent in futile eradication efforts. Now we can use continual doses of very high animal impact followed by well-planned recovery periods that enhance grass health and vigor, moving succession beyond the stage that suits the offending plants.

- Bare eroding ground that we once might have fenced off, ripped, and seeded at great expense can now be subjected to periodic heavy impact by giving a large herd a few bales of hay, which excites and concentrates them on the area. New plants then establish on the broken, litter-strewn ground, at no cost or lost production.

- Erosion gullies, whose steep-cutting banks grant no foothold to plants, spread across the land. Why pay for a bulldozer to slope the banks and chew up more land while consuming diesel and polluting, when a herd of livestock attracted to the gully can break down the sharp, cutting edges and create the conditions for plant growth to heal them. This high animal impact, while curing the gully, if also used in the catchment of the gully also tends to correct the noneffective water cycle that caused the damage in the first place.

- A mining company has millions of dollars tied up in legally mandated deposits until land it has stripped for coal mining is reclaimed and returned to productive use. Millions have already been wasted on reseeding and mechanical treatments that have failed. Now the company can use cheap hay full of weeds and seeds to attract and greatly concentrate animals on these sites and reclaim and maintain them for a fraction of the cost (as some companies have already demonstrated).

- Impenetrable brush clogs potential grassland. Although low densities of calm cattle will not touch it, a large herd attracted by the smell of molasses blocks, sawn in half and thrown deep into the thickets, will penetrate and break down the brush.

- Fish management often requires vegetated banks rather than steep, eroding ones, as shown in figure 22-1. Very high animal impact for very short periods can promote this as the well-vegetated river bank in figure 22-2 shows.

- Stock trails to water or down a hillside threaten to wash out. Although it may seem strange that damage caused by trampling can be cured by trampling, the treatment works because of the vast difference between the effect of prolonged, one-way trailing and the milling of bunched animals for very short periods.

- Coarse, fibrous grass has come to dominate a bottomland where low stocking rates and partial rest have prevailed. Traditionally fire has been used to "keep the grass palatable" but worsens the situation where high animal impact improves it.

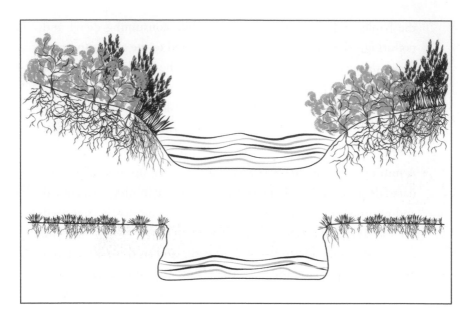

**Figure 22-1.** Well-vegetated stream banks are more stable and provide shade and cover for fish (top). Bare stream banks (bottom) are unstable, and their steep edges will keep cutting back, widening the stream and making it shallower.

- Humanmade desert soils have remained hard capped from lack of disturbance for over three thousand years. Pastoralists periodically shepherd their livestock over the land, overgrazing the few remaining plants while perpetuating the partial rest resulting in such capped soil. Occasionally a new grass plant establishes where a cow, donkey, sheep, or goat has broken the hard surface. However, once a herd is concentrated and moved to plan to prevent overgrazing, the desert starts to make its recovery at last.

All these examples are common situations where animal impact is the most practical tool available. Animal impact is indeed such a versatile tool that we run the danger of prescribing it as reflexively as some now hand out knee-jerk technology fixes. The context checks covered in upcoming chapters help prevent that.

The discovery that brittle environments need periodic disturbance to maintain stable soil cover makes animal impact the *only* practical tool that

**Figure 22-2.** Well-vegetated riverbank produced by using high animal impact with planned grazing to prevent overgrazing or overbrowsing. Zimbabwe (courtesy R. H. Vaughan-Evans).

can realistically halt the advance of deserts over millions of hectares. Here and there other tools can help, but what other way exists to treat millions of square kilometers of often arid and rugged country each year without consuming fossil fuel, without pollution, and by a means millions of even illiterate people can employ even while it feeds them?

> What other way exists to treat millions of square kilometers of often arid and rugged country each year without consuming fossil fuel, without pollution, and by a means millions of even illiterate people can employ even while it feeds them?

## The Role of Livestock

No other aspect of Holistic Management has caused such controversy as the suggested use of animal impact has. That trampling by livestock damages both plants and soils is a deeply held belief throughout the world, as mentioned in chapter 5. Some range scientists have for years rejected the one idea that has more promise of solving the riddle of desertification than any other. Meanwhile, they have supported the development of machines of extraordinary size and cost to break soil crusts and disturb vegetation through mechanical impact toward the same end. Because we have now lost most of the large herding wildlife species, and the predators that induced their movement, we are left only with livestock in most instances to simulate that role, which we do by bunching them (there is no need to panic or stampede them), using herding or fencing, and planning their moves. There is no other tool than animal impact, I believe, that can do more to regenerate the world's damaged soils and reverse desertification.

Unfortunately, livestock—cattle and goats in particular—are generally seen as an enemy of the land and wildlife, rather than its savior. Recent concern over the methane released by ruminating cattle has reinforced this view. Yet, as far as we know, all ruminants—buffalo, bison, antelope, sheep, goats, pronghorn, deer, giraffe, and the like—produce methane as a by-product of rumination. Moreover, atmospheric methane levels did not increase between 1999 and 2008, even though livestock numbers increased seventy percent over the same period.[1] Therefore, something other than the presence of cattle is most likely responsible for the high levels of methane in today's atmosphere.

The unnatural way most cattle are now fattened for slaughter—in crowded and unhealthy feedlots, where they are fed grains and other rations they are not genetically adapted to eat—is another strike against cattle. Now, not only do cattle destroy riparian areas, damage wildlife habitat, and create deserts, they also produce fatty and chemically tainted meat, and tons of manure and urine that are a liability in the feedlot, but an asset out on the land.

Is it the fault of the animal that we have taken a grazing creature, which developed a productive and mutually beneficial relationship with plants and soils over millions of years, and turned it into a meat factory? Rather than condemn the animal, we should be condemning our management for what *we* have done.

---

Rather than condemn the animal, we should be condemning our management for what we have done.

---

## Animal Impact and Biological Breakdown

*Grass plants play a significant,* often dominant, role in providing soil cover in brittle environments, which in turn promotes an effective water cycle and healthy soil life. In these environments, a few trees and shrubs are evergreen, but most drop their leaves in the dry or nongrowing time of the year. Shedding their leaves after having withdrawn the nutrients from them as they change color ensures adequate sunlight reaches new growing buds as the next growing season begins. Grasses, like those trees and shrubs, move their energy reserves down into the base of the plant or the roots. The aboveground leaves and stems change color and die off. Unlike trees and shrubs, however, grasses are unable to remove that dead material, having coevolved over millions of years with large grazing animals that did it for them. If not removed by trampling and grazing (or burning or mowing), this annually dying material oxidizes in sunlight and only breaks down over several years.

*This shift from rapid biological breakdown by animals to slow chemical and physical breakdown by oxidation and weathering has profoundly adverse effects in brittle environments.*

Commonly it leads to weakening and premature death of many grass species. In higher rainfall regions, communities tend to shift to ones dominated by woody plants with bare soil between them, and in lower rainfall regions to communities dominated by low-growing desert shrubs with bare, capped, and crusted soils between them. In either case available rainfall tends to become less effective, with more of it running off the land or evaporating out of bare soil surfaces, resulting in increasingly severe and frequent droughts and floods.

## What Animal Impact Does

Objectively speaking, few would question the salient aspects of animal impact—there are three things it does and for which we use it as a tool:

1. Hoofed animals tend to compact the soil at every step because their weight is concentrated on four small feet. The "sheep's foot rollers"* of modern civil engineering memorialize the herds used to compact roadbeds and earthworks not even a century ago.

2. When animals are excited or closely bunched, their trampling causes breaks and irregularities on the soil surface, as anyone who has tracked game, a cow, or a horse knows.

3. Such animals tend to speed the breakdown and reduce the volume of plant material returned to the soil surface through their dung and urine. They also speed the return of uneaten old plant material to the soil surface through the litter they trample down.

Whether any of these tendencies works for good or ill on the land depends entirely on management, *particularly of the time factor*.

### *Time and Trampling (Donkey Days)*

I myself did not overcome my old biases without considerable effort, doubt, and false beginnings. Early on I entertained the hypothesis that animal impact had some important function, but I could not articulate it well until I saw the distinction between brittle and nonbrittle environments and understood the importance of time and animal behavior.

The following example illustrates how timing may fundamentally change the quality of an event. Suppose you have a small house on a hill, and you and your donkey fetch water daily from the stream below. After one year of trampling the same path day after day a substantial gully forms, and the stream bank where you load the water cans becomes a trampled out bog. In this instance you could say that we had had 365 donkey-days (1 donkey × 365 days) of trampling. For centuries we observed such damage and in essence said that we had too many donkeys.

---

* Rollers with blunt spikes that cause more compaction than a smooth roller.

Now, suppose you took a herd of 365 donkeys down the hill and hauled a year's worth of water in one morning. In this instance you would again have 365 donkey-days of trampling. Though a passerby that afternoon would remark on severe trailing and trampling of the streambank, those "wounds" would have 364 days of plant growth and root development to heal before you had to come back. When you did, you could expect to find both the trail and the loading place completely overrun by new growth. In fact both might well be greener and healthier than before with the old grass removed and the dung and urine deposited, though they had still borne 365 donkey-days of traffic per year. Thus *time*, rather than animal numbers, was the critical factor in trampling. For thousands of years we simply overlooked the fact that

---

### Overtrampling

*Trampling carried out* for too long always causes damage to soil and plants. Trampling carried to extremes but over a very short time period can cause temporary damage to the ecosystem processes, but most environments have astounding resilience. The first time a rancher called me out to inspect the damage my advice had caused when his large herd concentrated during a violent storm, we photographed and discussed the resulting quagmire at great length. Finally, we decided to ignore it and keep on with the concentration and movement as planned. About eighteen months later we recalled the incident but could no longer determine exactly where it had happened, as all paddocks looked much the same. In many cases, the temporarily overtrampled area looks much better after a season than areas that were not overtrampled.

The *continuous* trampling we see so often around gates, water points, and feed troughs does not allow recovery to take place, but such examples often figure in arguments against trampling in general. Even sophisticated time management may not eliminate all such cases totally, but usually the area involved right at the water point, trough, or gate is insignificant and can be treated as a sacrificial area.

timing, rather than animal numbers, governs whether animal impact acted favorably or adversely on land.

### Stock Density and Herd Effect

One other observation has escaped many scholars of this subject, and poor appreciation of it continues to bias research at many levels. The herding animals that contribute most to the maintenance of the more brittle environments behave in a variety of ways that produce different effects. Two management guidelines—stock density and herd effect—show how to harness these effects when applying the tool of animal impact. *Stock density* refers to the concentration of animals within a subunit of land, or paddock, at any given time. *Herd effect* refers to the impact on soils and vegetation produced by a herd of bunched animals. We will look at both guidelines in detail in chapter 34.

My observations of how these different modes of behavior affect plants and soils led eventually to the definition of partial rest, mentioned in chapter 20. Land sustains partial rest when animals, either domestic or wild, are present but seldom have cause to produce herd effect. They may slightly disturb a recently capped soil, or a soil so long capped that it has become covered with algae, lichens, or mosses, but seldom can stimulate a successional shift to more complex communities and stability. The millions of hectares of America's more brittle environments that are deteriorating under the combination of partial rest and overgrazing in the presence of livestock, bison, or wildlife support this conclusion.

In the United States traditional range management favors protecting long-capped soils encrusted with a mix of algae, lichen, fungi, and bacteria, known as cryptogams, because they fix nitrogen and inhibit erosion to a degree and are said to promote a successional advance to grassland. Standard practice is to disparage any kind of trampling, despite abundant evidence that these crusts do not advance to grassland, as chapter 13 on community dynamics explained (see color plates 5 and 6). It takes a much deeper, long-term observation to see that heavy trampling over a short period leads to establishment of plants and litter that protect the soil much better than a cryptogamic crust ever can. These former grasslands developed with vast numbers of hoofed animals, which would have trampled any cryptogamic crusts in their path.

### *Kick Starting Land Recovery*

Color plate 13 shows a fairly dramatic response, over a period of two years, to animal impact on land that was deteriorating badly under partial rest and a little overgrazing. I owned this ranch before donating it to the Africa Centre for Holistic Management some years ago, and know it well. This part of it had been bare and very hard capped for more than thirty years at the time the first photo was taken, and the belief was that grass couldn't grow under the trees. The ranch lies in a very brittle 650-millimeter (26-inch) rainfall area, and even though many wildlife species are present, including predators, the wild grazers no longer function naturally and had not disturbed this area enough to break the capping so that new plants could germinate.

So we decided to concentrate our cattle herd of about five hundred animals on this area. We did so by placing our movable predator-proof kraal, or enclosure, on the site for about a week, concentrating the whole herd within it each night after they had spent the day grazing nearby. There was sufficient impact to break the capping, and that, plus the infusion of dung and urine, was enough to get grass growing again—in this case annual grasses. The next year, with no further treatments, although wild grazers now began to frequent the area, the additional soil cover provided by the new plants and plant litter helped to grow enough new grass, including a few perennials, to cover much of the bare ground.

The use of animal impact provided the kick start needed for this biological community to move forward again without any reseeding or other assistance. And we have repeated this treatment now on nearly five hundred additional sites. Annual grasses usually dominate the first successional advance, but perennial grasses begin to appear in about the third season and continue to increase in numbers in following years in a pattern that has become familiar to those using animal impact to reclaim bare ground.

## Impacting Croplands

We have learned much about how animal impact might serve in cropland management, particularly for small-scale communal farmers in Africa, where it has been used as a primary, or *biological*, form of tillage and also has the additional benefit of saving the farmer having to haul manure or fertilizer.

The concept of brittle and nonbrittle environments has not yet entered the thinking of most crop farmers, even after the American Dust Bowl of the 1930s showed how quickly cropland can be destroyed in the more brittle environments. Brittle environment croplands will require the removal and recycling of crop residue that does not decay fast, and animal impact may well provide the answer where machinery and fire have not.

Animal impact can be used to enhance crop production in other ways. Colin Seis in Australia sows annual grain crops directly into perennial grassland that his cattle have grazed, trampled, and fertilized, in what he calls "pasture cropping." In the United States Gabe Brown of North Dakota uses similar practices in an even lower rainfall brittle environment, and even though rainfall isn't sufficient to grow a grain crop every year he remains profitable even as his soil productivity soars. (See chap. 35 on cropping.)

## Impacting Forests

The management of forests that lie in the more brittle environments is a major challenge in Australia, Africa, and the United States in particular, where lightning or human fires burn vegetation readily. In many instances forests dependent on numerous small fires lit by humans have developed in response to both the burning and the protection such burning afforded from major conflagrations. In Australia and the United States such practices over thousands of years have favored tree species that require an occasional fire in order to reproduce and thrive. This is increasingly the case in Africa, though the change from fire-sensitive to more fire-dependent tree species appeared to begin only after European settlement, when the frequency of burning greatly increased.

Management in either case tends to alternate between using controlled fires to burn the understory, which, left in place, can fuel larger, more damaging fires, and doing nothing at all—leaving things to Nature. The pendulum generally swings from the latter to the former when doing nothing results in a dangerous and costly conflagration, including the mega fires of today. As pointed out in chapter 19, neither approach will maintain the few remaining forests populated by fire-sensitive tree species. But the well-planned use of animal impact and grazing, with no overgrazing, can. Animals used in this manner can clear the understory as effectively as fire, and can do so without damaging the soil surface or decreasing soil organic matter.

---

### Effects of Animal Impact in Brittle and Nonbrittle Environments

*The effects of animal impact at either end of the brittleness scale are as follows:*

*Nonbrittle Environments*

- **Community dynamics**—periodic high animal impact tends to help maintain high diversity in grassland communities and retard shifts to woody communities.

- **Water and mineral cycles**—high animal impact tends to improve both water and mineral cycles. However, it also tends to sustain grassland, and the mineral and water cycling in the grassland may not be as effective as they would be if the community advanced to mature forest.

- **Energy flow**—periodic high animal impact tends to increase energy flow, although, again, when used to maintain grassland in lieu of forest, energy flow may never reach that of mature forest.

*Very Brittle Environments*

- **Community dynamics**—periodic high animal impact promotes the advancement of biological communities on bare, crusted, gullied, and eroding ground, and retards a shift to woody communities.

- **Water and mineral cycles**—periodic high animal impact generally improves water and mineral cycles.

- **Energy flow**—periodic high impact tends to build community complexity, improve water and mineral cycles, and increase energy flow.

---

The African teak forests mentioned in chapter 19 did, in all probability, develop with a large animal component. It is difficult to imagine how else sand dunes eventually developed into mature teak forests that are so fire-sensitive. Today cattle, goats, and donkeys are allowed to graze in these forests along with wildlife, but they are few in number, rarely if ever concentrate, and remain in the same areas for prolonged periods of time. In effect the tools being applied are partial rest of the land combined with overgrazing of plants and fire and the result is entirely to be expected—these forests are simply desertifying gradually.

Forests that now contain mostly fire-dependent species may survive several millennia more under conventional management, but because of the atmospheric pollution stemming from the burning we cannot afford to let it continue. Animal impact can and probably should replace fire, but because the use of animal impact will eventually lead to a reduction of fire-dependent tree species, we must be clear on what we want, and that clarity can only come about in a well-developed holistic context.

## Conclusion

One of the greatest immediate benefits from animal impact can be seen in the restoration and maintenance of brittle environment water catchments, which store not only more water but also more carbon. While partial or total rest can sustain soil cover in the perennially moist nonbrittle environments, no technology exists that could replace animal impact on all the ranches, farms, pastoral lands, national parks, and forests that cover the bulk of most brittle environments, where either form of rest is so damaging to soil cover.

Those who remain opposed to livestock—and they are many, including scientists, environmental groups, vegetarians, governments, and international development agencies—remain unaware of the fact that no form of technology, nor burning, nor resting land can effectively address the desertification occurring in the world's grasslands while feeding people at the same time. The most troubled and violent region of the world, which extends right across North Africa, the Middle East, and up into China, consists of desertifying grasslands. And ninety-five percent of these lands cannot grow crops. They can only feed people from livestock, which will be essential for regenerating the grasslands.

Animal impact is a natural phenomenon that we choose to call a tool within the living organisms category because we can manipulate it to serve our ends. That distinction becomes equally obvious in regard to the next tool within that category—grazing—which, like animal impact, will be critical for restoring desertifying lands to health.

# 23

## Grazing
### *A Tool for Enhancing Plant and Animal Health and Productivity*

*IN CHAPTER 6 WE EXPLORED BRIEFLY* some of the old beliefs about overgrazing—what it is and what causes it—and why those old beliefs were confounding our efforts to reverse the desertification process. Now it is time to look at grazing in more detail to use this tool to beneficial effect.

Grazing ranks as a tool because management can manipulate the intensity and timing of it and the animal/plant relationships that govern it. But, like the other tools within the living organisms category, it has natural aspects in that humans did not design the mouths of livestock or wild grazers, teach them how and what to eat, or show them how to behave. And because the ways in which they do these things are crucial to the results of management, we need to better understand them.

In this chapter we will consider grazing as if animals float over the ground without dunging, urinating, salivating, or trampling as they feed. Though grazing never occurs apart from these things, separating the act of grazing from the other simultaneous influences of the animals on the land (animal impact) helps us to better understand the influence of each and thus the use of each as a tool.

Several examples from environments that lie closer to the very brittle end of the scale show how widely the effects of either grazing or animal impact can vary:

- Grazing might be maintaining the health of individual plants, while low animal impact (partial rest) is simultaneously exposing the soil between the plants as fewer new plants can establish.

- Grazing applied as overgrazing may be weakening or killing some plants in the community, while at the same time high animal impact is tending to increase the number of plants, the amount of soil cover, and the effectiveness of the rainfall.

- Grazing, together with adequate animal impact, can maintain soil cover, keep grass plants healthy and more productive, and in general enhance the functioning of all four ecosystem processes. But overgrazing combined with low animal impact (partial rest) produces the opposite effect. In fact this latter combination is the most commonly applied, and it is the one with the greatest tendency, in the more brittle environments, to lead to desertification.

The ability to analyze the influence of animal impact and grazing separately allows us to better interpret what we are seeing on the land. It also enables us to more easily unravel such questions as how ranchers and farmers damaged parts of Africa and the Americas more in three hundred years than pastoralists and their flocks managed to do in more than five thousand years in other parts of the world.

A thorough discussion of grazing requires a working definition of the term. Strictly speaking, *grazing* refers to the eating of grasses and herbaceous forbs, and not woody vegetation, such as brush and trees, which are technically *browsed*. The tool of grazing encompasses both. Browsing is covered in more detail later on.

Before we can meaningfully define grazing and overgrazing, however, we need to note the differences between annual and perennial plants, the types of perennial grasses, and their relationship with grazing animals:

- **Annual plants** include grasses and forbs (weeds\*) that germinate, grow, and die in one season. Their populations usually fluctuate widely in numbers from season to season. In some seasons they may fail to establish altogether, leaving the soil exposed. Annual grasses are less prone to overgrazing because they do not live long enough; they generally begin to die once they have produced seed.

- **Perennial plants**, be they grasses, forbs, shrubs, or trees, can be very long lived. Their populations fluctuate far less in numbers, helping to hold soil in place year-round and keeping more of it covered throughout the year. The lower the rainfall and the more brittle the environment, the greater the role played by perennial grasses in keeping soil covered. Their presence and health determine whether deserts advance or retreat. They are easily overgrazed, however, as you will see.

- **Perennial grasses** mainly grow in two forms: upright, or prone, with lateral running shoots above- or belowground. The more brittle the environment, the more likely the upright perennial grasses will appear as distinct bunches, the less brittle, the more likely they will appear as a sward or mat in which individual plants are hard to distinguish. This variation in the upright grasses can occur even within a species. *Themeda triandra*, a common grass in Africa, grows in a matted form in less brittle environments but in a bunched form in more brittle environments. The runner grasses do not appear to change form at different points on the brittleness scale, and they commonly become dominant where moisture is

---

\* I prefer to use the term *forbs* when referring to the smaller tap-rooted, or dicotyledonous, plants. The term *weeds* is almost always used in a negative sense and can apply to a tap-rooted or a fibrous-rooted (grass) plant.

adequate, many plants are overgrazed simultaneously and repeatedly, and animal impact is high. They are often planted in pastures and lawns, especially in the tropics, partly because most of them are fairly resilient to overgrazing (or very frequent mowing).

- *Severe grazers* and some perennial grasses are dependent on each other. The buds, or growth points, on perennial grasses of either form occur either very close to the ground near the plant base, or well aboveground, along or at the ends of the plant stems. The position of these growth points probably indicates the evolutionary development of the species. Those with growth points close to the ground probably evolved in close association with severe grazing animals that kept the plant clear of old stems so that light could reach the growth points. If they remain ungrazed, these plants can die prematurely. They have in effect become *animal dependent*. Those plants with growth points well aboveground probably evolved in places under little or no pressure from severe grazers (such as on the steep slopes of gorges), because they can be set back by severe grazing, but they thrive when rested. They are in effect *rest-tolerant* grasses.

## Grazing and Overgrazing

To grasp the difference between grazed and overgrazed plants, picture a healthy perennial bunchgrass plant with ground-level growth points, and imagine that a large animal bites the whole plant down to three or four centimeters (an inch or two) above the soil. That is severe grazing, but not unusual or bad in that most herding animals evolved to graze in such a manner. In the growing season, the plant receives a brief setback while it uses energy previously converted from sunlight that it now takes from its crown, stem bases, or roots to reestablish new growing leaf. But it also receives a long-term boost because the plant tends to end the season better off and less encumbered with old leaf and stem than its ungrazed neighbors. The growth points at the base remain intact, and no old growth of the previous year stands in the way of regeneration. The soil also receives a boost because anytime a plant is severely defoliated while growing, root systems die back and become feed for communities of bacteria, leaving porous passages and carbon-rich biomolecules that are aggregated into a sticky substance called humus.[1] When the animals move on, the root systems regrow along with the aboveground leaves. The process repeats itself when the regrown perennial grass is grazed again, increasing soil porosity and water and carbon content at the same time.

If the bite comes in the dormant season when the plant *has no further use for these leaves and stems of the past season,* which have become a potential liability, it loses nothing and gains the advantage of unimpeded growth at the start of the new growing season. Plants surrounding it that were not grazed are hampered by old material when the next growing season starts. This is why so many people use fire. It removes all the dead matter and allows the ungrazed plants to grow freely once again.

Overgrazing occurs when a plant bitten severely *in the growing season* gets bitten severely again while using energy it has taken from its crown, stem bases, or roots to reestablish leaf—something perennial grasses routinely do. Overgrazing of perennial grasses occurs

1. when the plant is exposed to animals for too many days and they are around to regraze it as it tries to regrow,
2. when animals move away but return too soon and graze the plant again while it is still using stored energy to reform leaf, or
3. immediately following dormancy when the plant is growing new leaf from stored energy.

Anytime a growing plant is severely defoliated, root growth ceases as energy is shunted from root growing to regrowing leaves. This movement of energy between leaves and roots and vice versa is important, not only to maintain the plant's ancient relationship with severe grazing animals but also to sustain the plant over dormant nongrowing periods. At the end of the growing season, most perennial grasses transfer reserves from leaves and stems to stem bases, crowns, and/or roots. This reserve carries the plant through the dormant period and supports the next year's first growth.

If bitten early in the growing season, however, when reserves have already been tapped to provide the initial growth of the season, perennial grasses then have to utilize what little energy remains and will severely deplete roots to provide that energy. If subsequent bites are taken before roots have reestablished, the plant may die. Some scientists argue that energy for new growth is taken from what leaves and stems remain on the grazed plant, and in the process some roots die to maintain root-to-leaf balance. Where the energy for new growth comes from, however, is not as important as what happens to the roots. No matter which theory you subscribe to, it is fairly evident that severe defoliation, repeated too frequently or repeated following inadequate recovery time, causes root mass to decrease so significantly that the plant weakens or dies. Thus a simple definition of overgrazing is any grazing that takes place on leaves growing from roots that have not yet recovered from a previous grazing.

---

Overgrazing is any grazing that takes place on leaves growing from roots that have not yet recovered from a previous grazing.

---

If the grass plant is of the runner type, rather than erect or bunched, there is less danger of overgrazing, even where animals linger or return too soon. As figure 23-1 shows, when individual plants are severely grazed, a lesser percentage of leaf is removed than is the case with an upright plant, because of the plant's horizontal spread. So much leaf, as well as stems with growing points, remain below the grazing height of most animals that fewer roots are affected. This helps explain why as upright grasses are killed by overgrazing there is a tendency for the space to be filled with a runner grass as long as there is sufficient moisture to sustain it; hence the runner grass mats so common close to water points and areas of very high animal concentration.

Although many perennial grasses can withstand high levels of overgrazing without actually dying, it is still damaging because it reduces the yield of the plant and reduces its root volume, which greatly affects soil life and carbon and water retention. If the aboveground part of the plant grows less, it provides less material to feed the animals and less leaf and stem to subsequently cover soil as litter and mulch. In brittle environments most soil cover comes from litter rather than the bases of living plants.

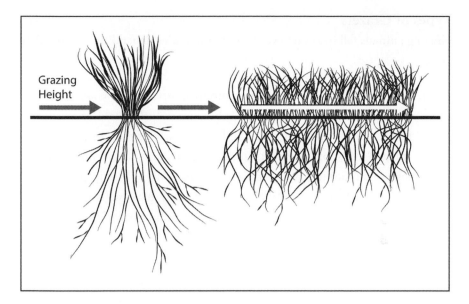

**Figure 23-1.** Because of its horizontal spread, less leaf is removed from an individual runner-grass plant when it is grazed severely than is the case with a bunched-grass plant.

---

### Overgrazing and Forage Utilization

*Many range scientists persist* in trying to prevent overgrazing by limiting forage utilization. They will specify various levels of utilization for individual plant species—thirty-five percent for bluebunch wheatgrass, for example—as though it really were possible for a manager to regulate how much an animal will take from plants of one species amidst many. Utilization levels are also specified for different types of communities (e.g., riparian areas twenty-five percent, and uplands fifty-five percent). But overgrazing continues no matter what the level of utilization.

---

Because time, not the number of animals, controls overgrazing and because preventing overgrazing is so critical for maintaining the productivity of the environments in which animals graze, this book devotes a whole chapter to managing the time that plants are exposed and reexposed to the animals (see chap. 33).

## Types of Grazers

Grazing animals fall into three very rough categories depending on how they graze:

1. Nibblers, endowed with narrow mouths, nip a leaf here and there off a plant, seldom overgrazing it because so little total leaf area is taken that the plant's growth, and particularly its roots, are hardly affected. Most of the grazers in this category are solitary, nonherding, small animals, such as the stembuck, dik-dik, klipspringer, duiker, and oribi of Africa. Generally they have self-regulating populations and thus never occur in high numbers. None have domestic relatives.

2. A second, broad-mouthed group, feeds by the mouthful. Buffalo, bison, zebra, horses, cattle, and hippo do this. Most are gregarious members of non-self-regulating populations capable of building up to high numbers and tend to defoliate grass plants severely. Elephants, which pull up grasses by the trunkful, also belong in this group.

3. Somewhere between these two extremes come animals capable of nipping an isolated leaf but habitually given to taking several at a time and concentrating on the same plant to such a degree that they have to be managed as severe grazers. Among them would be sheep, goats, deer, pronghorn, impala, and other herding antelope as well as perhaps some kangaroos. Again, they are gregarious and have non-self-regulating populations.

The distinction between self-regulating and non-self-regulating populations is an important one that we'll return to in chapter 37. We suspect that the solitary, self-regulating nibblers, which generally do not overgraze, control their population by some form of breeding inhibition allied to social stress above certain densities. The non-self-regulating severe grazers, who can and do overgraze, are heavily dependent for their survival on a high rate of annual loss to predation and, if predation is inadequate, to accidents, parasitism, and disease, which almost always take them down before starvation can. The predator-dependent species appear to herd as their main form of protection.

The presence of a wide diversity of different types of animals on a piece of land generally leads to more thorough utilization of available feed, as becomes evident in studies of complex wildlife situations in Africa, or where combinations of sheep, cattle, pigs, goats, and horses run together. It does not, however, change the basic dynamics of overgrazing or overbrowsing, and the grazing aspects of planning for most kinds of domestic stock are remarkably similar.

## The Relationship between Grass and Severe Grazers

Prior to human interference, the accumulation of old growth on plants in the more brittle environments was largely addressed by the grazing and social behavior of herding animals. But a prejudice against livestock has existed ever since because only recently have we considered regulating the timing of an animal's bite or tried to link the known behavior of animals in the wild to the problem of maintaining the vitality of grasslands.

The severe grazer, let's assume it is a cow, takes a mouthful from a particular plant, then moves on a step or two, leaving other plants of the same or different species untouched. The grazed plant should benefit from the bite, but frequently doesn't because its regrowth, offering more protein and energy, and less fiber, is very palatable and will attract a second bite some days later if the cow remains in the area. Thus one plant gets overgrazed while its neighbors rest ungrazed, and one cow may actually kill a few plants while a great many are rested far too long for their health. Increasing the number of animals increases the proportion of plants damaged from overgrazing and decreases the proportion damaged from overresting.

This process accounts for the apparent paradox that animals grazing continuously under the management most commonly practiced in the United States usually produce both overgrazed and overrested plants in the same area. Sometimes this may manifest itself as a startling mosaic of ungrazed and overgrazed patches. More often the overgrazed plants are dispersed among a whole population of plants and escape detection until some years later when a particular species disappears altogether.

Figure 23-2 gives an aerial view of a "conventionally well-run ranch" with a so-called correct (light) stocking rate. Over large areas nearly all plants have been overgrazed, as the ground-level close-up (fig. 23-3) makes painfully clear. The other areas contain a great quantity of overrested and dying grass plants. Changing the animal numbers will only alter the proportions of over-rested and overgrazed plants.

Very often the overrested sites will shift from grassland to herbaceous or woody plant communities. This underlies the invasion of problem woody species into millions of hectares of grassland. Vast sums have gone for research, chemicals, machinery, and publicity in the attempt to eradicate plants that never were a problem until our misunderstanding of plant/animal relationships resulted in them becoming one.

Before domestication no doubt cattle behaved like American bison or African buffalo in the wild. Even when feeding, those animals remained fairly close together for fear of predation, and they moved frequently off to new

**Figure 23-2.** Aerial view of a conventionally well-managed ranch, stocked lightly to avoid overgrazing. Light patches are severely overgrazed and darker patches overrested. Coahuila, Mexico.

**Figure 23-3.** Ground-level view of the overrested and overgrazed patches of figure 23-2. Coahuila, Mexico.

feeding grounds as the old became fouled. As mentioned in an earlier chapter, grazing animals do not like to feed over ground they have fouled. They keep moving to fresh ground and don't normally return until the dung has decomposed, usually long enough in the growing season for plants to regrow, thus avoiding overgrazing. Horses and cattle, especially in less brittle environments, often avoid dung sites long after the dung has decomposed, but this probably indicates that the extended recovery time given the plants has caused them to grow rank and fibrous. The fouling effect of the dung probably wore off much earlier.

When bison, pronghorn, springbok, kangaroos, wildebeest, or buffalo and other wild herding animals sense no danger from predators, including humans, who were a major predator, their behavior changes, and overgrazing of plants and overresting of soil and other plants increases. The herd remains spread for longer and longer periods, and even females with young will graze and lie well away from others. Then their dung and urine are scattered so

### Overgrazing Applies to Individual Plants, Not Whole Communities

*One should discuss overgrazing only in regard to individual plants. Applying the word to a whole area is irrelevant, as the following example illustrates.*

Near Albuquerque, New Mexico, the Sandia Indians have run livestock on a piece of land for over three hundred years, one of the oldest examples of continuous grazing of domestic stock in the New World. Despite the government restricting animal numbers, the U.S. Bureau of Indian Affairs has complained bitterly of catchment damage from "overgrazing and overstocking."

A cursory examination certainly bore out the validity of such allegations, but a closer inspection revealed the following:

- Many plants were matted with old, oxidizing leaves and stems and were dying prematurely from overrest.

- Other plants were thriving, as cattle had removed the obstructing matter.

- Other plants were weak from overgrazing.

- New plants were growing where physical trampling had created the right conditions.

- Previously trampled soil surfaces recapped following a rainstorm.

- Soil surfaces had rested so long from any disturbance that little grew on them.

- Generally litter and ground cover were scarce due to both the overgrazing of plants and the simultaneous partial rest of the whole.

Clearly, the blanket label "overgrazed land" means nothing and offers no guide to a solution. The presence of so many overrested plants belies the description "overstocked" altogether.

> On rangelands in Africa, the Americas, Australia, and the
> Middle East, I have seen and heard the same things. Even in
> those extreme cases where one hundred percent of the plants are
> overgrazed, it still does not represent overstocking because by
> simply planning the grazing to prevent overgrazing, the animal
> numbers can, and often must, be increased for the good of the land.
> As chapter 20 explained, partial rest leads to the same or a greater
> increase in bare ground as can be produced by overgrazing. The
> two influences compound each other, but the blame always falls on
> overgrazing, which is treated as synonymous with overstocking.

widely they no longer inhibit feeding nor induce movement, and the same
animals remain on the same ground day after day, overgrazing a great many
plants.

### Adaptations to Overgrazing

Although wild herds, at least under threat of predation, follow rather con-
structive patterns of grazing, they often do still overgraze plants. If they didn't,
plants would not have developed such ingenious defenses against it, and our
ignorance would have done more damage than it has.

On my own game reserve and research station in Zimbabwe it was a
common experience to see a buffalo herd concentrate for two or three days on
a site and then move off the fouled ground. Not uncommonly, a herd of zebra, wildebeest, eland, tsessebe, gemsbok, sable, roan, or some other species,
would follow them, apparently not bothered by the droppings of another
species. Although in complex wildlife populations, different animals may favor different plants and feed at different levels, from the rooting warthog to
the tree-nibbling giraffe, overlap is considerable. Plants severely grazed by
one herd often get grazed, or rather overgrazed, by another following close
behind. The overgrazing of some plants appears to do little damage as long
as animal impact is sufficient to ensure plants of that species still germinate

and establish. In fact, there are times when you might deliberately plan to overgraze with livestock. If you wanted to produce a mosaic of short and longer grass areas to improve wildlife habitat, for instance, you could plan the grazing so that your herd returns to an area before plants have fully recovered, as wild herds following each other will do, but at the same time increase animal impact to avoid creating bare soil between plants and to promote the germination and establishment of new ones. Or you might want to drastically reduce a dense stand of brush by deliberately holding animals long enough to overbrowse them severely, ring barking and killing them.

Some grass species cannot stand much overgrazing, and after a certain level of root reduction die out except where protected by thorny bushes and cracks in rocks. Others take evasive or defensive action. Some sacrifice the center of the clump but continue to hang onto life around the edges. Some distort their leaf and stem growth flat along the ground below the grazing height of the animals. Figure 23-4 provides an example of both these adaptations. Normally this plant would grow sixty to seventy centimeters (two to

**Figure 23-4.** Dying center and severe distortion of growing stems on an overgrazed plant. The plant would normally grow to sixty to seventy centimeters (two to three feet) high. The knife indicates how close and flat the leaves are to the ground. Zimbabwe.

**Figure 23-5.** Overgrazed for many years, this normally matted perennial grass plant has survived by forming a hedged ball. Baluchistan, Pakistan.

three feet) high. Other species develop a tight, round, spiny ball like a rolled-up hedgehog. The prickly aspect comes from old stem remains, among which small leaves persist (fig 23-5).

In all countries I have observed, perennial grasses appear remarkably resilient to overgrazing with one notable exception. For some reason not yet fully understood or explained by anyone, they seem particularly vulnerable in Mediterranean climates, such as predominate in southern Australia, the southern tip of Africa, and along the California coast. These places contain vast areas that have lost almost every perennial grass after years of overgrazing and partial rest. Fortunately, although we do not yet understand their extreme sensitivity, they do return when both overgrazing and partial rest are stopped.

## Browsing

So far we have considered grazing and grazing animals, but what of browsing and browsers? Among wild species, and among our domestic animals, we have many that browse woody plants more than they graze grasses, and some that purely browse, some that browse small shrubs in open grasslands, and some that subsist entirely on trees and shrubs. Do the same principles apply? In general, yes.

Though not nearly as much research has concentrated on browsing as on grazing, all results seem to point in the same direction. Woody plants can withstand heavy browsing that removes all of the green leaf as long as the plants get adequate time in which to recover afterward. They can also withstand continuous severe browsing as long as sufficient foliage remains out of reach of the animals. A few very sensitive species lacking an effective defense mechanism become conspicuous by their absence when under continuous browsing pressure.

### Adaptations to Overbrowsing

The most common response to overbrowsing by resilient plants is called hedging. Plants develop the look of a clipped garden hedge in which short, tightly spaced stems protect leaves crowded in among them. Figure 23-6 shows a heavily hedged plant growing in the biblical lands where heavy browsing goes back thousands of years. There is no knowing how old this plant is. Plants that do hedge can withstand overbrowsing for such prolonged periods that they may well live a normal lifespan, although I have seen elephants reach

**Figure 23-6.** Overbrowsed and hedged perennial shrub of great age. Yemen.

**Figure 23-7.** Browselines are evident on all trees—nothing is growing below the height animals can reach. Zimbabwe.

through the defenses and browse to death trees that had successfully hedged against lesser animals for years.

Other plants do not hedge, but larger individuals develop a browseline below which the animals take everything while the plant continues to grow as higher leaves trap sunlight. Figure 23-7 shows a browseline on the underside of the trees in a riparian area. Some species, when overbrowsed, develop very heavy root systems and straggly aboveground parts. This, however, can also result from frequent fire or, in the tropics, regular frosting, both of which remove leaf and damage stems routinely.

Unfortunately, these survival techniques are of no use to seedlings too small to hedge or develop browselines, and overbrowsing will eliminate them. Adult plants may hold on for centuries, but without replacements the population gradually declines. This gradual decline is apparent along many riparian areas in the western United States where livestock are encouraged to wander in low numbers in the belief that the low numbers will prevent such damage.

Browsing enhances the productivity of many woody plants, and thus, like the perennial grasses, they share an interdependence with the animals

**Figure 23-8.** Heavily overbrowsed stub of a winterfat bush exposed to a few cattle over prolonged time. Arizona.

that feed on them, which we don't understand very well at this point. Some woody plants also have elaborate chemical and physical defenses that provide protection against browsing. But even so, they still appear to share an essential relationship with the animals that promotes the spread and germination of their seed.

As in the case of grasses, overbrowsing bears no relationship to the number of animals, only to the proportion of leaf removed and the time that a plant has to regenerate.

Figures 23-8 and 23-9 show what can happen even when the animal numbers are doubled, but concentrated, and their grazing and browsing times planned. Figure 23-8 shows a highly nutritious but severely overbrowsed plant in Arizona called winterfat. Every single plant we found on this particular ranch had suffered to the extent of having no observable seed or seedling production.

Figure 23-9 shows this same plant in the same area of the ranch after two years of planned grazing with greatly increased and concentrated livestock.

**Figure 23-9.** Typical winterfat bush on the same ranch (fig. 23-8), growing out well and seeding with grazing/browsing planned and livestock numbers doubled. Arizona.

Far more leaf and many more stems have grown on this plant and many others, and seeds were produced. A year after, seedlings had established as well.

Where fouled ground induces movement in concentrated grazers it also does so with those browsers that feed on small shrubs near ground level. Those animals that do their browsing at higher levels of course are not to the same degree feeding on fouled ground because their noses and mouths are well above it. We may never know, now that we have lost natural populations

### Grazing and the Brittleness Scale

*A look at the impact* of grazing and overgrazing on the four ecosystem processes at the extreme ends of the brittleness scale gives a rough indication of the implications for management.

*Nonbrittle Environments*

**Community Dynamics**

In such environments where plant spacing is naturally close and soil cover hard to damage, grazing tends to maintain grass root vigor and soil life and structure. If grassland would normally progress toward woodland in that environment, grazing tends to impede this shift.

Overgrazing will reduce root mass but still not expose soil excessively due to the close plant spacing. Thus overgrazing tends to lead to a solid mat of grass. Species sensitive to overgrazing disappear, leading to a potentially more unstable community.

**Water and Mineral Cycles**

Grazing will probably enhance these cycles in grassland. In either natural or planted pastures, grass root reduction and compaction stemming from overgrazing adversely affect the cycling of both water and minerals.

**Energy Flow**

Grazing increases energy flow both above- and belowground in natural grassland or pastures. Where rest would produce woods or jungle, grazing, which holds the community at the grassland level, keeps energy flow down but available for our purposes.

Overgrazing reduces energy flow in grassland or pasture.

*Very Brittle Environments*

**Community Dynamics**

Grazing tends to maintain grassland communities, increase their diversity, cover soil, and retard shifts toward woody or herbaceous species.

Overgrazing, by reducing litter and soil cover even as it damages roots, fosters shifts away from grassland toward woody communities and herbaceous forbs (or "weeds"). It also leads to soil compaction and reduced numbers of grass species.

**Water and Mineral Cycles**

Grazing enhances both cycles through maintaining healthier and more stable root mass, increasing microorganism activity and aeration, and producing plants with more shoots and leaves that later provide more litter.

Overgrazing impairs water and mineral cycling by exposing soil and limiting the production of potential litter. By damaging and reducing roots, it also decreases soil structure, which increases compaction and reduces porosity, organic content, and soil life.

**Energy Flow**

Grazing increases energy flow by preventing old oxidizing blockages of material and promoting vigorous root and leaf growth. Healthier, more massive root systems also support millions of microorganisms and other life underground.

Overgrazing reduces energy flow because it damages plant roots, reduces plant litter, and exposes the soil surface.

to research in most parts of the world, but I strongly suspect that these species were heavily dependent on certain predators to induce movement. These predators would typically have hunted in bands or packs, such as humans once did and wolves, African wild dogs, and hyenas still do in some places.

## Conclusion

Because for centuries overgrazing was linked to overstocking, rather than the amount of time that plants were exposed and reexposed to grazing animals, society believes that healthy land turns to desert when it is overstocked. That totally rested land in national parks and research plots with no grazing at all was also turning to desert went unnoticed; we see what we believe.

Because no technology imaginable can ever prevent oxidation occurring on ungrazed perennial grasses or prevent the overgrazing of some grasses, livestock properly managed through holistic planned grazing (covered in chap. 41) will be essential to addressing desertification, as well as climate change. My hope is that this book will help society better understand that it is our management—not the livestock—that is the problem, and that we do have a way of successfully addressing it.

# PART 6

# HOLISTIC DECISION MAKING

**Holistic Management Framework**

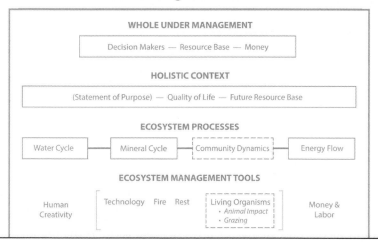

**WHOLE UNDER MANAGEMENT**

Decision Makers — Resource Base — Money

**HOLISTIC CONTEXT**

(Statement of Purpose) — Quality of Life — Future Resource Base

**ECOSYSTEM PROCESSES**

| Water Cycle | Mineral Cycle | Community Dynamics | Energy Flow |
|---|---|---|---|

**ECOSYSTEM MANAGEMENT TOOLS**

Human Creativity | Technology Fire Rest | Living Organisms<br>• *Animal Impact*<br>• *Grazing* | Money & Labor

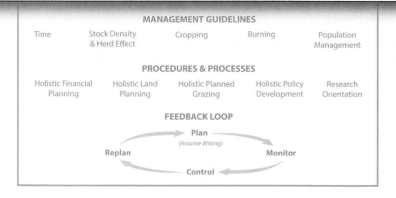

## ACTIONS & DECISION MAKING

Objectives, Goals, Tactics, Strategies, Policies
Customary Selection Criteria (past experience, expert advice, research, etc.)

### CONTEXT CHECKS

| Cause & Effect | Weak Link<br>• *Social*<br>• *Biological*<br>• *Financial* | Marginal Reaction | Gross Profit Analysis | Energy/ Money Source & Use | Sustainability | Gut Feel |
|---|---|---|---|---|---|---|

**MANAGEMENT GUIDELINES**

| Time | Stock Density & Herd Effect | Cropping | Burning | Population Management |
|---|---|---|---|---|

**PROCEDURES & PROCESSES**

| Holistic Financial Planning | Holistic Land Planning | Holistic Planned Grazing | Holistic Policy Development | Research Orientation |
|---|---|---|---|---|

**FEEDBACK LOOP**

Plan
*(Assume Wrong)*
Replan — Monitor
Control

# 24

## Introduction
### *Selecting Appropriate Actions*

UP TO THIS POINT I HAVE CONCENTRATED on the framework that guides Holistic Management: identification of the whole to be managed; creation of the holistic context; the four ecosystem processes that serve as the foundation for all human endeavor and on which the holistic context rests; and the tools for managing those ecosystem processes and the different effects some of those tools produce in brittle and nonbrittle environments. Now you are ready to begin making management decisions using your holistic context as a guide to ensure that you never lose sight of what is most meaningful to you in both the short and the long term.

As you move forward with management you will develop plans that include goals and objectives and the tactics, strategies, and policies for achieving them. Some of those goals and objectives will include *what you plan to do to create the quality of life your holistic context describes and the environment and behaviors that will sustain it and, if an organization, to meet your stated purpose.* The tactics, strategies, and policies that help you achieve these and any other goals and objectives will come down to actions, and those actions need to be checked to ensure they are aligned with your holistic context.

## Which Actions to Take?

Informed by your awareness of the four key insights, the ecosystem processes and their management, and the inspiration of your holistic context, you decide which tools to use and actions to take to achieve your goals and objectives. As always, you consider past experience, expert advice, research results, expediency, cost, cash flow, profitability, laws and regulations, and so on. In addition, you may consult the management guidelines, covered in chapters 32 through 37, before deciding how best to apply a tool or carry out an action.

## Selecting the Right Actions for Your Context

Now, before taking any action you run it through seven basic checks that consider the social, economic, and environmental impacts that may flow from that action relative to your holistic context. Your aim is to avoid actions that appear economically sound, but at the expense of the environment or the well-being of people, or that are environmentally sound, but hopelessly uneconomic or damaging to human welfare. Above all, you want to avoid actions that, while they achieve your objective, are likely to produce undesirable consequences down the road.

Each of the next seven chapters describes these context checks, which are condensed into one or two questions you ask yourself prior to taking action. Some checks will not apply to certain actions and can be skipped. Some checks will raise points you will again consider in other checks. The checking should take you minutes, rather than hours. Once you are more familiar with the checks and have internalized the questions, the checking will take you seconds and will become something you begin to do subconsciously.

## The Checking Process

When asked and answered in quick succession, the checking questions enable you to see the likely effect of any action on the whole you manage. You don't want to dwell on any one check to the point that you lose sight of the picture formed by scanning them all. Figure 3-2 in chapter 3 made this point. Spending hours examining every square in great detail would give no indication of what you are looking at. A quick glance at the whole picture from afar or with eyes squinted informs you immediately that it is a person and who that person is.

If the action passes most or all of the checks that apply, you should feel fairly confident in implementing it. If it fails one or more checks, you may want to modify how you implement the action (or apply the tool), abandon it altogether, or, in some cases, go ahead anyway—a subject we'll return to.

Table 24-1 summarizes the seven checks and the questions asked. There are no rules on the order in which to ask these questions, except one: the gut feel check should always be last. Your answers to the questions asked in this check should reflect the impression gained after passing through all the others.

As you become more familiar with the background information associated with each check you will automatically tend to order the checking

**Table 24-1.** The seven context checks

---

1. **Cause and effect.** Does this action address the root cause of the problem?

2. **Weak link**

   - **Social.** Could this action, due to prevailing attitudes or beliefs, create a weak link between us and those whose support we need?

   - **Biological.** Does this action address the weakest link in the life cycle of this organism?

   - **Financial.** Does this action strengthen the weakest link in the chain of production?

3. **Marginal reaction.** Which action provides the greatest return toward the goal for each additional unit of time or money invested?

4. **Gross profit analysis.** Which enterprises contribute the most to covering the overheads of the business?

5. **Energy/money source and use.** Is the energy or money to be used in this action derived from the most appropriate source in terms of our holistic context? Will the way in which the energy or money is used be in line with our holistic context?

6. **Sustainability.** If we take this action, will it lead toward or away from the future resource base described in our holistic context?

7. **Gut feel.** How do we feel about this action now? Will it lead to the quality of life we desire? Will it adversely affect the lives of others?

---

according to the nature of the action being checked. If the action addresses a problem, for instance, you will tend to go to the cause and effect check first. If you fail this check, it is often pointless to continue. If the action involves an organism whose numbers you want to increase or decrease, you might go to the weak link check first. Again, if you fail this check, it may be pointless to continue. Before long, you will instinctively know which checks apply to any action and which ones you can skip. Likewise, there will be times when your common sense tells you that, in light of your holistic context, a certain action is going to fail most of the checks, and you will modify it before going further.

As mentioned at the outset, speed is essential to the checking process. If you can't quickly answer yes or no to a question, simply bypass the check and move on to the next one. If you can't reach a conclusion after passing through all the other checks, then come back to the one or two you bypassed so you can give them more consideration. In most cases, you have bypassed a check because you don't have enough information to know whether the action passed or failed it. You may, for example, need to take time to diagnose the cause of a problem before you can answer the question asked in the cause and effect check. Or, you may need to gather actual figures, rather than estimates, for a gross profit analysis before you can pass this check with confidence. Once you have the needed information, run the action through the checks again and make a final judgment.

Don't worry that the speed of the checking will lead to an unsatisfactory result. You will be monitoring your actions to ensure they in fact achieve what you plan to achieve, and replan—and recheck—anytime you veer off track. When an action involves an attempt to alter ecosystem processes in any way, you assume at the outset that, because it is impossible to account for Nature's inherent complexity, the action, *even though it passes all of the checks*, could be wrong. On the assumption you are wrong, you determine the criteria you should monitor to give you the earliest possible warning of a need to replan. This monitoring process is explained in chapter 44, "Monitoring and Controlling Your Plans to Keep Management Proactive."

This process may sound daunting, but I can assure you that given a little practice it will become second nature. A description of how to ride a bicycle, which involves balance, steering, pedaling, braking, and so on, can be complicated enough to scare anyone off trying. But almost anyone learns quickly just by doing. One of the quickest ways to get familiar with the seven checks is to practice using them in your own home. Not only will it build your confidence in the checking process, it will show you how even small actions can affect the quality of your life. Question the detergents you use and the paper towels and light bulbs; how you dispose of waste products— garbage, batteries, computer parts, oil, paint thinners; how you deal with termites or cockroaches; your dietary habits or a new exercise routine, and so on. Two or more checks will apply to any one of these actions. If your holistic context is similar to my family's you may find that in the first year a lot of the things you are currently doing may fail a lot of the checks. By the next year not nearly so many will fail as you begin to modify your actions to bring them in line with how you want to live and what you want to accomplish.

Checking actions as a group also speeds learning, simply because some people will be more familiar with certain checks and the thinking behind them and can point out aspects others might have overlooked. You are more likely to be confident in your judgment on whether an action passes or fails a check when others come to the same conclusion.

Where the checking is likely to make the biggest difference, especially in the beginning, is with any actions that involve significant expenditures of money—for routine or emergency purchases, new business enterprises or products, and so on. Most of these sorts of decisions will be made when developing a financial plan for the year (chap. 39). Among other things, you are likely to find that actions you may never have questioned in the past, because the expenditures involved were so routine, will now have to be reconsidered. The context checks can also make an enormous difference in times of crisis because they will help ensure that the actions you take to resolve the crisis are not merely quick fixes undertaken in a panic.

## The Power Lies in the Holistic Context

On first exposure to the context checks, some people become so paralyzed by a fear they won't "do it right" that they avoid checking any action at all. However, it's no big deal if they don't get it right. Initially, *everyone* is a little fuzzy in asking the questions. Some people will accidentally skip an

### Transitioning to the New

*Often, in the beginning especially,* the context checks will show that some of the actions or policies you are already implementing are unsound. This does not mean you suddenly have to stop doing whatever it is you are doing. The checking merely warns you of the dangers you face over the long haul because this action is not in line with your holistic context. Now you can plan how to get out of the situation gracefully at low cost.

For example, the decision a farmer makes this year to grow monocultures of corn and soybeans may pass the gross profit analysis check as the enterprise likely to lead to the highest gross profit of several she is contemplating because the revenue they bring in is greater than the costs associated with their production. However, they are likely to fail most of the remaining six context checks, assuming the farmer's holistic context spells out prosperity that will extend to her grandchildren and beyond. Rather than go out of business or leave the land, the farmer wisely decides to make no change for now. But she does know that finding more regenerative ways to grow her crops—ways that cater for many of the hidden, but real, costs—has become a high priority, and that the checking of her future actions, coordinated through Holistic Financial Planning, will help decide when and how best to incorporate them.

Like the farmer, you may well find that you have to do something you know is damaging because, when viewed holistically, it is the correct thing to do *at the time.* In such situations, few of us, whether we are running a mining, manufacturing, or service-related business, can change what we are doing overnight. But all of us can begin to plan for change and how to regulate the pace of that change as rapidly as we can if we have ownership in the holistic context we create.

important question. Others will attempt to run actions through checks that don't really apply. In the end, however, they will still tend to arrive at the right decision *for them*—as long as their holistic context has meaning for them.

---

Ownership in your holistic context is more important to better decision making than an infinite understanding of each of the context checks will ever be.

---

Ownership in your holistic context is more important to better decision making than an infinite understanding of each of the context checks will ever be. If half the readers of this book were to learn the context checks to perfection and could run through them with one hundred percent accuracy, but had a holistic context to which they only paid lip service, they would fare no better than before. If the other half were committed to a holistic context in which they had a great deal of ownership, but could only perform the checking with ten percent accuracy, I would back their decisions every time.

Almost all the problems that beset humankind stem not from acts of Nature but from the way we decide which actions to take. I would stake my life on the premise that, if millions of humans in all walks of life would merely start making decisions holistically, in line with a holistic context they are genuinely committed to, most of the problems we face would evaporate.

# 25

## Cause and Effect
### *Stop the Blows to Your Head Before You Take the Aspirin*

THE CAUSE AND EFFECT CHECK is one that carries considerable weight when taking action to address a problem. It enables you to winnow out actions that only suppress symptoms when you need to correct the cause.

**The question you ask is:**
*Does this action address the root cause of the problem?*

The logic of going to the root cause of a problem is common sense, yet we have developed a culturally programmed habit of doing just the opposite. I don't apologize therefore for using the simplest metaphor to illustrate my point.

If I followed you around and periodically bashed your head with a hammer, you would develop a headache sooner or later. You could say the headache is a problem, but it is in fact a symptom produced by my hitting you on the head. You might take some aspirin to ease the symptom, and then even more powerful painkillers, adding yet more treatments for the side effects, or new symptoms, produced by the medicine. Or you might try to stop my hammering, and address the cause. Common sense dictates the latter but in practice we usually do the former.

Real life presents cause and effect situations that are less straightforward. Symptoms, or effects, can appear to result from multiple causes. Cause and effect is seldom a simple chain but a mesh extending infinitely in all directions (fig. 25-1). Nevertheless, the lesson of the aspirin and the hammer still holds, and as a practical matter we can usually see how cause A leads to result B without necessarily knowing why A happened or what will follow B. In other words, *why* I am hitting you on the head should not blind you to the fact that the blows are causing your headache. We can act effectively on that insight—to stop the blows rather than take the aspirin—without necessarily untangling the infinite ramifications that stem from it.

That we don't automatically address the cause of problems is probably allied to human nature. We generally tend to favor a quick fix over more permanent solutions because by nature we seek to avoid discomfort, which a quick fix alleviates right away. Since most of our quick fixes involve the use of some form of technology, it is tempting to believe that technology is responsible for this quick fix mentality, but it is more likely the other way around: in an attempt to deal with problems we develop technology.

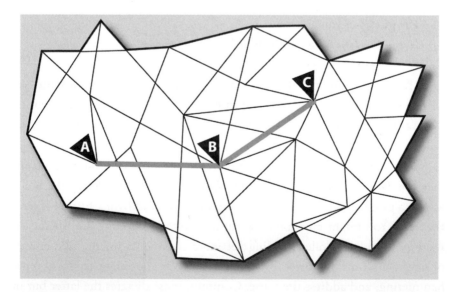

**Figure 25-1.** Cause and effect is never a simple chain of events leading from A to B to C, but a mesh extending infinitely in all directions.

Whichever it is, the fact remains that we are naturally inclined to resort to any one of the quick fixes modern science so often conjures up at the drop of a hat, rather than seek to address the underlying cause of the problem. Once the fix alleviates the symptoms, however, we tend to forget we even had a problem—until it recurs, as it surely will if the underlying cause is not rectified.

The seduction of the quick fix has weakened humanity's endeavors in all areas, from economics, to human and veterinary medicine, to the conduct of war and diplomacy, education, governance, and of course the management of natural resources. Instead of fixing what's really broken or finding a fundamentally different path, we print more money, invent a new drug, make a bigger bomb, suppress or buy off dissent, or build a dam. This check seeks to avoid nonsolutions by first asking you to think carefully about what might be causing your problem. If the action under consideration addresses that cause then it passes this check.

## Identifying the Cause

Identifying the cause of a problem can be fairly easy, but it can also require considerable probing. In most situations, this is a relatively unstructured exercise. You merely pose and answer the same questions over and over again: What is the cause of this? and when you have your answer, Well, what is the cause of that? You may have to ask this question three or four or more times, peeling away layers of symptoms, before you find the cause you should address. An incident from my farming days illustrates the principle.

I was developing my farm from raw bush and the cost of maintaining my three tractors was very high, enough to cause me considerable alarm. My men insisted that the reason for this was the age of the tractors, all of which had seen use for twenty years or more by the time I bought them. My accountant sided with my staff and tried to convince me that I would be on firmer financial ground if I took out a loan and bought new tractors, which had all kinds of tax advantages.

But I wasn't so sure. Was the age of my tractors really responsible for the high cost of maintaining them? My neighbors had all bought new tractors but their maintenance costs were high too, even though they only used their tractors for light work; they engaged contractors with heavy machinery to do the tough work. Mine were clearing virgin bush fourteen hours a day.

As I did all the repairs myself, I thought about the chain of events leading to each of my breakdowns and ultimately traced nearly every one back to having started with something becoming loose or dry. A few traced to other beginnings resulting from poor maintenance and almost none to unpredictable mechanical failure. Poor maintenance, rather than the age of the machinery, it appeared, was the cause of the problem. But why was maintenance so poor?

My initial response was to assume that my drivers, who were responsible for maintenance, were lazy and careless, and that I should think up ways I could penalize them for these faults. I found, however, that they often had a good explanation for a breakdown. They worked into the dark or started so early in the morning that they couldn't see as they checked their machines, for instance. So I continued to probe until I hit upon yet another underlying possibility: perhaps maintenance was poor because my drivers had no incentive to make it otherwise.

There was no need to push this line of questioning further. If they lacked incentive, then it was because I hadn't provided it. Accordingly, I did provide it by changing the nature of their employment. Each driver would become a private contractor who agreed to provide his services for a very small daily amount and a large daily bonus if the tractor he drove was maintained according to a specific list of criteria. However, one or more days of bonus could be lost if the oil level in the engine or gearbox was low, sludge appeared in the air filters, grease fittings were blocked, a fan belt was cracked, or a screw or bolt was loose, and so on.

Everyone received their bonuses for the first month while I made random inspections to ensure that each driver understood the criteria spelled out in his contract and what would constitute an infraction. I did have to deduct several days' bonus for a couple of drivers shortly afterward, but never again. My drivers earned far more than they had working for a wage and developed a sense of pride in being independent contract drivers; it became a game to see if I could ever catch them out. And my maintenance costs dropped to fifty percent of what my neighbors' averaged, despite my running three old tractors to their one new tractor. Had we been managing holistically in those days, the diagnosis of this problem might have led to a slightly different remedy. But the story does illustrate why it pays to keep probing for an answer until you can go no further.

Some Japanese companies are noted for insisting their people ask "why" at least five times in order to get to the root cause of a problem. Some years ago, one such company found that the water consumption in its office building was much higher than it should have been. They were advised to install low-flow toilets and water-saving faucets, but before spending any money, the company wanted to determine the cause of the high water consumption. By asking a series of "why" questions, they eventually discovered that when people used the toilets, they flushed twice: once to cover up the sound of urinating, and again when they were done. Having identified the cause, the company simply installed a small tape recorder inside each washroom cubicle with a button the person using the cubicle could push to produce the sound of a flush. The company's water bill plummeted.

There will be some instances, perhaps many, when you think you've got to the bottom of a problem and later find you haven't. If you've probed as deeply as you can and still not found the root cause, then you may need to look wider, rather than deeper. Sometimes outsiders, not necessarily experts, can readily diagnose the cause of a problem when the answer has eluded you simply because they can view the situation more objectively.

### Natural Resource Management Problems

If the problem concerns land or natural resource management (e.g., soil erosion, a plague of grasshoppers, increase in brush or weeds, or a decrease in the number of snow geese) look first to the four ecosystem processes for an answer. Then consider the tools (covered in the previous section) that may have been used in the past. How they have been applied will affect how the ecosystem processes are functioning now.

For example, if you have brush encroaching into a grassland, but your goal is open grassland with brush limited to fringes, you have a problem. In fact, when I first visited the United States on a lecture tour of eight western states in the late 1970s, everyone mentioned brush encroachment as the big problem of the hour. I saw where literally millions of dollars had gone toward both eradication research and actual control, to no avail. Nobody, however, discussed the cause, and when I asked the question a chorus answered in unison, "Overgrazing. And livestock spread the seeds." Yet brush *also* invades areas that have no livestock, and moves *less* rapidly into areas where nearly all plants are overgrazed.

Under pressure to do something about the brush and unable to really explain why the brush has come, most people just stop asking the question and look around for the best ways to kill the brush. John Deere, Caterpillar, Dow Chemical, and Monsanto accept the challenge. A whole new industry arises out of the ensuing competition, complete with research grants predicated on *not* asking the main question, and advertising to make it appear irrelevant in the light of apparent technological success.

Brush encroachment is in fact a symptom of an underlying problem resulting from prior management. As chapter 13 on community dynamics explained, a species can only establish in a place when conditions are favorable to its survival, and somewhere in the past management helped create the conditions favorable to the establishment of brush. As chapter 20 explained, partial rest in the more brittle environments tends to promote the establishment of woody plants or brush. All grasses have fibrous roots, but all brush species are tap rooted. For a shift from fibrous-rooted grassland to tap-rooted brush to occur, two factors must coincide: good germinating conditions, and porous or easily penetrated soil, which happens to be typical of rested areas where dead fibrous roots remain in the soil. These old and dying root systems assist the tap-rooted "invader."

Let's say you know that for years the land in question has been partially rested. Animals have been present but scattered and unexcited and thus have hardly disturbed the soil or plants. If partial rest has been the main tool applied to the land—either deliberately or accidentally—and you know it produces conditions ideal for tap-rooted brush plants, then it is reasonable to suppose that partial rest is likely to be the cause of your brush encroachment. That's what you should begin to rectify before killing any plants.

If instead you decide to attack the symptom (or *effect*) and poison, chain, or root out the brush, it won't be long before you have to repeat the treatment. The brush will keep coming back until you have remedied the cause. Clearing the brush with aspirin, or more potent remedies, is a costly nonsolution as long as the hammering from partial rest continues. Paradoxically, the likely solution, ending the partial rest, will probably make money rather than cost it.

Color plate 14 shows from the air an area that demonstrates the point. In the background beyond the fence we ignored the dense acacia encroachment and doubled the livestock numbers while increasing animal impact to combat the partial rest and stopping the overgrazing. The brush still dotting the landscape, which developed into grassland within a few years, became an asset because it provided feed, shade, and habitat diversity. The vast expanse of bare soil and brush in the foreground is where the government concerned enforced stock reduction, reseeding, and brush clearing—three costly forms of aspirin doomed to fail.

## When We Don't Identify the Cause

Going straight to the basic question of what is causing the problem demands courage, perseverance, and willingness to entertain new ideas, as everyone rapidly discovers who applies this check to the host of situations that arise in everyday land management. When you can actually hear the army worms stripping a crop field, can you stop and say to yourself, "I'm not going to spray until I know what I can do to cut the chance of army worms becoming so thick again?" In an emergency such as this you obviously go ahead with the application of a tool that only addresses symptoms, but in full knowledge of the dangers and only to buy time to rectify the cause. To knowingly repeat the application of a faulty tool is never wise.

The repeated spraying of grasshoppers on American rangelands, or the billions spent killing "invasive" brush and weeds, represent similar cases of continually attacking the symptom at ever-increasing cost without thought for the cause. The costs not only extend to the amount and price of the poison and many ecological side effects, they also show up in the price we pay for water to drink. Pesticides, together with other agricultural and industrial pollutants, account for much of the reason why bottled water has become a

### A Structured Diagnosis

*The Holistic Management framework* can be used in a structured way to diagnose the cause of a multitude of ecological problems (outbreaks of problem plants and insects, or disease; the disappearance of species; increased floods or droughts). Each of these problems stems from a malfunction in the ecosystem processes, and to determine the cause you have to consider how managers have applied the tools over large areas and for many years. Use the following steps to speed the diagnosis:

**Step 1.** *Which ecosystem process is the most appropriate to focus on to help you reason out what is happening?* If the problem involves an increase or decrease in a particular species (e.g., brush) look to community dynamics; if it involves a gain or loss of water (e.g., a falling water table or increased drought or flooding) look to the water cycle, and so on.

**Step 2.** *Has any natural disaster occurred that could have contributed to the problem?* Is there any evidence of major change in weather, volcanic eruptions, or anything else known to have occurred over the last century? If not then logically we can conclude the problem is a result of something *we* are doing. If we are causing it then it can only be because of the tools we use and how they influence the ecosystem processes, so the answer has to lie between the brackets in the tools row of the Holistic Management framework.

**Step 3.** *What is the position on the brittleness scale?* This will affect the results produced by some of the tools.

**Step 4.** *Which tools have been applied generally, for a prolonged period of time, and how?* Consider all of the tools separately: technology, fire, rest (partial or total), and living organisms (including animal impact and grazing). In the brush encroachment example the main tool applied extensively and for many years was rest—as partial rest. A few animals were present but animal impact was very low and grazing was applied as overgrazing.

**Step 5.** *How does the application of these tools tend to affect the ecosystem process at that level on the brittleness scale?* Ask this question with each tool applied over previous years and note its tendency either to create the problem or to have no effect. In the brush encroachment example, rest, applied as partial rest, tends to produce a shift to brush. The tool of grazing, applied as overgrazing, would have the same tendency, though more often produces a shift from grass to forbs (in many cases, weeds).

**Step 6.** *What can we do to remedy the cause?* After going through all tools applied and what the tendency or probable result would be at that level on the brittleness scale you will generally come to a conclusion as to how management is causing the problem and thus know what has to be remedied. If any doubt persists, test the opposite practices on a sample area to confirm your diagnosis. In our brush encroachment example, increased animal impact and grazing without overgrazing would be the most obvious remedy. A small test plot on which livestock were concentrated briefly to apply high animal impact without overgrazing would soon show whether the diagnosis was correct.

Commonly (almost universally) a simple diagnosis in cases of natural resource management problems shows they are humanmade and thus fortunately relatively easy to solve by simple changes in policy, education, and management in the field.

multi-billion-dollar growth industry in America. Homeowners, too, are part of the problem if they use pesticides routinely to rid their homes, gardens, and lawns of weeds and insects.

Massive monocultures in farming and industrial production of poultry and livestock have spawned a whole cluster of spiraling problems characterized by institutional unwillingness to question the root cause. Rather than admit to the inherent ecological instability of monocultures and factory-farmed livestock production, we try to keep them viable through chemistry, drugs, machinery, genetic engineering, and ultimately cash subsidy. More often than not, however, the side effects of these fixes exacerbate the problems.

Politicians, more than those in any other profession, have the most difficulty in overcoming the temptation to ignore cause and effect. Pork barrel legislation is only the most mundane example. The worldwide response to desertification shows how people may fall into the same trap without the slightest trace of cynicism. We've fed starving people, reduced livestock herds, settled nomads, imposed grazing systems, spent billions planting trees, installed mighty irrigation works, and done a host of other things time and again. Yet the deserts continue to grow because none of these actions tackles the cause.

In this sad tradition, a plan presented for signing at the first "Earth Summit," held in Rio de Janeiro in 1992, called for an annual expenditure of $6 billion on quick fixes for the symptoms of desertification.[1] To point out that none of the proposals made had ever reversed the decline of any land anywhere would have offended certain academic, diplomatic, and political sensibilities. Twenty years later, at the "Rio + 20" summit, participants admitted that desertification had only worsened and agreed to draft a list of Sustainable Development Goals. Eventually published in 2015,[2] these goals had one thing in common with the proposals of the first Earth Summit: they again focus on symptoms, which are likely to result in proposals and projects that do not address the underlying causes and thus fail to achieve the goals.

## Dealing with Short-Term and Long-Term Effects

Bear in mind that, especially in the beginning, you may have to take actions you now know are unsound in the long term. A farmer who suddenly stops fertilizing his crops with chemicals because of the groundwater pollution they create may have addressed the cause of the pollution but may also reduce his crop yields so significantly the first year or two that he does not survive financially. It makes more sense for him to *begin* to address the cause by weaning himself off the chemicals gradually.

Sometimes, actions that pass the cause and effect check result in short-term "problems." A rancher who uses animal impact to overcome the partial rest that has created a lot of bare ground and a moribund stand of a few rest-tolerant grass species may find that all she produces in the first year or two is a healthy crop of weeds. This was not what she had in mind when she described a future resource base requiring dense perennial grassland. Rather than consider the weeds as yet another problem to be tackled, she should view them as an intervening stage of succession that she is likely to move beyond. *Any* new plant is progress when you start with bare ground and dead or dying plants. The influx of weeds is just as likely to be an indication that progress is being made. Dense perennial grassland will not appear overnight, but given time—and planning that assures that animal impact is provided where needed and little or no overgrazing occurs—perennial grasses will appear and flourish.

## Summary

In general, the cause and effect check dictates that you not implement any action unless you feel sure that it addresses the cause of a problem, rather than its effects or symptoms. In an emergency you may proceed, but only in full knowledge of the dangers and only to buy time to rectify the cause.

It can be argued that it is sensible at times to remove a cause and treat the symptom simultaneously. Take the aspirin because your headache is painful, but make sure you receive no more blows to the head. Be wary though, because in practice this often results in draining resources from a more effective action. From years of practice I have found it wisest to remove the cause first and see what happens. Most often the symptom disappears at no additional cost.

When performing the cause and effect test, also bear these points in mind:

- Go to this test first, when you are dealing with a problem.
- When dealing with a resource management problem, look first to the four ecosystem processes for answers. How management tools have been applied in the past will be reflected in the condition of the ecosystem processes and give clues to the cause of your current problem.
- If the problem persists or returns, you haven't addressed the cause.

Remember that Holistic Management is driven by a holistic context, which leads people to pursue different strategies from the conventional ones. In conventional management, "getting rid of the weeds" would be the goal. In Holistic Management, "producing prosperous people on healthy land that is so rich in its diversity of plant and animal life that the weeds cease to be a problem" would be the comparable goal because that is what is reflected in the holistic context.

# 26

## Weak Link
### *The Strength of a Chain Is That of Its Weakest Link*

*A CHAIN STRETCHED TO BREAKING* will, by definition, fail at the weakest link. At any moment in time every chain has one, and only one, weakest link that alone accounts for the strength of the entire chain, regardless of how strong other links might be. To strengthen a chain when resources are limited, one must always attend first to the weakest link. Other links, no matter how frail they appear, are nonproblems until the weakest link is fixed. If $100 would correct the weakest link, and we spent $200 to make sure, we would have theoretically squandered $100 because, after the first $100 repair, the chain had a different weakest link on which the second $100 should have been spent.

The undetected weak link can cause mighty undertakings to fail outright or suffer continual setbacks. Thus we have a context check that compels us to ensure that our actions address the link that is weakest at any moment. The check applies in three different situations: social, biological, and financial.

**The questions you ask are:**
- **Social:** Could this action, due to prevailing attitudes or beliefs, create a weak link between us and those whose support we need?
- **Biological:** Does this action address the weakest link in the life cycle of this organism?
- **Financial:** Does this action strengthen the weakest link in the chain of production?

## Weak Link—Social

Consider the question: *Could this action, due to prevailing attitudes or beliefs, create a weak link between us and those whose support we need?*

You, and the future you envision for the whole under your management, are linked by a chain made up of all the actions you will take to get there. Any action that runs counter to prevailing attitudes and beliefs is likely to meet with resistance, creating a blockage that, if not addressed, will at some point become a weak link between you and the people whose support for your efforts is important, even vital. If such a potential weak link is identified, then it's important that you address it *before you take action.* Thus this question is not asking simply for a yes or no response but reminding you to address a potential social weak link before it becomes one. You want to ensure that as you move forward you do not inadvertently create disagreements that could have been avoided, or conflict or resistance that later hinders your progress.

When you suspect that an action is likely to result in a reaction that blocks further progress, you will fail the weak link check *if* your decision does not also include a plan for dealing with the blockage. An ounce of prevention is more than worth the pounds of cure it would take to undo the problems associated with the ensuing conflict. If there are no foreseeable obstacles to implementing the action, you of course pass this check.

## Weak Link—Biological

Consider the question: *Does this action address the weakest link in the life cycle of this organism?*

In the biological context, the weak link question applies when you are dealing with populations of plant or animal organisms that have become a problem, either because they are too many or too few in number: the parasites infesting the farmer's sheep, the loco weed that invades the range, the water hyacinths that choke the hydroelectric plant, the cockroaches that infest the kitchen, the rare aloe that needs protection, or the tortoise or owl threatened with extinction. Whether we see these organisms as friend or foe, the same question is asked and the same logic used. Before any action is taken to increase or decrease their numbers, we need first to ensure that it addresses the weakest link in the organism's life cycle. In doing so, we are likely to maximize the effectiveness of the action and to ensure the results will be lasting.

Every organism in its life cycle has a point of greatest vulnerability, a weakest link. Recognize this, and you have a good chance of inexpensively and effectively increasing or decreasing that species' ability to recruit new members to its population. When the action you are contemplating addresses that weak link, it passes this check.

Finding the weak link in the life cycle of any organism can be a challenge. Sometimes the answer is fairly obvious; other times it will require some research. Nature often provides clues that can help because all plants and animals have developed ways to ensure survival where they are the most vulnerable. Most plants, for instance, are most vulnerable during their initial establishment when the seed has germinated and the emerging root and leaf, dependent on energy in the seed, must find sustaining conditions in a limited time. If seeds, once sprouted, do not encounter the right soil, moisture, temperature, and sunlight for long enough to establish, they will not survive. Plants are able to overcome this vulnerability to some extent by the sheer number of seeds they produce, and the ability of those seeds to survive in the soil for many years awaiting the right conditions.

Insects and amphibians that produce a mass of eggs would appear to be most vulnerable while still in the egg or larval stages. Mammals, such as lions or dogs that produce several young at once would appear to be most vulnerable from birth to breeding age. Mammals such as humans that produce only one or two young at a time and nurture them for years, but remain fertile and sexually active throughout the year, would appear to be most vulnerable at conception, which is why humans try to limit family size at that point.

With some animal populations, careful observation should tell you fairly quickly at what stage they are most vulnerable. If, for instance, you have a population of antelope whose numbers you want to increase, you might observe that many fawns are being born, but that very few are surviving to adulthood. Thus, if the decision you planned to take was to purchase more adults and release them on your property, it would not pass this check. You first need to address why so few fawns are surviving, because that is the weakest link in their life cycle—from fawning to breeding age. Any action that enhanced fawn survival, such as the provision of more cover, if cover was lacking, *would* pass this check.

Far too often we make the mistake of concentrating on the adult members of the species when the adult stage of the life cycle is rarely the point of greatest vulnerability. Thus, while we bulldoze mature brush or poison well-established weeds we are creating favorable conditions for millions of their seeds to germinate and establish. Sometimes we use expensive and dangerous poisons to attack mature insect pests, and unwittingly select for new, unscathed and poison-resistant replacements. It is more sensible and more economical to address the invading brush, weeds, or insect pests by changing the environment that has become so favorable to their establishment. How will the tools available—rest, fire, technology, or living organisms—affect the four ecosystem processes relative to that organism's needs at its most vulnerable stage?

As mentioned in chapter 25, brush establishes easily where seedling taproots can take hold—on grasslands that have been overrested. Trees do too, wherever rainfall is high enough. Many insects flourish in sites where eggs are guaranteed a high rate of survival—on bare ground or on specific plants (made all the more attractive when a great number of the same plants are present, as in a monoculture). Animal impact can be used to help to cover bare ground through trampling down old standing plant material, and through breaking the capping, and compacting the soil to provide seed-to-soil contact so new plants can grow. Other living organisms can also be enlisted—increased variety of crops, hedgerows, and tree-belts—to increase the diversity of plants on croplands.

If you haven't identified the weak link in the organism's life cycle by the time you are ready to question an action this check will alert you to the need to do so. You will have to take the time to find out where that weak link is, through your own observations, by reading up on the basic biology of the organism, or by seeking expert help. Once you have your answer, you should know whether or not the proposed action treats the organism at its most vulnerable stage. If it doesn't, your research may suggest an alternative action—as my research did in the case of the bilharzia parasite—that is likely to pass.

---

### Addressing a Biological Weak Link Safely

*Identifying an organism's weak link* and tackling it at that point enables us at minimal cost to manage and control undesirable species and often leads to surprisingly simple ways to supplant destructive practices.

My family once had to draw water from an irrigation canal infested with parasites that cause bilharzia, a major scourge of Africa that leads to paralysis and even death in humans. Naturally, I did not want my family or my farmhands and their families to become infested. The government researchers recommended adding copper sulfate to the water as it killed the parasites. However, as there was no research to show the effects on children of drinking copper sulfate I decided against this remedy. Instead I asked to have access to the research so I could understand the life cycle of the parasites, which the researchers willingly granted.

Their research showed that the parasite in the water had to find a human host within 24 hours after leaving the host snail that carried it. So all I had to do was ensure that the delay in finding a human host was longer than the parasite could survive. By keeping our water in a holding tank for 48 hours before letting it into our main cistern, we imbibed only parasites that had suffered a natural death, and no chemicals.

---

## Weak Link—Financial

Consider the question: *Does this action strengthen the weakest link in the chain of production?*

Each year in conjunction with Holistic Financial Planning (chap. 39) you need to identify the weak link in the chain of production that stretches from the raw resources you work with to the money you receive for the products produced in each business enterprise. This chain has three links to which human creativity is applied: *resource conversion, product conversion,* and *marketing (or money conversion)*, as shown in figure 26-1.

The first link, resource conversion, involves the use of human creativity and money to convert resources that differ slightly depending on the type of business or enterprise, of which there are two broad categories:

1. **Sunlight harvesters.** This first category includes those businesses whose primary production is based on the conversion of sunlight energy (through plants) to a saleable or consumable product, such as food, fiber, lumber, or wildlife. The money their efforts reap represents solar dollars, as long as soils are regenerating in the process, and is the only form of wealth that can feed people. Thus, what they do is fundamental to our civilization's long-term survival.

2. **Resource enhancers.** The second category includes businesses that are one step removed from the sunlight-conversion business, such as a shoe store, bakery, software producer, or accounting firm. (This would include, incidentally, some pig and poultry producers or any others who buy, rather than grow, their own feed.) Their primary production is based on the conversion of raw materials and energy to a saleable product (goods or services). The money their efforts reap represents mineral or paper dollars. What they produce enhances our lives but cannot sustain a civilization.

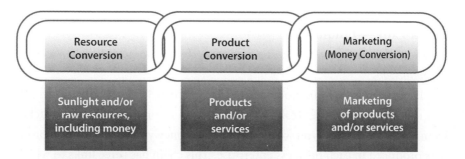

**Figure 26-1.** The chain of production. Human creativity first needs to use money and raw resources (including sunlight) to create a product or service. Then the product or service needs to be perfected and finally marketed to produce money. The chain is only as strong as its weakest link.

In the *product conversion* link, the sunlight harvesters turn the plants grown in the first link into a marketable form, commonly crops, livestock, wildlife, or fish. The resource enhancers convert the resources in their first link into a plethora of goods, services, or marketable skills.

In the *marketing (or money conversion)* link the products or services of the second link are marketed, and money is finally derived from the sunlight captured, or the raw materials and energy utilized, in the first link.

Obviously, whenever strengthening the chain of production requires money, and profit is a goal, the proposed investments should pass the weak link test. Increasing investment in advertising (marketing link) will not profit a business that turns out a poor product (product conversion link) as much as improving that product would do. Only investment in the weak link will result in more profit at the end.

In practice, we normally determine the weak link for each enterprise, or line of goods or services for sale, immediately prior to financial planning. Some businesses will only have one enterprise—such as the rancher who runs a yearling operation, or the professional income tax preparer—but many businesses will have several. In this case, each enterprise will have one link in its chain of production that is weakest at any moment. And that enterprise will only be as strong as that weakest link.

The aim in Holistic Financial Planning is to keep constantly strengthening the chain of production in each enterprise and in the business as a whole. Obviously, the entire year's budget cannot be spent on the one link that is weakest in any given year, but until money has been allocated to actions that do strengthen a weak link the business will not be as profitable as it could be.

### *At the Personal Level*

This thinking would also apply on a personal level where you are the product. Individuals working for a salary, for example, would be using their human creativity, time, and money to gain knowledge and skills to improve themselves (the product link) and marketing their abilities to derive money from the resources utilized in their first link. They might need to address the resource conversion link if they lacked skills, or the product link if, despite high skills, their attitude made them unemployable. They would have to market themselves better to get a raise or move to a new job.

### *Is It Product Conversion or Marketing?*

In virtually every kind of business, the product conversion and marketing links are closely related. Marketing is usually the weak link when producers fail to meet the needs of their market, such as the manufacturer who continues to build chrome and horsepower into automobiles when consumers want durability and fuel economy, or the cattle producer who raises chemically tainted beef in an environmentally destructive manner when buyers want it clean and environmentally enhancing.

Marketing is also the weak link whenever available markets remain untapped because they are not researched, because the product is poorly presented or badly promoted, or because the supply is erratic or out of synch with peak demand. One farmer I knew did exceptionally well year after year in a market where other onion farmers continually failed. He merely perfected his storage system and released his crop whenever supplies ran low in the local market.

There will always be gray areas where you are not quite sure which link applies. Is the bruised fruit you are trying to market a production problem (product conversion link) or a transport problem (marketing link)? In the end, it really doesn't matter as long as you detect it and address it. The power in this check is that it asks you to focus on the chain of production as a whole, and only then to determine where reinvestment in the business is needed most in any one year. The products you finally sell are not responsible for your profit: how you reinvest your money in the chain of production each year is.

---

The products you finally sell are not responsible for your profit: how you reinvest your money in the chain of production each year is.

---

## Conclusion

Identifying and addressing a social weak link before you take action helps prevent conflicts from arising. Identifying and addressing a biological weak link ensures that the actions you take are more effective and lasting. Once you have discovered a financial weak link, based on your chain of production, it *has* to be dealt with. It is not merely desirable or important to do so.

In discussing the financial weak link, I talked of misspending exactly $100, but real life never allows that kind of accuracy. Being consistently right with the vast majority of your dollars is what matters. Once you have identified the weak link, you look at all the possible actions you could take that would strengthen that particular weak link right away. When allocating funds for expenses, these actions will receive priority if they have also passed the other checks and are thus in line with your holistic context and likely to generate the most revenue. However, funds are often limited, so you will not be able to allocate dollars to every action that addresses the weak link; you will have to choose among them. The marginal reaction test, covered in the next chapter, helps you select those actions that do the most to propel you to your goal for the money and effort invested.

# 27

## Marginal Reaction

### *Getting the Biggest Bang for Your Buck*

THE MARGINAL REACTION CHECK only applies when you are considering two or more actions that have passed the other checks. What you want to check now is which of them will provide the maximum possible thrust toward a given goal or objective in terms of time and/or money invested.

**The question you ask is:**
*Which action will provide the greatest return toward the goal for each additional unit of money or time invested?*

Many people summarize the marginal reaction check with the phrase "getting the biggest bang for the buck," because the hypothetical example below involves just that. Although used when questioning two or more alternatives to achieve the same end, this check should ideally be one that we apply continually in many less tangible situations, including the mundane dilemmas of everyday life that consume so much of our time. I do not know of any management situation, right down to the family budget or the planning of your personal time, where the marginal reaction check would not be of great benefit.

In this check you are essentially asking yourself which of two or more actions will result in *each additional dollar or hour of labor* being invested where it provides the highest return in terms of your goal or objective. No two actions can possibly give you the same return for each unit of effort (money or time) invested at that moment. Thus, when resources are limited you want to select the one from which you gain the most. In doing so you will end up spending less time or money, and achieve what you want more quickly. The following hypothetical example illustrates this principle in dollars and cents.

Suppose you have $20,000 and must invest it in two banks under a peculiar set of rules. You may only open one account in each bank, and the interest earned with each additional deposit in that bank declines. Bank A pays 5 percent on the first $5,000 but on each additional $1,000 they give you 1 percent less (i.e., extra deposits up to $6,000 pay only 4 percent, the next $1,000 brings only 3 percent, etc.).

Bank B pays 4.5% on the first $7,000, but the rate declines 0.75 percent on each additional $1,000 deposited after that initial deposit. In practice such rules would discourage saving, but you can only get the best possible yield from your capital by determining the marginal reaction for each dollar invested. Think about it and then look at figure 27-1 to see how the investment would take place.

As you discover, you wind up investing $9,000 in Bank A and $11,000 in Bank B. No other combination except opening four accounts in Bank A—a violation of the rules—will yield more interest. Figure 27-1 was worked out by taking each dollar and asking where it would earn the highest interest. The first $5,000 earned 5 percent in Bank A, but Bank A would pay only 4 percent on the very next dollar instead of the 4.5 percent offered by Bank B. The next $7,000 would thus go to Bank B, but the $1,000 after that would go to Bank A because its 4 percent now beats Bank B's 3.75 percent. Similar thinking for each of the remaining $1,000 deposits determines the final outcome. Such neat examples do not occur in management, but the real-life situations that we cannot quantify are no less real.

## Marginal Reaction per Dollar Invested

This check should be used anytime expenditures are to be made and there are alternatives to choose from. In practice, this will most often occur in conjunction with financial planning. But emergencies do arise, and money may have to be allocated to address them.

| Bank A | Bank B | Interest (percent) | Balance | Interest Earned |
|--------|--------|--------------------|---------|-----------------|
| 5,000  |        | 5.00               | 15,000  | 250.00          |
|        | 7,000  | 4.50               | 8,000   | 315.00          |
| 1,000  |        | 4.00               | 7,000   | 40.00           |
|        | 1,000  | 3.75               | 6,000   | 37.50           |
| 1,000  | 1,000  | 3.00               | 4,000   | 60.00           |
|        | 1,000  | 2.25               | 3,000   | 22.50           |
| 1,000  |        | 2.00               | 2,000   | 20.00           |
|        | 1,000  | 1.25               | 1,000   | 12.50           |
| 1,000  |        | 1.00               | 0       | 10.00           |
| **9,000** | **11,000** |                | **20,000** | **767.50**    |

**Figure 27-1.** The investment of $20,000 in Bank A and Bank B illustrates the principle of marginal reaction per dollar invested. If you took each dollar and asked where it would earn the highest interest, the first $5,000 would go to Bank A, but the very next dollar, in fact the next $7,000, would go to Bank B. The next $1,000 would go to Bank A, and so on.

In Holistic Financial Planning this question is used in two ways. Initially it will help you prioritize the actions to be taken that address the weak link in your chain of production, as mentioned in chapter 26. Each action will provide a different marginal reaction. Those actions that provide the highest will get the most money allocated to them. Some may be dropped altogether because funds are limited and they cannot earn, dollar for dollar, what other actions would.

Let me use a few examples to illustrate these points. Before I do, I should remind you that this check, in most instances, never does boil down to the simplicity of the Bank A/Bank B case. Many times you cannot produce quantifiable figures, and your judgment will be highly subjective. That being the case, *your answer to the checking question will only be as good as your knowledge, common sense, and determination to achieve your objective.*

### Prioritizing Actions

Suppose you were creating a financial plan for the bakery you manage as a family business. A month or so before, you had determined that marketing was your weak link this year. Sales had fallen off in the previous six months while, to the best of your knowledge, the quality of your products had been maintained. In discussing the various actions you could take to address the marketing link, two appear most promising, enhancing your advertising program, and engaging a student from the local business college to conduct a customer survey. Redesigning and enlarging the ad you run in the local paper, and running it more frequently, will cost $3,000. The survey will cost somewhere between $500 and $1,000.

Either action is likely to address the falloff in sales, but you now use the marginal reaction check to see which one should receive priority in the allocation of funds. You find it difficult indeed to quantify the return from each alternative, but after some discussion you realize that knowing why customers have deserted you might enable you to bring them back and attract new customers as well, and that it could also influence the content of your advertising.

This discussion and your own intuition convince you that you will gain more per dollar from the survey. Increasing your advertising budget doesn't make much sense now, and for the time being you won't allocate additional dollars to it. If the survey indicates a need, you might adjust your plan later and add this expense. Let's say that the survey is done, and you learn that the falloff in sales was largely due to the temporary assistant you hired while some family members were away on vacation. At the counter, he had been rude and unhelpful to a number of customers and he had also neglected to fill several home delivery orders. Customers complained among themselves, rather than to you, so you hadn't been aware of the problem.

This example illustrates the parallels between this check and the weak link and cause and effect checks. Had this family noted the drop in sales earlier on, they could have diagnosed the cause of their problem then and addressed it sooner. Even so, they would still have used the marginal reaction check to help them choose among the alternative actions they could take to address the cause.

### When Your Business Is Sunlight Harvesting

In sunlight-harvesting businesses, where money is being generated directly from the sun's energy through plants, the effects of the marginal reaction check are profound indeed. The check also shows in many cases that, of the actions compared, only one will survive because no other comes close to yielding the return that action does this year. This can also be the case in other businesses but is less common.

Suppose you and your family manage a cattle ranch and have determined that this year energy conversion—that first link in your chain of production—is your weak link: you need more grass to feed your animals. There are a number of actions you could take, but you are leaning in favor of clearing the brush that still covers large portions of the ranch. Your goal is to have productive grassland with scattered patches of brush to provide cover for wildlife and livestock and diversity in the vegetation in line with the future resource base described in your holistic context. You have removed the cause of the brush encroachment and have halted its spread, but now you want to remove some of what's left to provide room for more grass to grow. So why not buy a used bulldozer at auction and root out the brush?

Before you do that, you need to compare the brush clearing to the other possible alternatives for increasing energy conversion to make sure you are getting the highest return possible on the dollars invested. Again, this comparison will not be quantifiable like the Bank A/Bank B example, so it demands careful thought.

One of the most obvious alternatives to brush clearing is to subdivide paddocks with fencing to increase forage production. What does brush clearing (Bank A) offer on the land compared with more fencing (Bank B) this year? If you clear the brush it will allow more grass to grow by letting in more light, and the disturbance created by the tractors and chains may increase rainfall effectiveness much as animal impact would do. Dead roots left underground will provide a mass of organic matter that will eventually enhance water retention, mineral cycling, and soil structure for some years. More forage in the cleared paddock may allow you to hold animals there a day or two longer and that means more recovery time elsewhere and thus more grass growth over a wider area. (Chapter 33 on time management explains this in more detail.)

The cost of brush clearing will be $50,000. On the other hand at $500 a mile, you could build ten miles of electric fence and split four large paddocks into eight for a cost of $5,000. From this you could anticipate the following benefits:

- Halving the size of the four paddocks would double the stock density in the divided areas during each use, thus increasing animal impact somewhat. Forage production will consequently improve steadily in these paddocks for many years to come.

- With four additional paddocks the grazing periods can now be decreased on average in every paddock on the ranch in every growing season over the fifty-year life of the fences. Thus this single investment will increase the amount of grass that grows in all paddocks over the next fifty years.

- Disease risks are reduced because animals receive a higher plane of nutrition and spend shorter times on fouled ground.

Although you cannot quantify perfectly the comparison of brush clearing and fence building as ways to increase energy conversion, clearly the fence beats the bulldozer in terms of what you gain. Not only do you grow more grass year after year, you have cleared another $45,000 to be added to the profit planned this year.

## Marginal Reaction per Hour of Effort

Nowhere does the marginal reaction check apply more than in our allocation of time. We have only a fixed amount, and it ticks by day and night. Constant awareness of the marginal reaction when it comes to investments of time frees time to do things we love, and the emergencies and crisis management we thereby avoid saves the money to pay for them.

Some years ago I visited a tobacco farm where near panic reigned, as reaping was to start in ten days, and the curing barns still had no roofs. Somewhere in the prior year, the owner had spent time in the coffeehouse or fixing a tractor when he might have worked out a construction schedule. Now he was paying heavily in extra labor, rushed transport, and high blood pressure, not to mention the probability of getting shoddy work and losing part of his crop anyway.

Naturally the marginal reaction per hour of effort figures in Holistic Financial Planning. The cost of most actions contemplated will be influenced by the amount of labor or hours of effort involved in implementing them. So will your judgments about how much each maintenance expense can be trimmed. If, for instance, you decided to save money by cleaning your offices yourself, rather than pay someone else to do it, you may be going too far if your time could reap higher gains when devoted elsewhere. Or, imagine spending a day of your time, at your salary, in the laundry business you own and manage, trying to fix one of your new computerized machines. How does that compare to bringing in a specialist who immediately knows what to do, has the right tools, and guarantees the repair?

In chapter 25 I mentioned how, as a struggling young sugarcane farmer, I managed to cut my machinery maintenance costs to a fraction of what they had been, simply by spending my time where it would provide the highest return at that moment, thinking through the chain of events leading to each breakage and then planning what to do about it. When my neighbors and I compared our costs, I realized that I had cut mine to half theirs. Had we compared the number of hours each of us spent thinking and planning with paper and pencil, I estimate I probably spent ten hours to their one. The marginal reaction achieved per hour of my time was so high, it saved my family and our farm at a time when the bottom had fallen out of the sugar market and my country's products were under sanctions imposed by the rest of the world.

## Conclusion

I have stressed that the marginal reaction check is always, in the end, a subjective one. It has to be, because you are comparing two or more actions that have passed the other context checks and selecting the one that yields the highest return per dollar or human hour invested in situations often difficult to quantify.

The next check is far less subjective and is focused entirely on generating profit. Having the potential to convert sunlight or raw materials into a marketable product is of little help if you cannot be sure which of many possible enterprises enable you to do that most effectively. The gross profit analysis will enable you to find out.

# Gross Profit Analysis
### *Bringing in the Most Money for the Least Additional Cost*

THE GROSS PROFIT ANALYSIS is used to select those enterprises (products or services from which you derive income) that, after associated costs and risks have been factored in, produce the most income. The income from these enterprises has to cover your overhead costs *and* generate some excess for there to be any profit.

**The question you ask is:**
*Which enterprises contribute the most to covering
the overheads of the business?*

This is one of the few checks that requires a pencil and paper, and I suspect that this may keep those with an allergy to paperwork from doing it. Yet the marginal reaction per hour of effort is hard to beat, especially if you plug along for a year putting hundreds of hours into an enterprise that, while providing massive income, produces a low gross profit, and thus a lower net profit for the business as a whole at year's end.

In most businesses a great deal of money is tied up in overheads, or fixed costs—land, buildings, machinery, equipment, salaries, and so on. While essential to the business, most fixed costs do not generate income, and thus the wherewithal to keep the business going. That is only done by the various activities that actually lead to the sale of a product or service. To be most profitable we need to find that enterprise or combination of enterprises that brings in the most income for the least additional nonoverhead costs each year. The greater the spread between income per year and additional nonoverhead costs, the greater the contribution of that enterprise or combination of enterprises to covering overheads and producing the surplus that becomes profit.

Various techniques exist to help you do this, but I find most too complex, confusing, and impractical for widespread use. Computer programs are also available for this purpose, but they generally do away with the need to *think through* the variables involved. It is this thinking that is so essential to the success of this analysis, particularly when it involves potential new enterprises.

The gross profit analysis check, derived from the work of a Cambridge University agricultural economist named David Wallace, has flaws when performed in isolation, in that it does not take into account the social or environmental costs associated with an enterprise. But when used along with the remaining six context checks this drawback is overcome, and the analysis provides a clear and simple way to determine which enterprises are likely to generate the most profit.

In the gross profit analysis you simply look at the income likely to be derived from each enterprise and deduct the additional money you will have to spend to bring in that income. *The difference between money in and money out is the gross profit.* The additional money to be spent is that money you would not spend *unless* you undertook the enterprise. Through the analysis and comparison of many possible enterprises with this check, you are selecting the best enterprise, or combination of enterprises, to create profit and minimize risk.

Wallace originally used the term *gross profit* in describing his analysis, but was later persuaded to change it to *gross margin*, a term that has no intrinsic meaning. Although Wallace realized that net profit, which factors in overhead costs, is quite a different animal from gross profit, struggling British farmers found it confusing to compute positive profits of any kind when their actual bottom line was bright red.[1] For many years I went along with Wallace's change, but I encountered a number of problems. First of all, I found that American businesses had "improved upon" the gross margin analysis in ways

I will cover shortly, but in doing so had largely destroyed its value. Secondly, I found that many people confused gross margin analysis with marginal reaction because both contained the word *margin*. To avoid both problems, I decided to revert back to Wallace's original name, *gross profit analysis*.

The key to Wallace's gross profit analysis, particularly when researching possible new enterprises, is the careful distinction of fixed (overhead) and variable (direct or running) costs *at a given moment in time*. Wallace divided all business costs into these two categories. Fixed costs exist no matter what or how much is produced. Variable costs are a function of volume of production—the more you produce the more these costs increase. However, when performing a gross profit analysis, the definition of what is fixed and what is variable changes depending on the current situation.

When you plan wheat production, for example, seed and diesel for the machinery during sowing and harvesting are variable costs. You incur them only if you grow wheat, and you compute the amount from the acreage you intend to plant. Payments on the harvesting combine you already own, however, are fixed costs because *even though you use it exclusively for wheat* you must make the payments whether or not you actually grow wheat. No matter how much wheat you grow, these fixed costs remain unchanged.

Many of the techniques used for analyzing enterprises, including the "improved" gross margin analysis, try to apportion the fixed costs among various enterprises. In the foregoing case, for example, all combine expenses would be charged against the wheat. Perhaps half of the tractor costs would be charged to wheat and half to something else. Labor costs might wind up apportioned under many headings. This practice, however, only clouds the picture and makes for a much poorer analysis, as figures 28-1 and 28-2 illustrate.

Figure 28-1 compares income and expense projections for two enterprises in a conventional gross margin analysis where fixed costs are apportioned. I have used A and B to represent any two alternatives. The gross margin for A is only slightly better, but if all other factors were equal you would, based on this analysis, probably favor enterprise A over B because it shows the highest gross margin.

In figure 28-2, using a gross profit analysis, fixed costs are not apportioned, and the story is very different. Since most of the expense allocated to enterprise B would have to be paid anyway, clearly enterprise B contributes

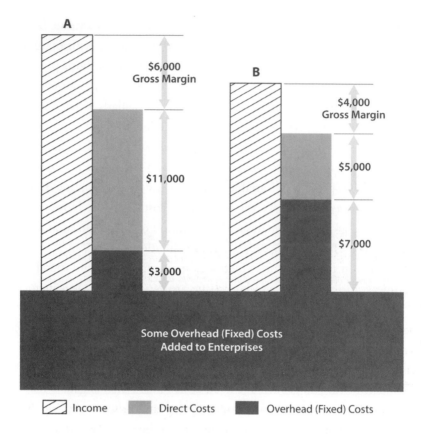

**Figure 28-1.** In a conventional gross margin analysis, fixed costs are apportioned. In this case enterprise A looks slightly better than B.

far more toward covering the overhead expenses of the business than enterprise A. And if you needed an operating loan to cover variable costs, obviously B would take far less than A.

Many find the matter of sorting out fixed and variable costs rather confusing, particularly when considering a new enterprise. No formula or list can assist you in this because whether an item is fixed or variable depends on the situation and the time frame under consideration—hence my apprehension concerning computer programs used for analyzing potential new enterprises. When contemplating the addition of a new enterprise, it helps to remember that *in the very long term all costs are variable* (you could sell the business) and *in the very short term all costs are fixed* (the new enterprise could be started

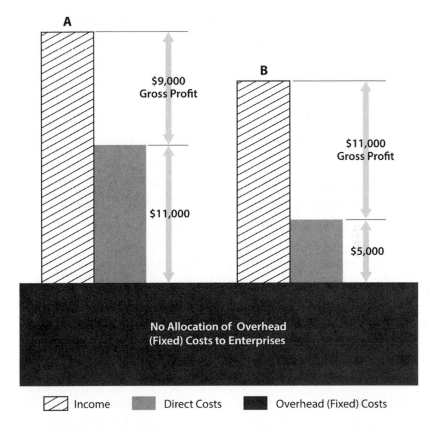

**Figure 28-2.** In a gross profit analysis, fixed costs are not apportioned, and the outcome is often different. In this case enterprise B looks far better than enterprise A.

using materials on hand and already paid for). To determine the fixed costs to be ignored in the *new enterprise,* and the variable costs to include, picture yourself standing on a bridge looking upstream. Any water (cost) that has passed under your feet is fixed and should be ignored, while any water (cost) upstream is still variable and should be included.

---

To determine the fixed costs to be ignored in a new enterprise, and the variable costs to include, picture yourself standing on a bridge looking upstream. Any water (cost) that has passed under your feet is fixed and should be ignored, while any water (cost) upstream is still variable and should be included.

---

If I were thinking of adding a new line of products to those I already manufacture, for instance, the factory I own but am still paying for would be a fixed cost and not included in calculating the gross profit for this new enterprise. The expense involved in the purchase of the new machinery required would be a variable cost, and that would have to be offset against the income I could expect to receive. However, because I would not use up the machinery in one year, this variable cost would have to be spread out over the estimated life span of the new machinery, and the average yearly cost used in my calculations. The actual payment for the machinery and how that will affect cash flow is planned later (see chap. 39 on Holistic Financial Planning) if the new enterprise passes the gross profit analysis check and any others that apply.

If, on the other hand, I did not need to purchase new machinery but could utilize some old machinery on hand, the machinery would be considered a fixed cost. However, it would take money to renovate and modernize the machinery, and this cost, annualized, as it was in the case of the new machinery, would be variable. It would be offset against the annual income anticipated to provide the gross profit.

Used in conjunction with Holistic Financial Planning, the gross profit analysis will enable you to weed out any enterprises that drain the business in that they contribute little or nothing to covering overheads. Surprisingly, it is not uncommon, at least in agricultural businesses, to find that the main line of business fails this check, and that it is the subsidiary enterprises that have kept the business afloat. This is something that more complicated techniques, which allocate fixed costs across enterprises, often conceal.

The gross profit analysis check is used at three different times in Holistic Financial Planning. Initially, when you have brainstormed a list of possible income sources, you use this test to narrow down the list. Very rough estimates of anticipated income and variable costs are all that is needed, and the exercise can be done in your head because you are only looking for major differences in enterprises at this point. A quick run through the rest of the checks will further narrow your list. Then you use the analysis a second time, but more formally. You will need to gather fairly accurate figures before you can pass the final list of enterprises through this check, and you will need to use pencil, paper, and calculator to get your answers. Those enterprises, both old and new, that pass this check by contributing the most to covering overheads, *and* pass the remaining checks, will be the ones you engage in. Finally, you analyze each enterprise again at the end of the year to determine how well it actually performed. You then take this information into account in planning for the next year.

## Refining Your Analysis

There are several refinements that can and often should be made in calculating gross profits that will further clarify the picture.

### Common Units of Measure

In agricultural businesses in particular, there will be occasions when you have to compare very dissimilar enterprises. How, for instance, does putting land into crops stand up against using the same land to graze livestock or for various forms of recreational, income-bearing activities? You can compare various crops, mixtures of crops, livestock, and recreation, or other enterprises to find the best strategy for covering those fixed overheads and making a profit at minimum risk. To do so effectively, you will have to find a common denominator on which to base your gross profit calculations. That common denominator in turn should be based on the factor that most limits production, generally the amount of land available, capital (in cash or assets), or labor. When land area is the most limiting factor, as it commonly is in agricultural businesses, gross profit per *hectare or acre* per year shows best how to put the land to use. In other cases, gross profit per dollar of capital or per human-hour of labor makes more sense. *The Holistic Management Handbook* (Third Edition) (and the Savory Institute e-book, *Fundamentals of Holistic Financial Planning*) explain in more detail how to calculate these figures.

### When Enterprises Do Not Overlap

There will be occasions when one or two enterprises do not overlap with others because they only utilize a portion of the production base (land, capital, or labor) and can stand on their own. An example would be a ranch that was considering the establishment of a bed and breakfast enterprise on one small corner of the property near a highway. The amount of land involved would be so insignificant it would not be a factor. The capital used to build the estab-lishment could be provided by a couple in town who wished to retire and run the enterprise. The only variable costs associated with the enterprise might be the money required to pay the lawyer for drawing up the agreement. That would be offset against the income received for the lease of the land to pro-vide a gross profit that could be fairly high. If the remaining checks showed the decision to be aligned with your holistic context, you would go ahead with it. A similar scenario would unfold if you wanted to lease out unused office space.

An increasingly common example that fits this case, but is generally missed, is when your business is game ranching, and you are operating in a more brittle environment, as most of these businesses do. You are likely to need to bring in livestock with Holistic Planned Grazing to keep the land vital and capable of producing adequate forage, because the more-difficult-to-control game on their own cannot do this. In this case, the livestock are not competing with the main enterprise (game), *but they are an essential tool for sustaining it.* The livestock could in fact be considered a variable cost, but because they generate income in their own right, they should be considered a separate enterprise, in this case, one that should stand on its own.

### Break Enterprises into Segments

Some enterprises will yield to analysis best if broken into subunits or seg-ments. For example, cattle production could be broken out into raising the calf, growing the calf, or finishing the animal for market. When you analyze each segment as though it were a separate enterprise, and you know what each segment's gross profit is, you can, if you choose, limit your effort to those seg-ments that contribute the highest gross profit. Thus a cow-calf operator may find it more profitable and less risky to reduce the cow herd and thus free up land to carry the progeny longer before marketing them. The same thinking would apply in many other production businesses.

Most manufacturing firms have already found that it does not pay to undertake all facets of production within their own factories. By contracting (outsourcing) certain segments of production they are able to achieve a higher gross profit overall.

## *Assessment of Risk*

While doing a gross profit analysis and calculating the anticipated income and the variable costs involved in any possible new enterprise, you could be far off the mark in your estimates due to a number of variables outside your control. To assess the risks involved, project the *worst, average, and best* scenarios. When doing so, keep most income or expense figures average, and in each scenario vary the figures for whatever is least under your control. For example, for a dryland farmer of pinto beans in the American Southwest, weather might be the most critical factor. For a clothing manufacturer, input costs (mainly the cost of fabric) might be. The dryland farmer would therefore pick an average price and compute his gross profit for low, average, and high yields. For the clothing manufacturer, a comparison of low, average, and high fabric costs would mean more. Obviously, one could add infinite levels of sophistication to this process, using a computerized spreadsheet and a bit of time. Your focus, however, should be on major differences, not minor ones. If, in comparing new enterprises, you find that the risks associated with any one of them are far higher because the spread between the low, average, and high gross profits is very great, this may influence your decision on which ones to adopt.

## Conclusion: Generating a Net Profit

Having selected an enterprise or a combination of enterprises based on the gross profit analysis, you still don't know if you can make a *net* profit. The Holistic Financial Planning process, which chapter 39 and the *Holistic Management Handbook* describe in some detail, will show whether your strategy does or does not add up to black ink.

In spite of all its benefits, the gross profit analysis has some serious shortcomings, some of which show up in other context checks and some in the light of common sense. In the theoretical case portrayed in figure 28-2, enterprise B turns out so far ahead of enterprise A that you could easily argue for committing the entire business to it. In practice, however, that could be very unwise. First of all, the advantages of B derived from being able to use assets already on hand and accounted for as fixed costs. If, however, doubling enterprise B meant paying for more equipment or labor, those would become variable costs and might lead to a different conclusion.

More important, gross profit analysis takes no account of ecosystem processes or many less tangible considerations. In farming, it quite frequently shows a complete and chemically enhanced monoculture as the most profitable strategy, and yet we know that this damages soils and the life in them and leads to a spiraling chemical dependency and rising costs. The chemical companies that manufacture the pesticides, pesticide-dependent GMO seed, and fertilizers no doubt find these enterprises yield a fairly high gross profit for them when they do not factor in the social and environmental costs that are borne by society.

While this check does not account for anything but money earned and spent, the technique still throws fresh light on many situations, and other context checks usually compensate for its limitations. The ideal is to find the best enterprise or combination in which all the technology and other tools pass all checks. At that point, to the best of your knowledge, you have enterprises that are economically, socially, and environmentally sound.

# 29

## Energy and Money
### *Using the Most Appropriate Forms in the Most Constructive Way*

CHAPTER 28 EXPLAINED HOW A GROSS PROFIT analysis helps determine the most profitable enterprises, but it ended on a note of warning. The means to profit may not be holistically sound. Not only the enterprise itself, but also the secondary inputs that support it, must pass the other checks that apply.

Rather late in the development of Holistic Management, it became evident that we had to press the checking one level deeper and to examine both the sources and the patterns of use of the energy and money involved in production. We lump money and energy together because any action contemplated usually requires one or the other, and often both.

**The questions you ask are:**
- *Is the energy or money to be used in this action derived from the most appropriate source in terms of our holistic context?*
- *Will the way in which the energy or money is used be in line with our holistic context?*

This check helps you avoid
- actions likely to lead you into an increasing dependency on, or addiction to, fossil fuels or any other inputs; and
- actions involving an addictive use of borrowed money on which you are paying compound interest.

## Sources of Energy

In terms of availability, energy sources fall into two categories: sources that are abundant or unlimited, and sources that are limited in supply. In terms of their effects on the environment, energy sources also fall into three categories: sources that are benign, damaging, or potentially damaging. This depends on the rate at which they are consumed and the methods used to harness and distribute them. Obviously, we stand a better chance of long-term success by favoring the energy sources in unlimited supply, but only if we can assure their effects on the environment are benign.

Most energy is derived either directly or indirectly from sunlight. When green plants convert sunlight directly, through photosynthesis, to a usable or edible form, they do no damage in the process. When those same plants are burned as a fuel, they produce polluting by-products; but if allowed to decompose first, in order to produce biogas, their effect on the environment is, as far as we know, benign. Solar panels or collectors can also be used to convert sunlight directly to a useful and benign form of energy.

Geothermal energy and energy derived indirectly from sunlight or gravity (wind, the falling action of water, the rising and falling action of ocean tides) is abundant or unlimited and generally considered benign, though the manner in which it is harnessed, such as a hydroelectric dam, may not be. Nuclear energy is virtually unlimited in supply, but its production, through nuclear fission, is potentially damaging and its radioactive by-products can be lethal.

Modern society is mainly powered by energy sources derived from sunlight trapped by ancient plants and converted to coal, oil, and natural gas. But all fossil fuels are finite, and we are consuming them at such a rapid rate that our ecosystem cannot reincorporate the residues of their consumption quickly enough to maintain a stable climate. Even if you are limited in your ability to select the source of energy you use in your home or business, you always have the option of using that energy sparingly, and of pressuring property owners and developers, utilities, and public institutions to convert to benign alternatives. In fact, public pressure is in part responsible for the astonishing growth in the transition to solar power in the United States. Use of solar energy grew 418 percent between 2010 and 2014,[1] and in 2015 another 30 percent.[2]

You are choosing a benign source of energy over a potentially damaging source of energy anytime you decide to walk or ride your bicycle rather than drive a car that burns fossil fuels. So is the farmer who uses livestock to break down the stubble on his cornfield rather than a tractor. What should concern you in this context check is whether or not the source of energy you plan to use in implementing any action is appropriate in your situation right now.

Ideally, governments should consider the long-term health of nations when forming energy policies, but their track record is dismal for reasons that, sooner or later (and hopefully the former), society needs to question.

## Sources of Money

The money used to implement any action can be derived either from internal or external sources. The source is internal if the money is taken from your own earnings, what the business or land generates. Anytime you can rely on an internal source you are likely to be better off, but there will be many occasions when money will have to come from outside the business in order for you to move forward. Economists sometimes argue that when you finance a venture with your own internal money, you should also consider the "opportunity cost," the interest you lose on your money by not investing it elsewhere. In managing holistically, opportunity cost is rarely a concern because the context checking helps determine where, among various alternatives, your money gives the highest return in terms of your holistic context, which is the more important consideration.

If you are in a sunlight-harvesting business, then you want most of the internal money you invest in any action or enterprise to come from solar dollars, not mineral dollars gained through mining soil in a nonrenewable manner.

When the money to be invested is derived from an external source, you need to be wary of the strings attached to it. If the outside source is a bank or other lending institution, compound interest will be included, meaning you could be using a greater sum of money to repay the loan.

External money can also be derived from the government in some form of cost-sharing or subsidy. This is a source of money that commonly becomes addictive, and its sudden withdrawal can spell financial ruin. There are strings attached as well, generally in the form of regulations that may force you to take actions not in line with your holistic context. It is also important to realize that governments have no money to give, unless they take it from you and your fellow citizens. Bear this in mind if you are considering an action purely because the government is providing half the money. If the action fails most of the checks and you go ahead anyway, it means that not only the half supplied by you but also the half supplied by the taxpayers is wasted. We'll return to ethical and moral considerations such as these in chapter 31 on the last of the checks—gut feel.

Philanthropic organizations are another source of external money, and a common one for nonprofit organizations. Only in relatively few instances are philanthropic grants or gifts made without strings attached, often in the form of influence or bureaucratic red tape or, on occasion, outright interference in management. Foreign aid programs are notorious for making money available to a developing country, then insisting the money be used to buy technology and expertise from the donor nation. If the influence or interference that comes with the money in such situations compels actions misaligned with your holistic context it will likely fail this check unless you can negotiate more favorable terms.

## Energy and Money Patterns of Use

In the second part of this check, you look at the specific way in which the energy or money will be used and whether it is in line with your holistic context. There are no rules that tell you what is right or wrong, but there are some questions that will help you decide:

- *Is the proposed use providing infrastructure that will assist in reaching your goals?*

*Infrastructure* refers to the sort of things that are essential to running your business more effectively: knowledge, skills, trained staff, buildings, roads, equipment, machinery, transport.

If energy or money is used to create infrastructure and all the materials or other aspects involved pass the other checks, you would tend to say the proposed infrastructure passes. If you are creating infrastructure that is not needed this year, and it does not pass the other checks, you would likely fail this check too. You might think this is so obvious it hardly needs stating, but you would be surprised. A number of ranchers have put money into fencing when it fails other checks, particularly the weak link (financial) check— which indicated product was the weakest link and they needed more livestock first. In this case, although the money would be building infrastructure, that infrastructure isn't needed now.

### • *Is the proposed use merely consumptive, with no lasting effect?*

A use of money or energy is consumptive if it is consumed in a one-time use. Many of the running costs involved in a business are consumptive uses, such as the fuel required for your vehicles, accounting fees, or salaries. So are many of the services you might purchase relating to a particular action, such as legal fees or consulting advice. If the source of the money or energy used is in line with your holistic context, and the action passes most of the other checks, a consumptive use automatically tends to pass this one.

### • *Is the proposed use cyclical in that once initiated, it would not require more money, or the purchase of more energy?*

A good example of cyclical use would be the single expenditure of money to install a hydraulic ram for pumping water. Since falling water provides the energy that drives the pump, all water is thereafter pumped at no cost, assuming you ignore the minimal maintenance it takes to operate the ram, which has few moving parts.

Using animal impact to break down crop residues is another. This might require an initial expenditure of money for the temporary polywire fencing or panels that confine the animals in any one place. But each year thereafter, it would merely require planning, implementation, and reuse of the movable fences or panels to have the animals do the job using only solar energy.

A cyclical use of money can be achieved in a number of ways. When money is used for a community revolving loan, for instance, it is cycling constantly and, if employed wisely, it is growing too.

Generally, a cyclical use that makes your money grow or enables you to forgo further purchases of energy is highly desirable, but again, the answer will depend on what you are trying to achieve and the holistic context guiding your actions.

> • *Is the proposed use addictive in that, once initiated, you risk an undesirable dependence on further inputs of energy or money?*

It is usually wise to avoid an addictive use of money or energy. An addictive use is one that obliges you to take the same action again and again, possibly with increasing frequency and/or increasing cost. Addictive uses are common in industrial agriculture and also in development projects, particularly when they seek to resolve a problem without addressing the underlying cause. An example of an addictive use of money and energy would be a parasite control program where livestock were dipped in a chemical (derived from the energy contained in fossil fuels) that not only killed the parasites but also their predators and thus generated a greater parasite problem needing more chemicals and more money.

## Conclusion

As mentioned, some individuals may have limited power to control the source of energy they use, though informed citizens can speak out for better governance, policies, planning, and regulation. You can also let conscience enter your own decisions to flip a switch, design a building, or raise a crop. The millions of little things Americans did, starting in the 1980s, to weatherize their homes and plug up leaks, plus the purchase of more fuel-efficient cars, yielded over seven times as many additional BTUs as the net increase in supply from all the new oil and gas wells, coal mines, and power plants built in the same period.[3] Without those improvements in efficiency, it would have required about fifty percent more energy to deliver America's GDP in 2013.[4]

Remember that your final decision will seldom be based on any one of the checks. You will be building a mental picture based on your answers to each of the questions. Only at the end of the questioning, when the picture is fully formed, will you finally decide whether or not to implement the action.

# 30

## Sustainability
### *Generating Lasting Wealth*

*THE SUSTAINABILITY CHECK ASKS YOU TO CONSIDER* the long-term environmental and social consequences of your actions relative to the future resource base described in your holistic context. In every case, there will be *people* who influence or are influenced by your decisions and *biological communities* on land and in the air or water that will be affected either directly or indirectly by your actions.

**The question you ask is:**
*If we take this action, will it lead toward or away from the future resource base described in our holistic context?*

This is one of the few checks that asks you to focus on a specific aspect of your holistic context. The future resource base describes the environment and behaviors that will be essential for sustaining the quality of life you desire for yourself and your descendants. Thus the sustainability check assures that actions you take to meet short-term needs also provide lasting gain—that they are socially and environmentally sound in terms of the future, as well as the present.

Far too many actions that prove correct in the short term lead to disastrous consequences in the long term. This check seeks to avoid that. In earlier chapters, I wrote of past civilizations that had been destroyed, along with their environment, as a direct result of actions people took to meet immediate needs, with the context for those actions generally confined to the problem at hand. The millions of environmental refugees caught in the mire of increasing civil unrest and violence, disease, and starvation today are a product of past management actions (and policies) that have exacerbated desertification. Their desperate plight will be shared by all of us if we do not begin to consider how our actions affect our future resource base. To do that we need to look at both aspects of the future resource base: *social*—our behavior as it will need to be perceived by the people whose support we need; and *environmental*—the land, as it will have to be to support our quality of life for many years to come.

## Social Considerations

No matter what type of business you are in, you need to consider how the perceptions of the people included in your future resource base (clients/customers and suppliers, extended family, advisers, and so on) are affected by the actions you take. If you have described yourselves as honest, reliable, and professional, you want to make sure that the actions you take reflect this behavior because it is your actions they will judge, not your words.

## Environmental Considerations

It is impossible to have a holistic context without some reference to the environment. There is not a citizen among us who does not eat, drink, produce bodily wastes that must be disposed of, or consume products that affect our environment, either in their manner of production or in their final resting place (usually the landfill).

Those specifically engaged in land or resource management will have described a future landscape in terms of how each of the four ecosystem processes will need to be functioning far into the future. Others may have only a fairly general reference to the environment in their future resource base that refers to the landscape surrounding their business or community. However, the civil engineer and the land use planner should attempt to find out how the layout and design of a new road or plans for a residential development, or for disposing of water runoff from roads and roofs, are going to affect the four ecosystem processes relative to the future resource base desired by the community and the level of brittleness that exists in that environment. As more urban communities begin to manage holistically and describe their resource base of the future, they will begin to appreciate the value of this check.

Most people, and many businesses, particularly those that provide services rather than manufactured products, may find it difficult to see how their actions affect land they are not directly responsible for. As I mentioned in chapter 9, all of them at some point, through consumption of raw materials manufactured into a product, and their use of technology, have a tie back to the land and affect the functioning of the four ecosystem processes. Thus any actions they take should pass through this check if they involve the consumption of products or the use of technology. I single out sunlight-harvesting and manufacturing businesses here because in both cases they involve fairly detailed considerations and demonstrate the considerable power of this check.

### Sunlight-Harvesting Businesses

If you are in a sunlight-harvesting business (e.g., farming, ranching, wildlife or forest management) most of the actions you take will relate more directly to the land and its necessary future condition. The actions you check will generally deal with one or more of the tools covered in chapters 15 through 23. In this check, you want to determine how the proposed action or tool is likely to affect the four ecosystem processes. Is the likely result going to take you toward or away from the future resource base described in your holistic context?

If your goal is to have open grassland with scattered trees, and certain areas are moving toward dense brush, you might consider applying the tool of fire, but before taking this action you would first review the effects of fire on the four ecosystem processes, as described in chapter 19. Anytime you are dealing with organisms that become a problem either because they are too few or too many in number, focus first on community dynamics.

As mentioned in chapter 13, when you manage for the health of the whole community, various species tend to take care of themselves and thrive within their community without becoming so numerous as to be classified as pests, or so few that we classify them as rare and endangered. Any actions that address a particular species in isolation, *with little regard for the whole*, can meet with only short-term success and will generally fail this check for long-term sustainability. Among the exceptions would be cases involving slow-breeding, rare, and endangered animals, such as rhinos or pandas, but they won't be saved in the wild unless you manage for the whole and preserve their habitat too.

Many people dedicated to saving a particular species find this thinking surprisingly difficult to translate into action, as numerous laws and programs for saving the ferret, the owl, the tortoise, the gorilla, or the local trout show. How many commit enormous resources into protection, and little into regeneration of the environment as a whole? No amount of captive breeding, plantings, culling of predators, poisoning of competitors, or other narrowly focused actions will bring back a creature that has lost its niche entirely or even a critical element, such as cover, food, or water, in its habitat.

The sustainability check is particularly important in assessing actions that still flow from policies created during the so-called Green Revolution, when we had supreme confidence that modern high-tech agriculture could feed the world. As a result, we don't question often enough the fertilizers, pesticides, genetic engineering, and extraordinary machines emerging under the heading of technology. These have encouraged monoculture plantings of annual crops, and the engineering of new plants and new pesticides to which these plants and no others are immune. However, abundant evidence now indicates the resultant damage to all four of the ecosystem processes.

When phasing farms from Green Revolution agriculture to regenerative practices based on healthy soil and soil life we often have to use measures initially that fail the sustainability check, just to stay solvent. However, this is no longer done in ignorance. Knowing that a measure is unsound allows you to start shifting your management, in the time bought by its use, and to find a way back to practices that regenerate soils and create healthy biological communities.

## Manufacturing Businesses

The majority of actions taken in a manufacturing business will involve the tool of technology, which, although it may not be used directly on the land, will impact the environment at various stages:

- When the raw materials to be used in manufacturing are extracted or produced
- During the manufacturing process itself (and the wastes it generates)
- In the final disposal of the manufactured products

In this context check, you are considering how actions relating to any one of those three stages will affect the ecosystem processes relative to the future resource base you have described. Few manufacturing businesses include a detailed description of the ecosystem processes in their future resource base, but some knowledge of them becomes necessary to weigh the effects of any product on the environment. Those effects are most obvious when you look at community dynamics. The waste produced in the manufacturing process, and many of the materials used in the products themselves, may not break down.

---

Few manufacturing businesses include a detailed description of the ecosystem processes in their future resource base, but some knowledge of them becomes necessary to weigh the effects of any product on the environment.

---

Very few products manufactured today are biodegradable, meaning that organisms of one form or another can consume them rapidly and completely. The cumulative effects of these nonbiodegradable and unnatural substances affects the web of interrelationships that exist in all biological communities, ultimately affecting their health and our own. Being "unnatural" the substances are frequently toxic to many organisms, adversely affecting their ability to function, impairing their ability to produce healthy offspring, and sometimes killing them and the organisms that feed on them. The horrendous amounts of plastic now polluting our oceans and predicted to outweigh the fish within them by 2050,[1] the industrial chemicals measured in Arctic wildlife living far from where those chemicals were released, and the amount of toxic substances showing up in human bodies, reveal the extent of the long-term damage. In one U.S. study, an average of ninety-one toxic, cancer-causing compounds was found in nine volunteers—none of whom worked with chemicals or lived near an industrial facility.[2] Since water and mineral cycles and energy flow are all dependent on living organisms, these are adversely affected as well.

Almost certainly, the future resource base of any manufacturing business will include customers, clients, and employee families who will be concerned with the business's impact on the environment. So there is no situation where these considerations and this check would not apply. In table 3-1 (chap. 3) I noted the many successes of everything we "make" and the genuine comfort that technology has afforded us. Each example, however, could be considered successful only if we ignored the effects on our environment stemming from its manufacture or disposal. This check is a reminder of that, and should encourage manufacturers to find ways to create products that, from their conception to the end of their useful lives, are environmentally benign, if not enhancing.

Nearly two decades ago, the International Organization for Standardization (ISO), headquartered in Geneva, produced a series of environmental management standards known as ISO 14000. By 2015 there were more than three hundred thousand certifications in 171 countries.[3] Such standards are a step forward. But compliance based on regulation will always fall short of the ideal. In the long run doing something because it is the morally right thing to do and is sensible, long-term business strategy needs to become the norm. Corporate boards and CEOs who align their strategies and actions with a holistic context will recognize early on that it is not in their self-interest to produce profit at the expense of society, our environment, and their own grandchildren.

### Service Providers, Households, and Consumers

Not only does the company manufacturing products need to concern itself with the pollution it creates, but so do the people who purchase, use, and then discard the products. If an action you plan to take involves the purchase or use of a product that, once used or discarded, will not be consumed by living organisms, you may want to substitute it for another or find ways to recycle it. In some cases, your options will be severely limited because most of what is manufactured and available today is not environmentally benign. But you will know that, if the future resource base you have described includes healthy land and healthy people on it, you will have to do something—and, fortunately, you will not be alone in your dilemma.

You can make a start by decreasing your consumption of nonbiodegradable products, by supporting with your purchasing power manufacturers who work to create products that do not damage the environment or who buy back spent products and reuse or recycle their raw materials, and by participating in recycling efforts in your own community.

## Conclusion

None of the changes you are trying to bring about on the land and in your behavior will be attained quickly and with only a few actions. Yet every action, however small, that takes you in the direction you want to go and is in line with your holistic context is progress, and cumulatively small actions add up to a big difference.

All of this may seem a lot to concern yourself with. But until you do, until we all do, we cannot hope to create a community, nation, or even a civilization, that is viable or sustainable.

# 31

## Gut Feel
## *Finalizing Your*
## *Decision*

*THE GUT FEEL CHECK IS DONE LAST* because it builds on the mental picture that has formed after passing through all the other checks. In each of the previous checks you have in effect been looking at one of the small squares that make up Lincoln's face (see fig. 3-2). Now you are asked to blur them together to see the face as a whole, and, based on that picture, make your decision. But where each of the other checks asked what you *think*, this one asks how you *feel*. And that is in large part going to be based on the values reflected in the quality of life statement embedded in your holistic context.

**The questions you ask are:**
- How do we feel about this action now?
- Will it lead to the quality of life we desire?
- Will it adversely affect the lives of others?

These values, in many respects, are a reflection of the traditions, customs, and culture shared by those who have created the holistic context. In our pursuit of progress no consideration deserves more reflection, and typically none gets less. Corporations concentrate on shareholder response to the next quarterly report, politicians on the next election. Ranchers concentrate on production of livestock, farmers on production of crops. Environmentalists concentrate on growing trees, loggers on cutting trees. Generals fixate on counting bodies and missionaries on counting souls. Few of us stop long enough to notice that, in our pursuit of progress, we shoot down our own dreams and those of others. Where the other checks have only touched on quality of life concerns, this one addresses them directly.

Occasionally, this check may cause you to question the value of certain customs or traditions. Are they really worth preserving? Have circumstances changed to such an extent that they have become counterproductive? If you didn't ask these questions when creating your holistic context, they may have to be addressed here. The answers may well cause you to revise your view of how you want your life, in this particular whole, to be, and to reflect this change in your holistic context. On the other hand, this check should also be used to ensure that customs and traditions you value are not lost.

Last, this check asks you to consider how an action could affect the lives of those outside your immediate whole—from the society you live in, to the greater society that comprises all humans. Pleasing everybody may seem impossible, but you can go a long way by embracing the holistic principle that the health of your particular interest is not distinct from the health of the greater whole. This is in effect a check for social consciousness and, more than any other, helps assure that a decision is socially sound.

The lack of attention to the quality of our lives in our national decision making has resulted in numerous tragedies, one of the most obvious being the state of American agriculture. The American government, with the acquiescence of many in the industry, undertook to increase production, solely in terms of quantity. From government and universities and industry leaders the message was put over powerfully—get big or get out. No one paid enough attention to the families that would be displaced as big farms swallowed smaller ones, and more powerful machinery and larger monoculture fields displaced labor.

Production boomed, but at the cost of a polluted environment, massive soil erosion, and enormous social dislocation. Hundreds of thousands of family farms vanished, dissipating generations of practical knowledge. The dispossessed drifted to urban centers and struggled to adjust to an alien culture.

Many churches, small businesses, and cultural centers in the small towns that served those people withered away, with suicide being the chief cause of death among the farmers that remained.

Hindsight gives us perfect vision. What if the American government had developed an agricultural policy guided by a national holistic context and used the seven checks to ensure each action described in that policy was aligned with that context? How different things would be. With its once vast and fertile prairie soils, the likes of which no nation had ever enjoyed, the United States would still most likely have become the world's greatest agricultural producer, while maintaining its healthy rural populations, the villages and towns that served them, and vast, diversified markets.

Most of the actions you consider will not be of this magnitude. Yet it is just as important that your holistic context spell out quality of life desires as clearly as possible and that they speak for all of the decision makers. Because your final decision is based on the picture that forms as you pass through the other six checks, the gut feel check may be one that those unfamiliar with the process fail to understand. These people may in fact exert considerable pressure, often out of genuine concern, to dissuade you from taking a particular action. If that happens remember that you are considering which actions to take based on your own cultural and social values, and it would be close to impossible for anyone else to understand them as well as you do.

## Context Checking Summary

We have now covered all of the context checks developed to date. If you have considered each of these checks in evaluating actions you plan to take, you will have gone far in preventing costly and unsound decisions. Ideally, all actions should pass all the checks that apply, and those that do almost certainly will give improved results economically, ecologically, and socially. Any actions that fail this year may pass later as your management takes effect and the whole situation changes.

Remember that speed is essential to the process. It is the speed that gives you the big-picture clarity you need. If you cannot quickly answer yes or no to a question, bypass the check. Most of the time you will only come back to that check if you are unable to reach a conclusion after passing through all the others. However, if you have to bypass the cause and effect or weak link checks, it may be pointless to continue the checking until you have answers for these two.

Given time and practice, the concepts underlying the seven checks will become so familiar that you will automatically take them into account before planning an action. You will start to look for the underlying cause of a problem before you even begin to consider potential remedies. Before you contemplate what to do about the borers eating your corn, you will look for the weak link in the borer's life cycle. Going into a preliminary financial planning session, you will already have calculated the gross profit on that enterprise you are so keen to develop.

When you've reached this stage, the checking really does go quickly, and it genuinely becomes nothing more than a final check to ensure the action you plan to take is sound and in line with your holistic context. In the meantime, bear in mind the following points:

- The tendency to slip back into making decisions in the context of any immediate need or problem will always be with you—especially in emergencies. Take actions to deal with the emergency, by all means, but check each of them to ensure they are in line with your holistic context while also dealing with the emergency.

- When you are dealing with a problem, go to the cause and effect check first. If an action does not address the underlying cause of the problem, you will not solve it.

- The gross profit analysis check applies only when two or more enterprises are compared; the marginal reaction check applies only when two or more actions are compared.

- The weak link check applies in three different situations: social, biological, and financial.

- The gut feel check is based on your feelings about the picture that emerges after passing through all the other checks that apply and should be done last.

Remember that there is no tyranny in Holistic Management, nothing that you must or must not do. You may decide to implement an action that fails one or more checks, simply because you have little option at present to do otherwise. You at least know that the decision is not in line with what you hope to achieve in the long run, and that you have to do something about it. Finally, even when a decision passes all the checks, it could still prove wrong. You can't be sure unless you monitor what you have planned, a subject we'll cover at length in chapter 44, "Monitoring and Controlling Your Plans to Keep Management Proactive."

# PART 7

# GUIDELINES FOR USING THE MANAGEMENT TOOLS

## Holistic Management Framework

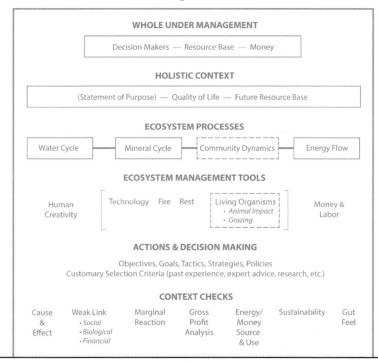

**WHOLE UNDER MANAGEMENT**

Decision Makers — Resource Base — Money

**HOLISTIC CONTEXT**

(Statement of Purpose) — Quality of Life — Future Resource Base

**ECOSYSTEM PROCESSES**

Water Cycle — Mineral Cycle — Community Dynamics — Energy Flow

**ECOSYSTEM MANAGEMENT TOOLS**

Human Creativity

Technology   Fire   Rest   Living Organisms
                              • Animal Impact
                              • Grazing

Money & Labor

**ACTIONS & DECISION MAKING**

Objectives, Goals, Tactics, Strategies, Policies
Customary Selection Criteria (past experience, expert advice, research, etc.)

**CONTEXT CHECKS**

| Cause & Effect | Weak Link • Social • Biological • Financial | Marginal Reaction | Gross Profit Analysis | Energy/ Money Source & Use | Sustainability | Gut Feel |

**MANAGEMENT GUIDELINES**

Time   Stock Density & Herd Effect   Cropping   Burning   Population Management

**PROCEDURES & PROCESSES**

Holistic Financial Planning   Holistic Land Planning   Holistic Planned Grazing   Holistic Policy Development   Research Orientation

**FEEDBACK LOOP**

Plan
(Assume Wrong)
Replan                    Monitor
Control

# 32

## Introduction
### *Lessons Learned in Practice*

THE MANAGEMENT GUIDELINES covered in the following chapters have crystallized out of a struggle to connect what is possible in theory with what is practical in real life. They reflect years of experience in a variety of situations, and the contributions and criticisms of many people, but these guidelines are chiefly the result of what we have learned through continually making mistakes.

Whether or not you are managing land, study each of the next five chapters thoroughly. They are relevant to anyone making or supporting decisions that will affect the land in any way. The management guidelines will influence many of the decisions you run through the context checks because they in fact help shape those decisions, providing definition and detail that might otherwise be lacking. This is certainly the case for the oldest of the guidelines, those that arose when we discovered the significance of herding animals to the health of brittle environments. Before we could utilize livestock to restore deteriorating land, we had to develop guidelines for managing their grazing and trampling to ensure that the animals, the land, and the people involved all benefited.

Anytime we attempt to alter ecosystem processes we do so through the use of a particular tool. The management guidelines are specific to the tools of living organisms (including grazing and animal impact) and fire. In each case, the guidelines attempt to work with rather than against Nature to ensure that when you use a particular tool it will achieve what you want it to achieve. In brief, the key guidelines for managing ecosystem processes through these tools include the following:

- **Time** (Tool: living organisms—animal impact and grazing). When the whole you are managing includes grazing animals for any reason, you need to make sure that their presence enhances all four ecosystem processes. Timing the exposure and reexposure of the plants and soils to the animals will be critical to ensuring plants are not overgrazed or soils overtrampled. Chapter 33 gives guidelines for managing grazing and trampling time depending on the level of brittleness, the climate, the season, the types of plants, and the needs of the animals.

- **Stock density and herd effect** (Tool: living organisms—animal impact and grazing). The stock density and herd effect guidelines, covered in chapter 34, apply when you are using the tools of grazing and animal impact to alter soil conditions or vegetation. The biggest challenge in inducing herd effect, which requires that you produce behavior change in the animals, is inducing it often enough and over a large enough area, particularly in the more brittle environments. Chapter 34 summarizes the techniques developed to date.

- **Cropping** (Tool: living organisms). If we are to sustain our present civilization and its enormous population, we must strive to create an agriculture that more closely mimics Nature, one that enhances rather than diminishes water and mineral cycles, energy flow, and community dynamics. Toward that end, chapter 35 gives some fundamental guidelines that apply in any cropping situation.

- **Burning** (Tool: fire). While fire is a tool that has a definite and useful role to play in land management, we always need to question its use. Chapter 36 reminds us of the environmental dangers associated with burning while providing appropriate safeguards.

- **Population management** (Tool: living organisms). The population management guideline, covered in chapter 37, bears on the tool of living organisms, but more generally on the management of community dynamics. It applies anytime we want to encourage or discourage the success of a species. Guidelines are given for assessing the health of a species' population, for determining the environmental factors that will enhance or limit that population's success, and for dealing with predators that become a problem.

The management guidelines are merely a set of principles that help you to determine a course of action. These chapters include only the most obvious principles and the relevant guidelines developed to date. As new challenges emerge and knowledge is gained further guidelines may well be added.

# 33

## Time

### *When to Expose and Reexpose Plants and Soils to Animals*

*Most management situations that involve grazing animals* for any reason require maximum functioning of all four ecosystem processes. To meet this requirement, overgrazing and overbrowsing of plants need to be avoided or minimized. This chapter explains the principles to follow if you are managing livestock to enhance forage production, regenerate soils, and reverse desertification.

Earlier chapters have mentioned the finding of André Voisin that overgrazing is linked to the time plants are exposed to animals rather than to the number of animals, but how in practice do we time the exposure and reexposure of plants to animals? Should we monitor "key indicator" plants, set arbitrary grazing or recovery periods, or follow some aspect of animal performance? Should timing in planning reflect the growth rates of plants, and if so, which of the millions in the community? Do we choose individuals of a particular species or a random selection? What about the wildlife on the same land? Should animals be allowed to select their diets, or should they be forced to eat everything in a nonselective manner over a short time?

Voisin's work with European pastures answered only a few of these questions for me when, in the 1960s, I first began to see that timing mattered on the savannas of Africa and began to look for ways to successfully manipulate it. Healthy savannas include a mind-boggling diversity of plant species, from the simplest algae-like forms to a variety of trees. Animal life ranges from trillions of microorganisms to a vast complexity of birds and animals, small and large. In such a complex biological community, any change produced by management in one area inevitably changes everything to some degree. So when we decide that a particular plant species can be sacrificed to overgrazing, as some do, we unleash consequences beyond human ability to even understand, let alone manage. Other species depend on species that depend on those we have sacrificed, and on and on.

If your holistic context includes healthy biological communities, you need to minimize overgrazing on every plant you possibly can. To achieve this you base the time of exposure on the most severely grazed plants, wherever they are and whatever species they are.

The Reverend Martin Niemöller, imprisoned by the Nazis at the Dachau concentration camp, captured the idea precisely when he wrote, "In Germany, the Nazis first came for the Communists, and I didn't speak up because I wasn't a Communist. Then they came for the Jews, and I didn't speak up because I wasn't a Jew. Then they came for the trade unionists, and I didn't speak up because I wasn't a trade unionist. Then they came for the Catholics, and I didn't speak up because I was a Protestant. Then they came for me, and by that time no one was left to speak up for me."

I will return to this famous saying as a constant reminder that, once any plant has been severely grazed you need to pay attention, or as the Reverend Niemöller would say, you should protest when the first person of any persuasion is taken.

## Monitor the Perennial Grasses

Because of the many variables involved, a systematic accounting of time is nearly impossible. Grazing animals select different plants and different parts of plants in different seasons. Different plants recover at different rates. And plants on different parts of the land are experiencing very different growth conditions daily. At the beginning of my wrestling with this problem I chose to watch the perennial grass plants as the group most vital to the stability of the whole community.

We did not then recognize the distinction between brittle and nonbrittle environments, but experience has borne out the hypothesis that, particularly

---

### Time and Trampling

*Adverse consequences of trampling* are a function of time rather than animal numbers. Prolonged trampling has adverse effects— pulverization of the soil surface, excessive underground compaction, and injury to plants.

Chapter 22, on animal impact, gave the example of 365 successive donkey-days of traffic producing a beaten track between a house and a water hole. On the other hand, the same traffic produced by 365 donkeys on a single day, followed by 364 days of recovery time, would produce a different result. The plants and the whole soil community could recover from any damage due to trampling and benefit from the intense deposition of dung and urine. Time, rather than numbers, governs the ultimate impact.

Maximum impact *over minimum time* followed by a sufficient recovery period makes trampling an extremely effective tool for regenerating brittle grasslands and water catchments as well as crop-land soils. Specific guidelines for trampling are given in chapter 34.

---

in low-rainfall, brittle environments, perennial grass stability in fact contributes to the stability of the whole biological community more than any other plant group because the grasses provide most of the soil cover. Since well over half the world's land surface leans to the more brittle end of the scale, and little of that enjoys enough rainfall to support full tree cover, the health of perennial grass acquires enormous significance.

---

To reach the richest level of biological diversity in any predominantly grassland environment, time grazings according to the needs of perennial grasses.

---

In less brittle environments where the future landscape includes the maintenance of grassland, perennial grasses are more prevalent than annuals, and their health will affect your ability to prevent or slow the community from moving to woodland or forest. The overgrazing of perennial grasses in these environments causes little or no soil exposure. Some pastures in England have been overgrazed for centuries yet still remain completely covered. However, forage volume, and thus energy flow, is greatly decreased when plants are overgrazed, and diversity is invariably reduced.

In choosing to consider perennial grass first, we risk overlooking two important factors. A tree or shrub species might suffer severe defoliation before animals start on the grass. Also, time allowed for recovery of a severely bitten grass plant might not suffice for a severely browsed shrub or tree. This potential problem had to be worked out, and I will return to it further on in a way that still justifies the practical guideline: to reach the richest level of biological diversity in any predominantly grassland environment, time grazings according to the needs of perennial grasses.

## Monitor Plant Growth Rates

Overgrazing, remember, occurs when a severely bitten plant is bitten severely again while using energy it has taken from its stem bases, crowns, or roots to reestablish leaf. This can happen in the *grazing period*, when the plant is exposed to the animals for too many days and they are around to regraze it as it tries to regrow. It can also occur following an inadequate *recovery period*, when animals have moved away but returned too soon and grazed the plant again while it is still reforming leaf and has not yet reestablished its roots.

So how long is too long a grazing period? How short is too short a recovery period? No matter what the perennial grass species, this depends on two things: the proportion of leaf removed by the grazing, and the daily growth rate of the plant. The less leaf removed, the quicker the subsequent regrowth and the faster the recovery.

To be safe, we assume that the grazing has been severe, because some plants are always grazed severely, and thus we focus on plant growth rate. If growing conditions are favorable, and a severely grazed plant can thus grow one centimeter (half an inch) or more per day, you can expect plants to be overgrazed after about three days' exposure to animals. If growing conditions are poor and a severely grazed plant can only grow one centimeter every four or five days, overgrazing will occur after about ten days. Thus the faster the growth rate, the shorter the grazing period *needs to be.* The slower the growth rate, the longer the grazing period *can be*. The reasons for this become clear with an understanding of the link between grazing and recovery periods, which I'll cover shortly.

---

Thus the faster the growth rate, the shorter the grazing period needs to be. The slower the growth rate, the longer the grazing period can be.

---

When it comes to the recovery time needed for a plant to reestablish its roots, the situation is similar. When the daily growth rate is fast, because growing conditions are good, the plant reestablishes its roots quickly. Thus the faster the growth rate, the shorter the recovery time needed. With runner-type grasses, where a smaller proportion of leaf is always removed, the recovery time needed can be as short as twelve to fifteen days. With bunched grasses, where a higher proportion of leaf is usually removed, recovery time can be as short as twenty-five to thirty days. When the daily growth rate is slow, the plants need longer to restore their roots following a severe grazing. Recovery times for runner grasses can stretch to thirty to fifty days, for bunchgrasses sixty days or more. Figure 33-1 illustrates the principle using a bunchgrass as an example.

## Grazing and Recovery Periods Are Always Linked

As long as a herd of livestock moves through a series of subdivisions, the grazing periods will be inextricably linked to the recovery periods. The dynamics of this relationship are simple, but easy to overlook.

Assume that the top diagram (A) in figure 33-2 represents a piece of land divided into six areas that animals will graze for four days each. From the time they leave an area until they return to it will then take twenty days—four days in each of five areas (six grazing areas minus the one they are in). Plants in each area will get four days of exposure to grazing and twenty days to recover.

If, on leaving area 1 you decided it will require forty days for a severely grazed plant to recover, you will have to add twenty more days somewhere in the other five areas, as the middle diagram (B) shows.

Therein lies the rub. *Any change in recovery time in one area will change the grazing times in the remaining areas to be grazed.* In the bottom diagram

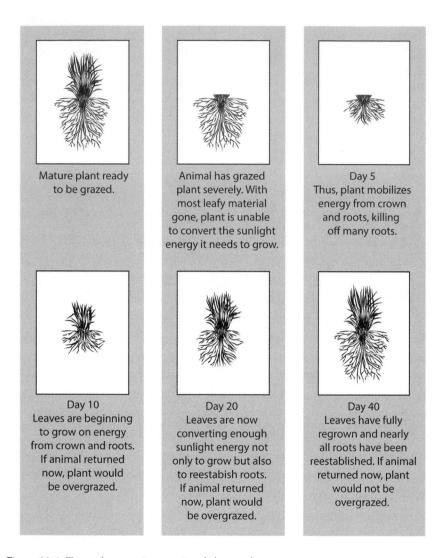

**Figure 33-1.** To avoid overgrazing, monitor daily growth rates.

(C), the operator planned a forty-day recovery period for the plants in each area and thus eight-day grazing periods. But after five days of grazing, area 3 looked a bit sparse, so he moved on. He thereby cut the recovery time in all areas back to thirty-seven days. Each area that is grazed for fewer days than planned *reduces recovery times in all areas.* Conversely each day that stock are held longer in an area adds a day of recovery to all remaining areas.

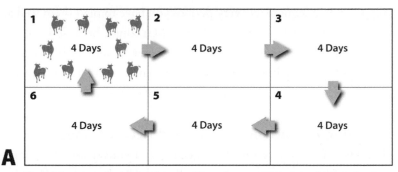

If animals are to spend 4 days in each of 6 grazing areas, then from the time they leave an area until they return to it will take 20 days—4 days in each of 5 areas (6 grazing areas minus the one they are in.) The plants in each area thus have 20 days to recover.

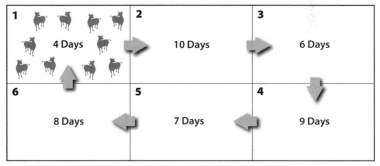

If, on leaving area 1 you decided it would require 40 days for a severely grazed plant to recover, you would have to add 20 more days somewhere in the other five areas (10 + 6 + 9 + 7 + 8 = 40 days.)

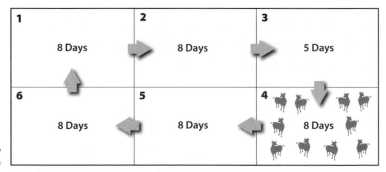

Here the manager planned a 40-day recovery period and thus 8-day grazing periods. But after 5 days of grazing, area 3 looked a bit sparse, so he moved on. He thereby cut the recovery time of the other areas back to 37 days (8 + 8 + 8 + 8 + 5 = 37.)

**Figure 33-2.** Grazing periods and recovery periods are linked. Any change in grazing time in one area will change the recovery times in all remaining areas.

To maintain adequate recovery periods you have to plan them well ahead (as covered in chap. 41 on Holistic Planned Grazing), because it takes time to build them up. Grazing periods, on the other hand, can be changed on impulse by simply moving the animals, *but* remember that anytime you do this, it will have a cumulative effect on recovery periods.

## Base Grazing Periods on a Preselected Recovery Period

I refer to land managed as a unit for grazing as a *grazing cell.* The timing of herd moves within the grazing unit, or cell, naturally depends on the number of subdivisions within it, which I refer to as *paddocks,* a term that varies country by country. Domestic stock, being severe grazers (as are most of their wild cousins), will predictably defoliate some plants severely soon after entering a paddock, regardless of how few animals there are. Color plate 15 shows what one horse did in one hour to one plant among hundreds of thousands of plants. Remembering Niemöller's mistake, "They came for the Communists, but I wasn't a Communist . . . ," we generally try to minimize the chance that the first severely grazed plant will get overgrazed, regardless of its species.

Perennial grass plants in the Rio Grande Valley of New Mexico, where this photograph was taken, might need a recovery period of sixty days in slow growth. If, in our simple case, the grazing cell contains nine equal paddocks, then a sixty-day recovery period will require a seven-and-a-half-day grazing period in each paddock. The reasoning goes thus: After leaving any one of the nine paddocks, the horse can pass through the other eight before coming back. A sixty-day recovery period divided by eight paddocks yields seven and a half days of grazing in each paddock.

In contrast to this method, seat-of-the-pants management would tend to eyeball each paddock after stock had been in it awhile and then decide when they should move, based on the amount of forage left as opposed to plant growth rates. It might work, but it more likely wouldn't, because it leaves to chance the really crucial time, the recovery period. As the number of paddocks increases, naturally the length of the grazing period decreases, because the same recovery period gets divided by a larger number. As it turns out, there are a number of other advantages that flow from having many paddocks in a cell.

## The Advantages of Many Paddocks

Increasing paddock numbers, by subdividing a cell further, either by fencing or by strip grazing within a paddock, decreases time in each paddock (or strip), and thus increases your ability to minimize overgrazing. What's more, as paddock size decreases, stock density increases, causing better distribution of dung, urine, and trampling and a number of other benefits, including more even grazing, increased energy flow, and improved animal nutrition.

### More Even Grazing

As paddock size decreases, given a constant herd size and constant recovery period, the proportion of plants grazed increases. This does not, however, mean animals are any less able to select and balance their diets. Because the time they spend in the paddock also decreases, the same volume of forage essentially is taken. In general, only a change in the number of animals or in the time they spend in a paddock will change the amount of forage they will harvest.

What we do tend to find is that, as the animals select a diet balanced for levels of protein, energy, fiber, and other nutrients, they tend to feed over a higher proportion of the plants available. This has the marked tendency to keep a higher proportion of the leaf and stems on more of the plants fresh and young. When grazing at lower densities animals generally feed off a smaller proportion of the available plants and thus allow a higher proportion to become cluttered with old stems and leaves of low nutritive value. I have also found that the longer we hold any number of animals in any paddock, the higher the proportion of plants that get severely grazed.

### Increased Energy Flow

Much research in several countries has shown that, during the growing season, the amount of green leaf removed greatly affects the rate at which plants regrow after being grazed. Figure 33-3 shows two equal perennial grass plants. Both had almost all their old leaf and stem material removed in the previous year and began growth as equals in this season. Early on, some animal severely defoliates A, removing ninety percent of the leaf, but takes only about forty percent of the leaf on B. The two plants then recover at very different rates. B

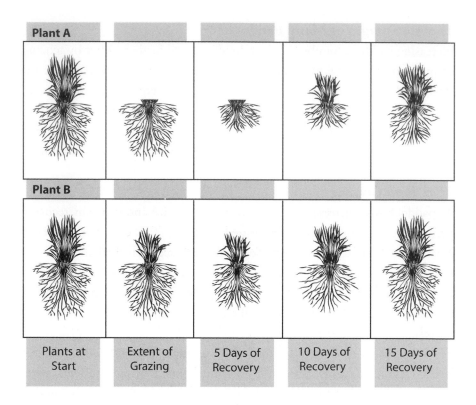

Plant A

Plant B

| Plants at Start | Extent of Grazing | 5 Days of Recovery | 10 Days of Recovery | 15 Days of Recovery |

**Figure 33-3.** The amount of leaf removed in a grazing affects the rate at which the plant regrows. Plant B loses far less leaf than plant A and thus draws less energy from roots, stem bases, and crowns. Less root is killed, and it begins to regrow almost immediately.

draws less stored energy and loses less root and starts regrowth almost immediately. Over the next two weeks, it produces much more volume of leaf and stem than A. At some point, however, B slows down, and A will catch up. Plant A, on the other hand, suffers a severe setback to root growth following the severe grazing because so much leaf has been lost. It will be awhile before enough new leaf has grown to concentrate sufficient energy to regrow the lost roots. The higher the proportion of less severely grazed plants to severely grazed plants, the more total forage produced in a given recovery period.

The higher the proportion of less severely grazed plants to severely grazed plants, the more total forage produced in a given recovery period.

Thus, in practice, the more paddocks per cell, the better the distribution of the grazing on the plants, the fewer severely grazed plants, and the greater the proportion of plants able to recover quickly from grazing, all of which results in increased energy flow. Unfortunately, many people in such situations try to prevent animals from selecting their diets by forcing them to eat everything. This is generally unwise.

### Improved Animal Nutrition

The fact that animals move more frequently onto fresh, unfouled ground means they receive a better plane of nutrition and reduced danger of parasite infection and buildup. Animals moving into a fresh paddock are free to graze almost everywhere, as they encounter no fouling from dung and urine. They easily take in a good volume of a well-balanced diet.

During the first day in a paddock, the animals tend to select what is readily available, while at the same time dunging, trampling, and urinating over much of the paddock. They do the same the following day, but do not find it quite so easy, because less of the most desirable leafy material remains, and they try to avoid grazing on their own fouling of the previous day. Consequently, the second day the animals may experience a lower-quality diet. Quality drops again the third day, and so on. The pattern of declining quality and/or volume of diet continues until the animals go to a new paddock. If they ever stay in a paddock until all forage is depleted, the consequent severe drop in nutrition inevitably results in poor performance.

The next chapter will further explain how to manage these generally beneficial aspects of smaller paddocks. Nevertheless, reducing paddock size has one other effect that falls under the heading of time management. Because animals are grazing at higher density, if they are left in a paddock long enough to overgraze plants, they will overgraze many more plants than they would if grazing at a lower density. Even a twenty-four-hour mistake in timing could mean extreme depletion of forage in a given paddock, loss of selectivity, and a drop in animal performance, so *time management and planning become more critical as paddock sizes decrease.*

Some, on considering this possibility, have decided that more paddocks mean more *risk.* This word implies chance beyond management control. I would rather say that more paddocks decrease risk from weather problems but increase the *penalty* for poor management. Using many paddocks with good management reduces the risk of drought and disease, although poor planning will result in poor animal performance.

## Time and Overbrowsing

Earlier I promised to return to the question of possible discrimination against woody plants when grazing is timed to the needs of perennial grasses. A couple of real-life examples illustrate how this apparent dilemma generally works out in practice. The Karroo area of South Africa has seen some three hundred years of overgrazing of plants and partial rest of soils over a vast expanse of low-rainfall, very brittle environment. Some of the ranches where I worked had once been grassland under herding wildlife populations, but had long ago declined to bare soil capped with an algae and lichen crust with a few scattered desert bushes. In the absence of any alternative, these desert bushes had become the main feed for all livestock and were consequently highly prized by ranchers and range management professionals alike.

Given a holistic context requiring the reestablishment of perennial grassland, I advised the use of recovery periods that would promote perennial grass plants. To many academics this seemed illogical, as for all practical purposes perennial grasses no longer existed. They claimed, as did many ranchers, that severely defoliated desert shrubs could not recover in the short recovery periods my recommendation suggested. Available research showed that the desert shrubs in question required anything from a year to eighteen months to recover after severe browsing. Pressure grew to make recovery periods reflect this finding.

The ranchers generally had between five and sixteen paddocks available. To plan on an eighteen-month recovery period in a nine-paddock cell means grazing periods of about sixty-seven days. Even in a sixteen-paddock cell, eighteen-month recovery periods would require average grazing periods of thirty-six days. Such grazing periods would guarantee overgrazing of any perennial grass plant that might try to establish. In addition, such grazing pressure at the increased stock density of the smaller paddocks would severely defoliate the desert shrubs, thus probably causing them to require the very long recovery period. At the same time, it would put extraordinary nutritional stress on the livestock and could lead to damage from overtrampling in some areas.

It would, in other words, make the regeneration of perennial grassland extremely difficult, if not impossible. Also it would make it difficult to achieve good animal performance without heavy and costly supplementation.

With nine paddocks, sixty-day recovery periods would result in seven-and-a-half-day average grazing periods. Sixteen paddocks lowers this to four days. Such a regime reduces stress on animals, cuts trampling time, and ensures a good chance of achieving a grassland landscape. Also, it does not in fact expose bushes to the kind of heavy and prolonged browsing that necessitates long recovery. Bushes do not regrow from energy in crowns or roots like grasses, but more from remaining leaf. Thus the species that had survived under the conditions described were not likely to die out from a little over-browsing. Even their seedlings, once complexity returned, would have more chance of survival.

Color plate 16 is a photo taken on the first ranch on which I put this into practice. Over the entire ranch there was not a single perennial grass and almost no annual grasses. I had the rancher grade a small airstrip on which I could land my plane and immediately annual grasses appeared on the disturbed ground, as the photo shows. One perennial grass plant is visible under the wing of my plane. The rancher took the photo in color plate 17 some thirty years later. The old airstrip is now overgrown with grass. In fact the entire ranch is now grassland in which the desert bushes are still present.

All this argument, that woody plants can also thrive when the length of recovery periods is designed for the benefit of perennial grass, may seem paradoxical in light of many statements about shifting the composition of the biological community away from woody plants and toward grassland. The length of recovery periods, and thus grazing periods, and the degree and nature of animal impact, however, can be manipulated to move the composition of biological communities in either direction. By grazing grasses without overgrazing them and by using high animal impact, for instance, most grasslands remain grassland without excessive weeds or shrubs.

## Grazing in the Nongrowing Season

So far we have looked at managing time to minimize overgrazing, overbrowsing and trampling during the growing season. What about manipulating recovery periods and grazing periods while perennial grasses are dormant, and thus not as susceptible to overgrazing?

Even then, of course, animals eat and have a physical impact on soil and plants, so many of the same considerations govern the situation as before. Livestock will avoid ground fouled by their own dung and urine, and parasites and infection will usually increase when herds linger in the same area. Hooves will continue to trample. The timing must still assure that these factors contribute to the health of the soil surface microenvironment and good performance of animals, both wild and domestic.

Nongrowing periods are the most critical times of year for most wildlife, and their food and cover requirements are heavily influenced by the grazing plan for the livestock. If livestock are merely rotated through the paddocks on an arbitrary schedule, as is tragically so common, it can devastate wildlife. And that influenced my thinking greatly when developing Holistic Planned Grazing as a way to overcome the problems inherent in rotational grazing and to ensure integration of the livestock with wildlife, crops, orchards, and other land uses.

### *Limit the Number of Selections*

From what we currently know, animals select their diet in the same way, regardless of season. Each time animals enter a fresh paddock they balance their diets as best they can in ways we may never fully understand. If they enter the same paddock a second time during dormancy plants will not have regrown any new leaf. However, the effect on animal nutrition will not be the same as a single prolonged stay. An intervening recovery period will have allowed fouling to wear off and the mere act of moving onto fresh forage, even when depleted, seems to stimulate livestock in ways we do not properly understand. Yet undoubtedly the selection from the forage remaining the second, third, and fourth time through will certainly contain less protein and energy and more fiber than before, because of no regrowth between each of these grazings.

Given a high enough number of paddocks, as becomes possible with herding or strip grazing, the stock will constantly move to fresh ground throughout the nongrowing season, and even on the last day will still enjoy a reasonable plane of nutrition and a quick move to fresh ground.

### Plan a Drought Reserve

It used to be that ranchers would plan for drought by withholding certain areas from grazing during the growing months. The forage that accumulated would be kept in reserve in case the following nongrowing season lasted longer than expected. This practice, however, always decreased the productivity of both forage and animals. Now, we plan for drought by *reserving time* as described in chapter 41 on Holistic Planned Grazing.

## Time and the Management of Wild Grazers and Browsers

The same principles that apply to livestock govern wild grazing and herding animals. Trampled litter and soil do not distinguish between buffalo and cow. The health of the community in the more brittle environments demands some trampling, but any number of animal species can provide it either in a beneficial way, or too long and too often.

In the case of livestock, we can easily distinguish between time and numbers, as no matter what the numbers are, we can control the time through fencing or herding. Wild animals do not submit to the same kind of control, and the distinction blurs, especially *because in their case, numbers can influence time.* Where heavy predation, accident, and disease control numbers, the size of a herd's territory or home range tends to regulate the frequency of return to feeding areas. The concentrated fouling of animals bunched for self-protection will ensure short periods of grazing on the same ground. If lack of predation or disease allows numbers to rise, home ranges and territories appear to become smaller as more herds occupy the same area, and herds return to past feeding grounds sooner. This starts a snowballing breakdown of the ecosystem processes, including the loss of many nonherding wildlife species.

I thought through the logic of all this years ago when doing my early work in the Luangwa and Zambezi Valleys of Zambia and Zimbabwe. However, one bit of evidence did not fit and appeared to disprove the theory. Both the areas in question still had naturally functioning predator populations that should have provided population control and reinforced herding instincts.

The Luangwa in particular had the highest known concentrations of lions in the world. Why were the ecosystem processes breaking down so badly after we formed game reserves?

One thing both areas had in common was the removal of humans as a predator. A closer look at both hunted and protected areas revealed that where African villagers continued to hunt, the game was wild, grass was abundant, and burning was frequent. Inland, where the burning was taking place, much of the soil was bare, but the riverbanks remained covered in lush vegetation and were stable. In the protected areas, game was tame and the soil was bare right to the water's edge, clearly a result of prolonged grazing and browsing at pressures no vegetation could withstand. The evidence was striking in the 1950s and early '60s, when the Zambian side of the Zambezi had predation from humans and the Zimbabwean side didn't.

Chapter 23 described the destruction of vegetation and soil communities on the Zimbabwean side despite years of heavy elephant culling, initiated as a result of my own faulty research. In order to keep the elephants tame for tourists, park managers shot them well away from the river, and the elephants soon learned that the river was the safer place to be. As more elephants crowded into the riparian areas, park managers increased their culling efforts, but, again, they shot them well away from the river and tourists. As a result, the vegetation along the river to this day remains badly overgrazed and overbrowsed and much biodiversity has been lost.

Where we once, as principal predator, ensured constant movement of the herds we preyed upon and had limited ability to light fires, we have now taken the part of protector and friend, stopping movement and reducing other predators while increasing the frequency of fire. Successional communities that evolved over millions of years could compensate no better if wolves and lions donned business suits, moved to the suburbs, and sent their agents out to burn the forage and expose soil.

---

Where we once ensured constant movement of the herds we preyed upon and had limited ability to light fires, we have now become protector and friend, stopping movement and reducing other predators while increasing the frequency of fire.

---

Our concept of conservation in setting national parks aside for large game needs to change, and fortunately circumstances are compelling scientists in many disciplines to rethink old concepts. In Africa in particular, home of most of the large national game parks, the sense of urgency to increase understanding has grown dramatically as the decline of the parks in brittle environments is so obvious.

Much new work will be needed to find ways to induce movement again in wild herding populations and to maintain concentrations in brittle environments. Management schemes now commonly call for cutting off water periodically to force herds to move to other sources. To an extent, this causes movement. However, it does not cause concentration. It does not cause frequent enough movement. It may let nonmobile species die of thirst, thus hindering the buildup of complex communities. I have seen attempts to use this technique but have never seen it work as a realistic means of managing the crucial time factor.

At this stage, you may feel that even a superhuman time manager could not simultaneously restore the most severely bitten plant, defend the "Communists," and think about game tagging along behind her cattle. Fortunately time factors do not stand alone, and perfection is not necessary. Other

---

### Using Livestock Grazing to Enhance Wildlife Diversity

*Where we used to use fire* as the best way to create a mosaic of different habitats that will host more species, we can now use livestock to achieve the same end. Traditionally, we burned patches of ground, which exposed soil and polluted the atmosphere. Now, we can simply return the cattle or sheep or mixed herd to chosen sites, keeping them there long enough to overgraze virtually all plants while applying high animal impact (herd effect) at the same time.

As the next chapter explains, this strategy promotes close plant spacing, much as the regular mowing (or overgrazing) of a lawn does.

influences, such as animal impact, go on simultaneously in the community and generally override the main cause of degradation, which is partial rest associated with overgrazing.

You should do your best, through careful planning of time, to minimize overgrazing, but between the livestock and the wild grazers, some will occur nonetheless. Yet, even as overgrazing tends to push biological succession backward, at the same time high animal impact can overwhelm that tendency and keep it moving forward. Today's alarming rate of desertification in the more brittle environments came about under low animal impact, which in turn provided less than ideal conditions for the establishment of young plants. If seeds and sprouts can establish, losses to overgrazing matter far less, as eons of evolutionary history show.

---

Even as overgrazing tends to push biological succession backward, at the same time high animal impact can overwhelm that tendency and keep it moving forward.

---

## Conclusion

Because it is so new to us, it has been necessary to devote considerable space to the concept of time as it relates to our management of plants, soils, livestock, and wildlife. Experience in many contexts in many countries is now showing clearly that planning grazing to manage recovery time will be important in maintaining the health of grasslands and water catchments and in halting the advance of deserts throughout the world. I am sure we have by no means yet seen the full implications.

Let's now proceed to those management guidelines that focus on animal impact, the other aspect of livestock grazing that has to be understood and planned.

# 34

## Stock Density and Herd Effect
### *Using Animals to Enliven Soils and Enhance Landscapes*

STOCK DENSITY AND HERD EFFECT are the two management guidelines that apply when the tool of animal impact (chap. 22) is used to alter soil conditions or vegetation. Because for thousands of years people have observed the trampling effects of their animals over many days or months and the damage this causes to the land, recognizing the beneficial effects of even higher amounts of trampling, dunging, and urinating for short times initially requires a significant change in thinking.

*Stock density* refers to the density of animals on a defined, usually fenced, area of land at a given time. Thus, if one hundred animals are in a hundred-hectare/hundred-acre paddock today, the stock density is 1:1 (one animal to one hectare or one acre). If tomorrow these hundred animals are moved to a two hundred–hectare or –acre paddock, the stock density would then be 1:2 (one animal to two hectares or two acres). Stock density does not indicate whether the animals are spread and feeding placidly or bunched and behaving differently.

*Herd effect*, on the other hand, concerns animal behavior and cannot be quantified. It is merely the effect on soils and plants that a large number of animals have if they periodically bunch so closely that *their behavior changes.* When the animals are spread out and calm, their hooves leave few signs of disturbance on the soil surface, apart from when it is wet. When they are bunched and milling around on the land or are excited, they tread down old coarse plant material, raising dust at times, and chipping the soil surface. The larger the herd, the greater the effect.

## Stock Density

As chapter 23 made clear, stock density has a strong relationship to the management of grazing, browsing, and trampling time, but it deserves some discussion in its own right. Because of the prejudice that hooves in any context damage soils and plants, low stock density has characterized management of livestock on grasslands. Unfortunately, grazing at low stock density almost universally causes problems—chief among them partial rest of soils and excessive rest for most plants. Our traditional bias, however, made us attribute the land degradation and other side effects of grazing at low stock density to other causes.

Low stock density and the partial rest allied with it, not overgrazing or overstocking, should bear the blame for many serious land and production problems, including severe trailing, shifts toward brush and weeds, grasshopper and other pest outbreaks, excessive use of fire to even out grazing and suppress brush, and the development of mosaics of grazed-out patches with decreased water cycle effectiveness and thus an increase in both the frequency and severity of floods and droughts. Because it can lead to so many problems it is useful to be able to recognize the signs of low stock density.

A high degree of patchiness and trailing is a hallmark of low density, the grazed patches commonly ending sharply where ungrazed (and thus overrested) plants, often of the same species, begin. Some people refer to this as patch grazing or spot grazing, but the term *low-density grazing* serves better because it describes the process and suggests a solution.

### *Stock Density and Animal Performance*

Poor animal performance plagued my early work with ranchers. Though careful monitoring clearly documented the improvement of plant communities and soils when we started managing grazing time, no class of livestock performed as well as the same animals continuously grazing on deteriorating control areas. On stable irrigated pastures, André Voisin's work guided us to success, but animals at higher density on fenced rangelands did not thrive.

For eight years, I carried the albatross of almost continuous poor animal performance while my many critics rubbed their hands and snickered. As it turned out, a major part of the problem was my failure to question the conventional wisdom that patchy grazing was due to animals selecting certain species of more palatable grasses and rejecting others less palatable. Academic papers and textbooks had belabored the subject ad nauseam and allayed any doubt.

One day, while discussing the problem of poor performance with a ranch manager in Swaziland as we walked over his land, a pair of grass plants of the same species caught my eye, and the pieces of the puzzle began to fall in place. Range scientists considered this species (of the genus *Cymbopogon*) undesirable, believing that its strong turpentine aroma made it unpalatable, and indeed one plant stood untouched in a rank clump. But another, right next to it, had been eaten right down. I had noticed such things before but had never paid them much attention, so I just sat down and thought for a long time.

I asked myself, "Why would two plants of the same 'unpalatable' kind, enjoying the same weather, soil, moisture, and exposure to cattle, come to such different ends?" After a while, I startled the already bewildered manager by blurting out the observation that "cattle don't select species, they don't even know the Latin name of this plant." What they are selecting on any given day is the freshest and leafiest forage on *any species*.

---

Cattle don't select species. What they are selecting on any given day is the freshest and leafiest forage on any species.

---

It had taken me years to register that, although cattle, sheep, and goats carefully and intelligently select their diet, they do it by what they actually sense in front of them, not by choosing from a Linnean menu of desirable and undesirable species. They will eat fresh tender leaves of undesirable brand X and leave old, stale leaves of desirable brand Y.

I immediately determined to approach my old dilemma from a new tack. I had already noted that, in smaller paddocks, where animals grazed at higher density, the plant community tended to have more leaf and less fiber. If we increased stock density all over, by subdividing large paddocks and combining several herds into one or two larger ones, we could shorten grazing periods while increasing density and should be able to improve animal performance generally. To convince the ranchers I was then working with to try it, I used the following analogy.

Assume I asked you to visit for a year. As a good host, I would ask you for a list of your favorite foods (your most desired species), and on your arrival you find a smorgasbord of every one of your selections, from which you choose a substantial meal but of course leave many items untouched or only nibbled. While you rest, I replace exactly what you ate, leaving everything else as before. At the next meal you choose again, and I replace only what you actually ate.

After a few months, you will only dare eat things I replaced in the last day or so, despite the fact that everything on the table started out as a "desirable species." Some of the most delectable dishes now reek from mold and decay. If I suddenly stopped replacing your daily selection, your performance would take a nasty drop as you spent eating time picking through that old garbage. The problem was, of course, low-density feeding! Had I invited enough people to sit around the table, and replaced everything in the same way, every meal would have been as good as the last.

In the days before I understood the full implications of stock density and time, I cost my long-suffering clients on smaller properties in high-rainfall areas thousands of dollars in poor animal performance by recommending they start off with as few as eight to ten paddocks. Now, in order to attain good animal performance in that situation I would, as many farmers are doing, either start with many more paddocks or strip-graze within the existing paddocks using portable fencing.

---

All of these ranchers had started out with a fairly high percentage of overrested or stale plants because of low-density grazing in the past, and this was a factor in their poor animal performance in every case.

---

As soon as the ranchers increased stock density, animal performance did in fact improve, but the degree of improvement varied on the different properties and with different managers. By this time I had well over a hundred clients in five countries, and ample evidence led to a diagnosis that helped clarify the spotty results. All of these ranchers had started out with a fairly high percentage of overrested or stale plants because of low-density grazing in the past, and this was a factor in their poor animal performance in every case. However, rainfall and soil type and the manager's ability to plan and monitor the grazings were largely responsible for the variation in animal performance once stock density was increased.

### Rainfall and Soil Type Make a Difference

Animal performance improved the most in low-rainfall areas with highly mineralized and more alkaline soils. These areas supported forage that had less fiber, shorter height, and better curing properties. Higher mineralization in the plants from these soils, I surmised, kept rumenal microbe populations high in the animals, thus maximizing digestive efficiency and leading to better performance, *even on old forage*. Higher-rainfall areas, on the other hand, characterized by leached and more acidic soils, produced generally taller, tougher forage of much higher fiber content. Older perennial grass plants had little or no feed value compared to plants of similar age in low-rainfall, highly mineralized soils.

It was apparent that we could increase stock density immediately in the low-rainfall/highly mineralized soil situation and experience little or no initial drop in performance. From there on, the situation would only get better. In the high-rainfall/leached soil situation, however, we had to make some decisions about how to deal with the old forage and the inevitable drop in performance: we could burn the forage or perhaps mow it, at considerable expense; we could provide a high level of supplementary feed to the animals, or we could bite the bullet and accept the performance loss during the first few times through the paddocks, recognizing it as a legacy of the past. The latter could be mitigated somewhat if we used cows when dry and pregnant to graze and trample the old stale forage because at that time they are at their hardiest. If they dropped in weight slightly this not only made calving easier but then put them on a rising plane of nutrition at breeding time, which in turn led to higher conception rates.

### Grazing Planning Makes a Difference

Once we'd sorted out the variations occurring because of rainfall and soil types, we still had to contend with the differences that appeared to be attributable to management. With the old stale forage no longer a problem, the poorest results were now occurring among managers using relatively few paddocks per herd and who failed to monitor plant growth rates and adjust grazing and recovery periods. The problems generally arose when fast growth slowed down, but grazing and recovery periods were not changed to reflect this.

Rapid moves, and thus shorter grazing periods, generally led to a short-term benefit to the animals as they moved onto new ground. However, the quicker moves meant that recovery times were shortened as well, and in slow growth periods that meant the animals would return before plants had had time to recover from a previous grazing. Those plants grazed severely in the previous grazing period would now be overgrazed. Those plants not grazed at all or only lightly, would tend to be left and to grow somewhat staler.

Using the smorgasbord analogy once more, it is as though we have more people to feed at the table but they don't have time to eat all they actually can because their mealtime has been shortened. The butler moves them on halfway through the main course. That food eaten gets replaced according to standard practice, but the remaining food grows a little staler. When again the next day they have only ten minutes to eat, the stale gets staler. Once again, matters progress until they have to eat stale food, and performance drops. Technically we have enough people (density) at the table to keep the food fresh, but without enough time to clean their plates, much of the food grows stale anyway.

In this case, the animals are obliged to eat the stale food almost immediately because the plants they overgrazed the last time have been unable to produce enough fresh forage to feed them all. In a low-rainfall area with highly mineralized soils (those of volcanic origin), performance in this case may drop too little to draw immediate attention. In high-rainfall areas, the old grass will have so little nutritional value that stock performance drops almost at once.

### Brittleness Makes a Difference

My early attempts to solve the challenge of managing livestock in a way that could enhance rather than degrade an environment were made before I understood the concept of brittle and nonbrittle environments. I knew that high animal impact was key to solving the desertification problem, but did not fully understand that *the degree of brittleness influenced whether animal impact should be created through stock density or herd effect.*

Generally, in nonbrittle environments, where humidity is relatively well distributed throughout the year, farms and livestock herds are smaller, and fencing controls animals, animal impact is most easily provided through stock density. Desertification does not occur in these environments, although they do deteriorate under bad grazing practices. Grazing livestock at high stock densities with frequent moves—*as long as recovery periods govern such moves*—will help maintain grass communities that would otherwise tend to shift to tap-rooted plants, including trees, and will lead to better animal performance.

In the more brittle environments, where there is a prolonged period of very slow or no growth, whether or not rainfall is high or low, properties (ranches and pastoral grazing areas) are larger. Even where fencing is used to control animals, fenced subdivisions (or paddocks) are generally also very large for both practical and financial reasons. In such situations animal impact is best provided through herd effect to maintain perennial grass communities that would otherwise tend to shift to annuals, weeds, and bare ground.

These environments, of course, are where herd effect was applied routinely by the wild herds of grazing animals that helped create them. Long ago the world's most productive brittle grasslands, such as those in North America, Australia, Mongolia, or Africa, had extremely low stock density, as the "paddock" was a whole continent. However, as herds were so vast and pack-hunting predators so abundant, herd effect was great and occurring somewhere most of the time. On large ranches and pastoral lands today herd effect will be high wherever animals are made to concentrate, but the stock density overall will be very low. If, for instance, you herded a thousand animals daily on five thousand hectares (twelve thousand acres) of land, despite the large herd the stock density would be very low at one animal to five hectares (twelve acres). Thus increasing stock density can increase animal impact on smaller areas of land, but it has negligible effect as land units become larger, which is when herd effect becomes increasingly important.

## The Research on Animal Impact

*Chapter 22 explained that the tool* of animal impact is applied through the guidelines of stock density and herd effect. A situation may call for one or the other or both for maximum effect. A number of researchers have published papers concluding that animal impact does not produce the beneficial changes that I describe in this book. In fact, they designed their projects without indicating an understanding of the concept of herd effect and made no effort to apply it. As a result, they effectively proved that *low stock density* does not do what I claim *high herd effect* does.

Stock density was applied to brittle environment communities that really needed herd effect. The herd in one case consisted of two steers that could not have done much even if the researchers had excited or bunched them.[1] Two steers enclosed in a one-hectare/one-acre paddock will not have anywhere near the same effect as two hundred steers bunched for a time within a hundred-hectare/hundred-acre paddock, though the stock density is the same.

Researchers studying animal impact have consistently ignored herd effect, and thus a great deal of money and effort has been wasted over the years studying low animal impact over prolonged time instead of studying high animal impact over brief periods of time. While it cannot easily be isolated for research, herd effect can readily be observed and monitored in the field. For example, color plate 18 shows two pieces of land on which grazing is taking place. In the background there is a high level of partial rest and no herd effect, and in the foreground cattle crowded into an enclosure for handling have provided herd effect periodically. Common sense tells us that the soil in the foreground is healthier, and that more carbon and water are being stored here than in the soil in the background.

## Herd Effect

Herd effect is the main management guideline we use to achieve the high animal impact needed for restoring desertifying grasslands to health. As mentioned, herd effect is produced as the result of a change in animal behavior, and herd size influences the extent of it. Although this bunching, milling, and sometimes excited behavior strongly affects the entire biological community, and wild herds in truly wild conditions exhibit such behavior, neither herding wildlife nor livestock produce much herd effect without outside stimulus. Inducing adequate herd effect thus represents a challenge that must be addressed in the management of brittle environment grasslands and forests.

It is clear from the study of herding animals (and some fish and birds), that large numbers and bunched, milling behavior was the most effective protection developed against predators until the advent of modern humanity. Many excellent documentaries show how hard predators must work to beat that defense and to isolate an animal in order to bring it down. While the carnivores do their job, the animals in the herd, concentrated for protection, do not respect the grass and brush beneath their hooves as they do when grazing unmolested. Free from fear of fang, claw, or spear, even instinctively timid wild animals soon lose the habit of vigilance and scatter widely when grazing.

Starting in my game department days, I gradually built on the observation that, wherever predators caused bunching and the formation of large herds, the concentrated dung and urine of the herd also induced movement, and this in turn regulated the overgrazing of plants by governing their time of exposure and reexposure to animals. Wherever the pack-hunting predators and their large prey were reduced or absent altogether over prolonged periods, the grassland began to deteriorate, plant spacing widened, and algae/lichen crusts flourished on ground that had become bare between plants.

For several years, I lived close to large buffalo herds, as well as elephants and many other game animals, followed by a varied host of predators in high numbers. Buffalo gathered into herds running to thousands, and even elephants on good grassland gathered in loose herds that I estimated at six hundred or more. At times, I would pick up tracks several days old and follow them to find a herd. In places, even an inexperienced tracker could follow the spoor at a trot. Elsewhere the trail would dissipate almost entirely though the country did not change. That happened whenever the animals spread to feed. At such times, their hooves avoided coarse plants and did not break soil surfaces as much, nor did they trample old plant material, as they did when bunched.

Since then I have observed the same differences among other herding animals and even humans. When tracking men, as I did often during Zimbabwe's long civil war, one learns much about the mental and physical state of the quarry by noting the way he places his feet. Individuals in an excited group, walking and talking, leave a very different trail than an individual walking quietly alone. A starving, thirsty, wounded, or panicking person will not place his or her feet the way a calmer person would.

In the absence of pack-hunting predators, most herding animals break into smaller herds and remain spread out most of the time, resulting in land degradation through partial rest as we see in many of Africa's national parks today. For several million years, predator-induced herd effect was a feature of evolving grasslands. But in the last instant of the last million years, human activity changed that.

The vast scale of desertification and climate change in the world today attests to the enormous impact of these human-induced changes. We would instinctively understand devastation caused by withholding rain showers that had occurred for eons, but the damage to water cycles, in particular, caused by eliminating herd effect, and replacing it with partial rest and fire, has in reality done that very thing.

---

> We would instinctively understand devastation caused by withholding rain showers, but the damage to water cycles caused by eliminating herd effect and replacing it with partial rest and fire has in reality done that very thing.

---

### Practical Demonstrations of Herd Effect

To help people overcome their fear of animal impact and trampling, and therefore of herd effect, I frequently encourage them to conduct simple demonstrations for their own benefit. These usually take the form of placing hundreds of animals very briefly—a few hours at most—into small enclosures to see what happens to the land over time.

Color plate 19 shows before and after photos of one of these demonstrations. In this case, the ranch manager and his staff, as well as local environmentalists, wanted to see for themselves if what I was saying about herd effect applied in their case. The day before four hundred cattle were to be enclosed on about two hectares (five acres) of land, it rained heavily, and they phoned me to see if they should wait until the soil dried out. I saw no need as wet soils have been trampled billions of times over the eons. They went ahead, and, as you can see in the top photo of color plate 19, nearly every plant was grazed or trampled down. But one year later, as shown in the lower photo, the growth inside the enclosure was lusher than any outside it. Though the demonstration was convincing, its greater value was the confidence it built in those who participated in the exercise.

### Inducing Herd Effect Routinely

When our ancestors first domesticated livestock and protected them from predators by herding them, we removed much of the tendency to produce herd effect. This holds true for the American rancher as well as the Andalusian shepherd or African pastoralist today. Fencing and grazing systems designed to spread livestock evenly over the land in a totally unnatural manner have exacerbated the problem, severely disrupting the evolutionary interdependence of animals, plants, and soils.

Unfortunately, even though we now see the need to produce herd effect with livestock over large areas, generating it routinely remains a problem. Many people, however, do not see that the same problem arises in the management of wild animals, even in our national parks. In many parks predators are few. Frequently the land base is so limited it does not allow herd sizes large enough to provide adequate trampling, or sufficient to sustain enough predators.

Now we must learn how to simulate the predator-induced behavior, and we have learned much over the past forty years about doing so with livestock. The most successful methods so far involve simply herding livestock to a grazing plan without the use of fencing, which is fortunate when having to deal with vast areas, traditional pastoralists, and places where fencing is undesirable, such as national parks. Figure 34-1 shows cattle and goats being herded on an unfenced ranch surrounded by large wildlife populations and national parks, which make fencing not only undesirable but impractical. Because predators, lions in particular, are abundant here and the dense cover conceals them, herders keep the animals bunched much of the time, but also keep them moving onto fresh forage so they can still select their diet.

On fenced properties, other methods for inducing herd effect include the following:

- Using a movable electric wire to strip-graze small areas of land within larger paddocks, as shown in Figure 34-2
- Bunching the animals for even part of the day using horses and dogs
- Attracting the herd to anything that excites them, such as supplemental feed cubes, a bale of hay, or a few handfuls of granular salt, for animals purposely deprived of salt

Occasionally situations arise that allow use of herd effect without a concern for keeping the time short. Say, for example, you needed a firebreak through very dense brush. In this case attractants can be used that hold animals for longer periods, such as a dilute molasses or saline spray over the vegetation, or even supplement blocks.

### The Type of Livestock Matters

While herd effect provided by any type of animal is better than none, the type of animal may make a difference. Almost any livestock—sheep, goats, camels, cattle, or horses—can produce adequate impact on sandy soil. However, on soils that produce a hard surface cap, sheep and goats have limited effect and progress can be slow. Cattle, horses, or donkeys are capable of breaking hard capped soil surfaces better than small stock or camels. Horses generally have a higher soil surface breaking ability than cattle (that is why they are easier to track), but unfortunately we seldom have very large horse herds, so cattle must suffice.

**Figure 34-1.** Cattle and goats herded under Holistic Planned Grazing to produce continuous herd effect where, because of large paddocks, stock density is low. Zimbabwe (courtesy C. J. Hadley/Range).

**Figure 34-2.** Cattle strip-grazed at ultra-high density (three thousand per hectare/twelve hundred per acre) using a one-strand portable fence that moves every few hours, according to a grazing plan. Zimbabwe.

### Using Attractants to Create Herd Effect

*You can easily train animals* to come to a piece of ground on which you have scattered an attractant if you blow a whistle each time you do. The animals soon associate the sound of the whistle with a reward. Once they learn to respond, a herd can be drawn in an excited bunch to any spot on the land where herd effect is needed to build toward the future landscape you intend.

Figure 34-3 illustrates animal impact being applied on an area of land over a short period on a large Texas ranch. First I photographed the typical soil surface between plants (A). I took the next picture (B) as the cattle arrived, drawn by a whistle and a half bag of feed cubes. Then I photographed roughly the same spot as the first photo a few minutes later (C). The contrast is great indeed. Any gardener wanting to grow seeds on that soil would appreciate the difference immediately.

This technique has one major drawback—very little ground is impacted over time. Two thousand cattle will beneficially affect an area only about fifty meters across each time they are attracted, though somewhat lesser impact grades out from there. Very few managers are able, nor is it practical, to induce herd effect with attractants more frequently than once a day, but doing so, or simply herding the animals together in part of a paddock for a few hours, has a significant effect.

**Figure 34-3.** Induced herd effect.

(A) Close-up view of the soil surface between grass plants before herd effect is applied. The soil is hard-capped, covered with algae, and barely able to breathe, making it difficult for new plants to establish.

(B) To induce herd effect, a herd of two thousand cattle has been attracted to the spot shown in (A) with supplementary feed cubes. The bunched and excited animals place their hooves carelessly, breaking the soil cap and raising dust.

(C) Close-up view of the same spot shown in (B) three minutes later. The soil looks as if a gardener had hoed it, and it can now breathe. Water can also penetrate faster, and new plants can germinate and establish more quickly.

## *Herd Size Matters*

Deeply rooted emotion and myth surround the question of herd sizes, especially among cattle producers. Prominent cattlemen in the 1970s heavily condemned me for even suggesting that herds of 200 cows could be run and still breed well. Beyond 140 cows lay the edge of the world and a long fall to disaster. Nevertheless, we have gradually increased breeding herd sizes without encountering serious problems. To date, we have not yet had any evidence that conception rate or weaning weight in breeding herds depends on anything outside quality of handling, health, and nutrition. We have no evidence yet of any serious drop in performance in any large herd, given good handling facilities, calm handling, and well-planned grazing.

Having worked with vast buffalo herds, as well as cattle herds of up to four thousand or more, I have no doubt in my own mind that, to manage brittle environment grasslands, the larger the herd the better. Herds of two thousand to five thousand head produce better results than herds of two hundred to five hundred *when Holistic Planned Grazing is practiced.* I qualify this statement because people tend to forget that pastoralists for centuries herded their animals in herds both large and small, but produced the great manmade deserts of antiquity. There are two factors, I believe, that might explain why this deterioration occurred. First, families and small groups ran too many herds, which is still common among some pastoralists today, making it impossible to achieve adequate recovery times between grazings—after one herd leaves an area, another enters it a few days or weeks later, attracted by the fresh regrowth. And second, herders protect their animals from predators, which allows the herd to spread out and graze calmly without fear. Most of the herd effect I've observed in these situations is where animals are moved to and from an overnight enclosure to the grazing areas day after day, which leads to trailing and pulverized soil.

## Conclusion

The guidelines for stock density and herd effect concern a tool of great power—animal impact. An understanding of herd effect is essential for any land manager operating in a brittle environment, but also for those who will never manage livestock, but as informed citizens can influence government or nongovernmental-organization policies relating to desertification and climate change. Bear in mind that many such institutions currently promote reductions in livestock numbers, the creation of more water points to keep animals spread out, or pen-feeding to keep animals off the land—practices that only exacerbate desertification. In the future, as knowledge increases and attitudes change, I believe that fewer, larger herds will become the principal tool in watershed management on the public lands of America and in efforts to reverse desertification all over the world.

# 35

## Cropping
### *Practices That More Closely Mimic Nature*

*THE GROWING OF CROPS OVER THE AGES* has increasingly involved the creation of ever more artificial conditions and the loss of natural balances and stability within the original biological community. Instead of a variety of plants and perennial ground cover, a small number of crops make only part-time use of the space available. The soil is exposed to sun, wind, and rain to a greater extent than before, particularly where fields are left bare for part of the year, leading to loss of soil life and higher rates of erosion and a less effective water cycle than under natural conditions. Mineral cycling is also disrupted, and extra inputs in the form of manures or fertilizers are required to keep soils productive.

The history of agriculture is in effect the story of how various societies have attempted to deal with the inevitable problems linked with growing crops to feed ourselves: soil erosion, loss of fertility and biodiversity, and the instability associated when substituting simplicity for nature's complexity. Many civilizations—starting with those in Mesopotamia, the Indus valley, China, and later the Americas—have collapsed, largely because these problems overwhelmed them. Where past civilizations failed regionally we now face the same situation globally.

There can be little doubt that a change in direction is needed for agriculture if we are to sustain our present civilization and its enormous population.

With the rise of industrial agriculture at the end of World War II we have tended to treat our soils merely as a medium in which to hold crops upright while we pour chemicals over them. In reality, soil is alive and respiring, as most living organisms do, and it has to be nurtured. To do so we must strive to create a truly regenerative agriculture that is based mainly on the biological sciences rather than chemistry and the smart marketing of ever-advancing technology. A truly regenerative agriculture more closely mimics nature; it enhances, rather than diminishes, water and mineral cycles, energy flow, and community dynamics. In my own experience, and that of others seeking to find better ways, this focus leads to some fundamental guidelines.

## Keep Soil Covered throughout the Year

All living organisms have a skin or protective covering of some sort. If a significant proportion of it is removed, the organism dies whether the organism be living soil or human burn victim. A soil's skin is made up primarily of plant material, some of it living, but most of it dead or decaying at the soil surface. This covering, made up of both growing plants and dead, decaying plant litter, insulates the life in the soil from temperature and moisture extremes and protects the soil crumb structure so essential to water penetration and aeration from destruction by raindrops. It also provides a hospitable surface environment for billions of organisms that break nutrients down so they can return underground for recycling. An exposed soil is a dying soil at the mercy of wind, rain, and sun. As mentioned in chapter 1, we are losing more than seventy-five billion tons of eroding soil *each year* from our agricultural lands.

On conventionally managed croplands, the majority of the soil surface between the plants is exposed. After harvest, even more soil is exposed. Many croplands remain bare over winter. Others are deliberately kept bare and harrowed over a fallow year in order to grow a crop on two years' rainfall. Tragically, even small-scale farmers in many developing countries burn crop residues, leaving soil bare and baked. Yet there are many ways to keep cropland soils covered *throughout the year*, including the use of conservation tillage, the planting of cover crops over winter, intercropping low-growing shade-tolerant crops among taller ones, using animals rather than fire, plow, or harrow to remove crop residues (while leaving enough material to provide soil cover).

---

**Pasture Cropping**

*One of the most innovative strategies* for keeping soil covered year-round is pasture cropping—the simultaneous use of land for growing a perennial pasture for grazing and a cereal crop for harvest. Developed by Australian farmer Colin Seis in the mid-1990s, it is spreading rapidly among regenerative farmers throughout the world because, not only does it keep soils covered, it also leads to increased soil fertility and water-holding capacity, very respectable grain yields, and higher profits.

Colin drills the crop seed directly into his perennial native pasture. His sheep graze off the pasture growth through the winter and early spring and then move to another area, according to his grazing plan. The cereal crop then shoots up and he harvests the grain, as shown in color plate 20. Following harvest, the sheep return to graze the growing pasture, fertilizing the field prior to its being planted again.

---

## Do Not Turn Soil Over

To mimic nature, soil should be worked from the surface just as for millions of years the hooves of herding animals have done, or the claws of turkeys, pheasants, guinea fowl, and other birds, and many small creatures have done. Yet, dating from the invention of the first plow, we have generally done just the opposite. In the 1940s Edward Faulkner, in his book, *Plowman's Folly*, alerted farmers to the damage done by deep plowing in which topsoil is turned right under and subsoil brought to the surface en masse. In turning over the protective surface mulch, he explained, plowing not only exposes soil to the elements, it also compresses the turned under "trash" into a narrow layer deeper down that is unable to decay and thus hampers the growth of plant roots.

When you think of soil in terms of the complex biological community that it is, Faulkner's argument makes sense. If a mass of plant material grows and only a small proportion of it is harvested, corn for instance, what happens to all the rest of those plants, the crop residue? Generally it is plowed under "so the organic matter can be returned to the soil." But this is *raw* organic matter, not the mature humus that develops gradually with the help of plant roots and billions of microorganisms. Raw organic matter, together with the mass aeration caused by the plowing, leads to problems, including an increase in bacteria that consume the much-needed humus.

---

The raw organic matter, together with the mass aeration caused by the plowing, leads to an increase in bacteria that consume the much-needed humus.

---

Anytime you turn soil over, organisms that have established in a certain microenvironment, either deeper underground or close to the surface, are suddenly placed in a different one. The result is not unlike what would occur if you dumped all the residents of a European city into the middle of the Sahara and the Sahara's inhabitants into the city overnight. While we can imagine the chaos, suffering, and death caused by such an action, we fail to see its parallel in the trillions of soil organisms that are displaced anytime we turn soil over.

After a field is plowed, soil communities will start to rebuild, but they are immediately set back once the field is plowed again. In the 1960s, French pasture specialist André Voisin recorded the "years of depression" that followed the plowing of a field. Production steadily dropped to a low in about the seventh year. If the field was not plowed again, it would steadily improve, eventually surpassing its initial productivity.[1]

Fortunately, there are alternatives to plowing. Crop residues can be dealt with by animals, livestock in most cases, that reduce them to dung, urine, and mulch, leaving dead roots in the soil to further enhance fertility. Conservation tillage can be used in preparing fields. In fact, just over a third of all U.S. cropland is now under some form of conservation tillage. In most cases, however, it is linked with heavy herbicide usage, which tends to counteract the benefits. The solution to the problem of unwanted plants, which often increase when tillage is minimized, lies again in approaches that try to mimic nature. In this case, the answer may lie in combatting diversity (which weeds represent) with even more diversity, which progressive farmers already do.

## Endeavor to Maintain Diversity and Complexity in the Community

What we call weeds should not be blamed for stealing water and nutrients from our crops, but valued, as long as they do not dominate the community, for the diversity they contribute, including the insects and microorganisms they attract. This added complexity offers protection against the few insects or microorganisms that actually damage crops or cause disease.

All too often we reduce this complexity in our croplands for no good reason. Figure 35-1 portrays just such an example. The grass growing beneath the apple trees in this organically farmed orchard has been mowed short to keep it looking tidy, but the mowing has removed a host of potential habitat niches for the insects and microorganisms that help to control apple pests. Despite this, and at some expense in nonrenewable resources, far too many people continue to engage in similar practices throughout the world, when there are alternatives. Sheep, under Holistic Planned Grazing, could be used

**Figure 35-1.** Organic apple orchard that is regularly mowed to keep it looking neat, but the mowing has removed ideal habitat for a host of insects and microorganisms that help control apple pests. Pennsylvania.

instead to keep the grass down, as some managers are doing in orchards and vineyards to maintain a similar appearance while maintaining greater soil cover and biodiversity.

### *Avoid Monocultures*

You should avoid planting monocultures, particularly over large areas, as much as possible, if for no other reason than to avoid large-scale damage from insects and plant diseases. Unfortunately, the majority of farms in the United States and many other countries today are monoculture deserts. The recently planted bean field in figure 35-2 is a typical example. Such large fields of one crop are attractive to insects that only feed or lay eggs on that one crop, and they will congregate in great numbers. Unless some form of pesticide is used, the farmer will reap little from this field.

Farmers in some of the most heavily insect-infested areas of the world—tropical rain forest environments—have for thousands of years grown crops in complex polycultures without using pesticides. Although cultivated

**Figure 35-2.** View of a recently planted monoculture of beans. If you were an insect living in this field what would you eat? Iowa.

on a smaller scale, these polyculture fields yield far more per hectare than high-input monocultures. However, many farmers in developing nations are being persuaded to abandon polyculture fields in favor of machinery and chemical-dependent monocultures of cash crops, with unfortunate results for both the land and the people.

Monoculture fields cannot be converted to polyculture fields overnight. While that might be desirable, it would be highly impractical and financially risky. It makes more sense to move progressively away from monocultures, or rotations of monocultures, to intercropping, alley cropping, and ultimately to more complex polycultures. The oldest form of crop production, agroforestry, which grows crops among certain trees, has recently been reborn. Intercropping involves the planting of low-growing, or early-maturing crops (which may or may not be harvested), among taller, later-maturing crops. In alley cropping, mixed crops are grown between rows of trees. In some cases, tree branches are lopped off to provide a green manure, or mulch, and added sunlight for the crops below. As the crops mature and no longer require full sun, the trees regrow the limbs and leaves they sacrificed.

By moving progressively toward a more complex community, you can learn as you go, minimizing the risk of financial setbacks, and steadily improving practices that increase the complexity of your croplands. *There will be much to learn.* Insect damage may continue until the right combination of crops is worked out, or until small or large livestock, birds, bats, and other wildlife are brought in, either to feed on insects or to render the environment less conducive to their reproduction. The role of livestock in cropping programs is discussed later.

### Create "Edges" to Increase Diversity

One way to increase the diversity of species in any environment, whether you are managing a garden, a ranch or farm, a stream or ocean inlet, is to increase the amount of *edge*—where two or more habitats join.

Aldo Leopold, the father of modern game management, was the first to note the phenomenon of *edge effect*, particularly as it related to wildlife populations. Most wildlife occurs, he said, where the types of food and cover they need come together, that is, where their edges meet. "Every grouse hunter knows this when he selects the edge of a woods, with its grape tangles, haw-bushes, and little grassy bays, as the likely place to look for birds. The quail hunter follows the common edge between the brushy draw and the weedy corn, the snipe hunter the edge between the marsh and the pasture."[2] Where forest meets meadow, animals find cover in the wood, visibility across the open land, and feed from two types of environment.

It is of course possible to have too much edge, as can be seen in forests where timber harvesting has produced a patchwork of trees and clearcut areas, and thus eliminated plant and animal species that required a larger expanse of forest in order to thrive. In the artificial environment of the crop field, however, too much edge is rarely a problem; creating enough edge more commonly is.

In figure 35-3, one very large field provides minimum edge. Two different crops provide two different habitats, and the one water point provides a third for the many species of birds, insects, and small animals. The greatest diversity of species of all types will be along the edge where the three habitats meet, followed by the edges where the two crops meet. Assume that a certain species of insect-eating bird required food, cover from its predators, and proximity to water. If one of the habitat types provided cover and the other food, but no water, there would be no birds of this species. Even with intercropping in this field there is little diversity.

In figure 35-4, hedgerows and trees have been used to divide the large field into six smaller ones, and water has been dispersed. What a difference! The trees and hedgerows add another dimension of complexity, and the proportion of edge is many times greater. Many creatures can obtain food, cover, and water. Where an insect-eating bird was restricted to the cover at the edge of the field in figure 35-3, it can now range over the entire crop area.

Once upon a time most farmers planted hedgerows and trees around their fields, which were much smaller than they generally are today, creating edge and habitat for numerous species, including insects (ninety percent of which are beneficial to crops), insect predators, such as birds and bats, and larger animals as well.

Modern-day farmers, of course, have the option of returning to smaller fields bordered with natural or planted vegetation. They benefit even more if a

**Figure 35-3.** Crop field with limited edge—two different crops provide two different habitats, and the one water point provides a third.

planted border produces a harvestable crop and additional income, or fodder for livestock, or serves other functions, such as willows or eucalypts that help drain boggy areas.

In designing your fields also give thought to the needs of nocturnal creatures, such as bats. The tonnage of insects eaten by bats each night is

Water

**Figure 35-4.** In this crop field, the proportion of edge is many times greater.

staggering. A single little brown bat can catch six hundred or more mosquitos in an hour, a colony of thirty could easily catch more than thirty thousand insects in an evening's feeding. Unfortunately, many bat populations have been seriously decimated or destroyed altogether by poisons used in our struggle to sustain monocultures in defiance of nature. You can encourage their return by providing habitat for them. Most colonies of bats choose roosts within a quarter mile of water in areas of diverse habitat, especially where there is a mix of croplands and natural vegetation.

Creating edges of course also affects the species you cannot see, and they exist in uncountable numbers. All animal species are in one way or another struggling for the best assortment of places to feed, drink, hide, rest, sleep, play, and breed. By creating edges, you can assist them all in their quest.

### *Preserve Genetic Diversity*

Maintaining diversity refers not only to species diversity but also the genetic diversity within each species. Genetic diversity is just as important in domesticated plants as it is in wild ones. We only have to remember what a lack of it did to the potato crops in nineteenth-century Ireland.

It used to be that seed was the farmer's most valued possession, which reflected generations of selection for adaptation to a particular environment. However, in many countries today, most crops are grown from hybrid seed that must be purchased each year because seeds from the resulting crop are infertile or don't breed true. When farmers no longer raise their own seed, they are at the mercy of the giant corporations that patent and market hybrid seed and control both their price and the supply.

The genetic material we have lost as a result of the move to manufactured hybrids has led to the development of seed banks to preserve what remains. But this is little different than placing endangered animals in zoos in an effort to preserve their species. Our chances of sustaining them are slim if they aren't at some point returned to the environment that originally helped breed and shape them.

### Incorporate Livestock

Up until recent times livestock played an essential role in maintaining cropland fertility, and no farm was without them. Only in the earliest forms of agriculture, known as slash and burn, or swidden, were crops grown in the absence of domestic stock. Forests were cleared and burned and crops planted in soil that was rich in nutrients and organic matter. But within five or so years, much of the organic matter was lost and nutrients leached by rainfall down to layers that shallow-rooted crops could not reach, and the people would be forced to move on. When the forest had regrown and the soil was rejuvenated, twenty years or more later, the land could be cultivated once more.

When rising populations made swidden agriculture impossible, people were forced to settle and continually crop the same land. Fertility could then only be maintained through the use of animal manures, but the number of animals that could be fed throughout the year was often quite small because of the lack of fodder crops. Manure was removed from pastures to fertilize crop fields or for use as fuel, and this, combined with plant overgrazing, which even then was believed to be associated with animal numbers rather than time, reduced hay and grass yields on the pastures. In colder climates, many animals had to be slaughtered in the autumn because of a shortage of winter feed. When food was short, more land would be put under crops as a short-term measure to try to increase food production, thus setting up a vicious cycle. When more land was put into crops animal numbers had to be reduced. Less manure was produced and crop yields decreased.

Today, with Holistic Planned Grazing we can grow much more forage on less land and can better integrate livestock with complicated mixes of crops, including orchards and vineyards, in a way not previously possible. And progressive farmers are already doing this and learning that soil life can be regenerated amazingly quickly. Forages grown on terraces and in grassed waterways, or interseeded as cover crops, also provide stock feed while stabilizing eroding soil. Livestock can also utilize many so-called wastes, such as damaged grains, a drought-failed corn crop, food processing by-products, grain screenings, and especially weeds. The combination of crops and livestock makes the waste of one enterprise a valuable resource for another.

One of the most practical and important uses of livestock (cattle, sheep, goats, pigs, or poultry) is for breaking down after-harvest crop residues. As mentioned earlier, plowing raw organic matter into the soil does more damage than good. Burning residues is equally destructive because of the soil exposure and atmospheric pollution it causes. Animals, on the other hand, will reduce the residues to dung and urine, and still leave a mulch to cover the soil. However, the time the animals spend on any unit of land must be carefully

## Using Livestock to Regenerate Soils

*North Dakota farmer Gabe Brown* no-till plants more than twenty-five different crops—cash crops and a constant rotation of cover crops—to enhance soil fertility and boost yields on an average of four hundred millimeters (sixteen inches) of rainfall in a fairly brittle environment. His corn averages 159 bushels per acre, compared to the county average of 100 bushels per acre, and his cover crops pay for themselves as forage crops.

Cattle and sheep are integral parts of his operation, spending their time grazing this diverse sward of plants and helping to regenerate his soils, as shown in color plate 21. Their grazing, he says, stimulates plants to release root exudates that feed soil life, which in turn increase fertility and enhance soil structure. The animals are allotted fresh forage every day throughout the growing season. Every morning portable fencing is set up to subdivide larger paddocks into smaller ones to increase stock density sufficiently to create the behavior change that produces herd effect. The animals move themselves with the help of solar-powered automatic gate openers preset to open at specific times throughout each day.

Chickens—both laying hens that run free and less nimble broilers in mobile predator-proof pens—follow the cattle during the day feeding on the insects, including flies, that also follow the cattle. When bison ruled the plains, says Gabe, they were followed by vast flocks of various birds. That's what he's imitating with two more profitable enterprises.

planned. Concentrating the animals on small areas for very short time periods usually achieves the best results and can easily be done using portable electric fencing, tight herding, or, for poultry, mobile, bottomless cages. When they are allowed to spread over the whole field for an extended time, the animals are likely to consume too much material, leaving the soil exposed and highly susceptible to erosion.

When livestock of any type are incorporated into a cropping plan, you want to keep them on the land and out of buildings as much as possible. A number of farmers have already demonstrated that even small stock, such as chickens, turkeys, and rabbits, can spend a good portion of the year out on the land, given adequate protection from predators and frequent moves. However, the trend on most farms has long been in the other direction. Confining animals to pens, stalls, or barns not only promotes ill health, it also requires extra work in moving manure out onto fields. The nutrients contained in urine may be lost altogether. The beneficial trampling, digging, and scratching that help loosen mulch and aerate soils are certainly lost.

Some farmers who would otherwise keep their animals out on the land find it necessary to confine them during the winter months. Many, and most notably Virginia farmer Joel Salatin, have developed ingenious ways to overcome problems created by the animal wastes that accumulate over the months of confinement. Salatin uses what he calls a "carbonaceous diaper" to absorb manure and urine under a simple shed roof constructed with poles and undressed lumber. Every few days, he adds carbon (primarily woody material like chips, bark, sawdust, but also moldy hay, leaves, or straw) and as the bedding pack builds, the cows tromp out the oxygen, creating an anaerobic environment. About thirty-four kilograms (seventy-five pounds) of whole shelled corn per cubic meter (or yard) of carbon ferments in the pack. After winter, when the cows exit the shelter to go back out to pasture, pigs enter, seeking the fermented corn, and aerate the pack, creating compost, which provides the fertilizer for the farm's fields.

Having to transport manure back onto fields is never as convenient as having the animals deposit it there in the first place—something farmers may yet overcome by developing portable "barns" that can be moved with minimal effort to new sites through the winter. Some Canadian farmers soon learned they could keep cattle on the land all winter by piling dried forage in large rows that animals can easily find under the snow.

## Minimize Irrigation

Irrigation renders the cropland environment even more artificial than dryland farming, which relies on rainfall only. To date we have been unable to sustain any irrigation-based civilization over time. Overwatering, through the most common form of irrigation, flood irrigation, is largely the reason why. In most soils an excess of water leaches nutrients, carrying them below the crop root zone. Yields are reduced and food value is diminished through plants picking up fewer nutrients. In a poorly drained soil (one that is mostly clay or contains an impervious clay layer) overwatering leads to waterlogging, which prevents plants from absorbing needed nutrients. It also alters the mineral content of the soil and may eventually, especially in hot areas with high evaporation rates, produce a thick layer of salt on the surface that makes further cropping difficult, if not impossible.

The use of drip irrigation, or porous piping, or any other technology that enables water to be rationed, currently offers the most promise for sustaining land under irrigation. That these alternatives aren't used more widely is largely due to the higher capital outlay required to purchase the materials, the difficulty of maintaining them, and, in some cases, the energy for pumping the water. Unlike most flood irrigation systems, which rely on gravity to carry water to fields, water in a drip irrigation system usually has to be delivered under pressure, and energy is required for pumps that can do this.

## Manage the Water Catchments

Up to this point, our focus has been on the health of the soil on the croplands themselves, but no crop field can be sustained for long if the land surrounding it is degrading. To ensure the long-term viability of crop production, the water cycle must remain effective.

Unable to soak in where it falls, rainwater moves downhill, gathering silt as it goes, more often than not creating a flood at some point. Floods not only destroy crops, but the silt they carry also fills irrigation canals and drainage channels. The loss of the first great civilization based on irrigated agriculture in Mesopotamia was largely due to the silt from eroding catchments that filled canals and dams.[3] The demise of the Mayan civilization in the jungles of Central America was largely due to silt from deforested catchments that filled channels draining the marshes in which their raised-bed crops were grown. [4]

The only civilization we know of that has survived for many thousands of years is that of the Egyptians of the lower Nile Valley. Deforestation and soil erosion occurring over *three thousand kilometers away* in the highlands of Ethiopia and Uganda combined with eroding soil from vast regions lower down provided nutrient-rich silt that annual floods deposited in the delta. Using camel-powered pumps to move water from the river to fields, farmers grew a plentiful supply of crops. The sandy soil ensured good drainage and thus waterlogging and a buildup of salts did not become problems. By exploiting a natural process, *and someone else's environmental problems*, the people of lower Egypt managed to sustain their fields over seven thousand years. This security ended when the Aswan High Dam was built in the 1960s, and the silt that had kept the fields fertile was trapped behind it.

Obviously, if we are to create a regenerative agriculture, which depends on healthy water catchments, we cannot put all the responsibility on individual farmers. They can do their best to stabilize the catchments on the land under their care, but even if every farmer managed to stabilize the catchments on every farm, we could not guarantee future food supplies unless the remaining catchments (the mountains, forests, prairies, savanna-woodlands, and rangelands) that cover a far more extensive area, are stabilized and regenerating. With government agencies, and even major environmental organizations, today supporting policies that result in less effective rainfall over so many of these catchment areas, this will require a much larger effort than I can even begin to describe in this book. Nonetheless, it is one that must be undertaken. In the meantime, you as an individual farmer can make a start by working to ensure that the rain you receive, anywhere on your farm, soaks in where it falls.

## Minimize Energy Consumption

Most so-called primitive agricultural systems are highly energy efficient, producing about twenty times as much energy as they use. Paddy fields in China and Southeast Asia produce fifty percent more energy than they use. Modern industrialized grain farming, on the other hand, produces about twice as much energy as it consumes in the form of synthetic fertilizers, pesticides, and machinery, and it is becoming steadily less energy efficient. However, when processing and distribution are taken into account, then all food production in the industrialized nations uses more energy than it creates.[5]

Even farmers who feel no moral obligation to curtail energy inputs will be forced to do so due to the mounting social, environmental, and economic costs associated with the use of synthetic, fossil-fuel-based fertilizers, pesticides, and machinery.

## Feed Soil Rather than Plants

We can reduce the need for pesticides simply by increasing diversity in our croplands, as mentioned earlier. But reducing the need for fertilizer is a slightly different matter. With the development of synthetic fertilizers, we changed our focus from feeding the soil to feeding the plants, with catastrophic consequences. We need to shift our focus back to where it belongs—on the soil. Soil fertility involves much more than mere nutrients. As I found on my own farm, you can have all the necessary nutrients, but many may be unavailable to plants simply because the soil is poorly aerated. Poor drainage, excessive acidity, and exposed soil all affect the soil life that keeps soils fertile. It is just as important to identify and to correct these limiting factors as it is to supply the appropriate nutrients.

When you do supply nutrients, supplement only what is lacking. Sometimes an otherwise healthy soil will produce poor yields merely because a minor element or micronutrient is deficient. Long ago, some farmers found they could correct a copper deficiency merely by dragging a piece of copper wire behind a tractor. All too often, when we base fertilizer applications on the needs of plants rather than the needs of the soil, we overfertilize, creating a chemical imbalance in the soil that damages organic material, soil structure, and soil life. Groundwater also becomes polluted by excess chemicals.

Animal manures and composts can provide most of what the synthetic fertilizers do and, as long as they are not applied in excess, without killing microorganisms. However, they do not replace all that is lacking or has been lost, a challenge that still requires research, both into what may be lacking and into benign methods for providing it.

Biodynamic agriculture uses minute amounts of various preparations that enhance soil microorganisms and thus soil structure, with impressive results. As a general rule, *any* nutrients added to a soil should enhance the life within it. Otherwise, we cannot keep soil alive nor benefit from the "interest" such biological capital could provide.

## Conclusion

Throughout this chapter I have referred often to the agriculture of the past in an attempt to show that, good as much of it was, it is not something we should emulate altogether to solve today's problems. Slash-and-burn agriculture managed to maintain small human populations but quickly broke down when those populations increased. Settled populations managed to prolong the useful life of continuously cropped fields by adding animal manures and composts, and later crop rotations and soil-building cover crops. But soils continued to deteriorate.

What we do on the world's croplands to sustain ourselves cannot be divorced from the far greater areas of land that constitute the world's water catchments, which if not well managed can render crop fields unproductive or destroy them altogether. Crop production is likely to require far greater integration with livestock, properly managed, to keep both water catchments and cropfields healthy and productive. Fortunately, over the past few years some very creative farmers working with the principles outlined in this book and collaborating with exceptional soil scientists are already showing what is possible at scale. Farmers such as Colin Seis, Gabe Brown, and Joel Salatin are, I hope, but forerunners of millions of such progressive farmers who will show what is possible through regenerative agriculture.

# 36

## Burning
### *When and How to Burn and What to Do Before and After*

*FIRE IS AT TIMES THE ONLY TOOL FOR THE JOB,* but as we learned in chapter 19, there are dangers when it is used excessively, as it undoubtedly is today. Most people have no idea of the amount of burning that takes place each year on croplands, grasslands, and savanna-woodlands, nor of its consequences, and would not be aware of the amount occurring in wet tropical forests if the media had not exposed it. While fire is a tool that has a definite and useful role to play in land management, we need to question its use more rigorously. The burning guideline serves to remind us of the dangers while providing appropriate safeguards.

Use fire only when the context checks show it to be the most appropriate tool. Particularly avoid burning purely for the sake of tradition, accepted practice, or what other people say. Because results may vary considerably even in the best of circumstances, you must, as with any action you take that attempts to modify an environment, monitor closely after a burn on the assumption your decision to burn was wrong.

If your holistic context describes a future resource base that includes great diversity, then maintaining species that depend on periodic burns may well require limited burning to prevent a fire-dependent species disappearing entirely. When you are managing holistically, the most common justifications for burning include the following:

- To invigorate and freshen mature or aging perennial grass plants if, for some reason, animals cannot be used, or in cases where you want to sustain fire-dependent vegetation
- To expose soil in patches in order to create a mosaic of different communities that can support a greater diversity of plant and animal species
- To reduce selected woody species that are fire-sensitive at certain stages of their lives
- To provide intense disturbance to a community in which many dead plants are hindering growth

## Before You Burn

To burn or not to burn is an action that must always be checked to ensure it is in line with your holistic context. Some of the concerns the checks should address follow.

### *Cause and Effect*

If you intend to burn to overcome a problem produced by past management, you must at least ensure that you simultaneously act to rectify the problem's cause. Not doing this commonly results in people using fire to fight the effects produced by past fires. Burning very rank, fibrous grass to make it more palatable is a good example seen in the management of many ranches and national parks. The fire will freshen individual plants by clearing the old growth, but, because it exposes the soil between plants, it tends to increase the spacing between plants. This results in fewer, larger plants that in turn become coarser and more fibrous, thus requiring further burning to be edible at all.

Burning forbs considered weeds that spring from the cracks in bare, exposed soil is equally self-defeating. Such tap-rooted plants, which take hold easily in the cracks left by a previous fire, are the beginning of an advance in biological succession. The material of their stems and leaves will provide the soil cover for other species as the community becomes more diverse, dynamic, and stable. Burning them sets the process back, lengthening the time it will take to reach the landscape you require in the future.

### Social Weak Link

Because the prevailing belief of the public, and thus large environmental organizations and governments, concerning the vast grasslands is that fire is natural and essential for grassland health and reducing woody plants, it is not likely that your burning will offend or confuse the people whose support is critical to your success. However, this belief is beginning to change in the United States due to the catastrophic fires in the western states where goats and sheep are now being used to reduce the fuel load in fire season.

### Biological Weak Link

If you were attempting to reduce the population of a fire-sensitive woody species by burning, you would need to be aware of the weakest point in the life cycle of that species. If it was a species that established best in cracked bare soil you might, by burning, kill adult plants, but you would also be likely to enhance the success of the next generation. If it were a species that established best in long-rested clumps of perennial grass with weakened root systems, you would then be likely to kill the adult plants and to reduce the ability of their seedlings to establish.

### Financial Weak Link

The weak link in the chain of production that stretches from raw resources to money should always be considered before you burn. Burning to reduce a fire-sensitive woody species becomes most tempting in years of low forage production because the brush stands out amid the poor grass. But when forage production is low, this often indicates that energy conversion is the current weak link, in a livestock or wildlife management situation, and if that is the case, it could be a mistake to burn. You need all the forage you've got this year, and you don't want to risk losing any to fire. Burning would best be left to a year in which forage is abundant and energy conversion is not the weak link.

## *Marginal Reaction*

We often view burning as cheap because the only investment involved is a box of matches. However, when burning either forage or crop residues, factoring in lost forage and grazing time and the reduced effectiveness of rainfall shows the true cost to be high. With the contribution of biomass burning to atmospheric pollution factored in, the cost becomes higher still. All else being equal, alternatives, such as animal impact, would probably reap a higher marginal reaction per dollar than burning.

## *Sustainability*

In this check, you look specifically at the future resource base described in your holistic context. In terms of the land and the four ecosystem processes, you need to consider a number of questions: What degree of soil exposure do you have now? What litter will you lose? What might fire do to the mineral and water cycles? What will happen to the microenvironment at the soil surface, and how will that influence the biological community you are attempting to create?

In considering these questions, remember to think about the entire future community and the age structure of its populations, not just adult plants and animals that will be present only for the short term. In chapter 19, I cited the example of the annual burning of teak forests early in the dry season to save the trees from more damaging fires later. The practice doomed the forest because teak seedlings don't establish easily on the bare, inorganic sand left by the frequent low-intensity prescribed fires set to save the forest.

Burning to eradicate brush, a common reason given by extension services, almost always fails to pass the sustainability check. Fire, rather than killing woody species, invigorates many of them, causing them to thicken up and send out multiple stems. Because it exposes soil, it also tends to produce long-term damage to the grassland you might hope to enhance.

## *Gut Feel*

After passing through all the context checks, how do you feel about burning now? Should you go ahead or seek alternatives? If you, or any others making this decision with you, have had a strong disposition toward burning in the past, based on long-held beliefs or custom, I hope the context checks will help you to question that stance. You should bear in mind that, anytime you burn, you are releasing carbon and other pollutants into the atmosphere that are adversely affecting all of humanity. You will contribute to this worsening situation if you use fire reflexively when alternative treatments are available.

## Planning Considerations

There are several factors involved in planning for a burn that will affect the outcome of your context checks. *How* you plan to burn is just as important as *why* in considering each of the checking questions.

### *Types of Burn*

Burns may be either hot or cool, depending on the amount of fuel, its moisture content, and atmospheric humidity. Hot burns occur when large amounts of combustible material and dry conditions produce large flames that persist for a long time and can seriously damage the aboveground parts of woody plants. In the tropics where the year divides into wet and dry seasons, the best time for hot burns comes toward the end of the dry season. In temperate zones, opportunities may depend on several factors.

Cool burns are done when forage is still partially green or damp and difficult to ignite. In this case, the fire trickles along, barely scorching the woody plants. When cool burns come at the beginning of a dry (dormant) season, as they often do in the tropics, where they are also called early burns, the soil will remain exposed longer than it would be with a hot burn that comes later in the season. Where hot fires, being more fierce, tend to burn uniformly, cool burns are generally patchy and broken, as not all the material burns equally well.

If you decide your situation calls for a hot burn for high heat, more uniform burning, and shorter soil exposure, factor this into your grazing plan (covered in chap. 41) to ensure the presence of sufficient fuel by not grazing it down just before you need it. If you intend the hot burn to kill a fire-sensitive plant, the burn should coincide with the plant's most vulnerable stage, be it during growth, dormancy, or just before dormancy. A cool burn requires less fuel, which may mean taking some grazing out before the burn. Again, the grazing plan should assure this well before the event.

## Tools to Associate with Burning

When using fire, remind yourself that the decision to do so is not complete until one other tool is selected to be applied following the burn. Fire on its own—by default—becomes fire followed by rest. In brittle environments, as most are where burning is done, both fire and rest have the tendency to produce bare ground. Unfortunately, fire followed by rest is standard practice in most parts of the world where government extension services advocate prescribed burning. On public lands in the United States, regulations generally require it.

---

Fire on its own—by default—becomes fire followed by rest.

---

Animal impact will of course offset the rest and is the tool you would most often want to employ in conjunction with fire. Wild grazers often concentrate on burned areas, which provide ash and good visibility (of predators), and they soon flush green. In Africa, some animals will move onto a burn even before the ground cools off. However, in most cases, we are dependent on livestock to provide the animal impact. We can induce them to provide it by creating herd effect on the burned area using an attractant, or simply by herding the animals onto the burned ground.

If you are ranching in the United States on public lands and are prohibited from using animal impact following a burn, you might consider establishing some test plots for the benefit of the government agency people, who may in fact be persuaded to help change the regulations.

## Burning to Enhance Wildlife Habitat or Maintain Biodiversity

As some plants and animals are to varying degrees fire dependent, it can be a mistake in some environments, such as national parks, to suppress fire altogether if the holistic context includes maintenance of biological diversity. Such cases may require periodic burning guided by basic biological research on the life cycles of the fire-dependent species. There will be many years in which to conduct such research because shifting the community back to one dominated by animal-dependent species in stable grasslands will take many years, as I have seen on my own land, which, after nearly a decade without fire, is still dominated by the fire and rest-tolerant species produced by past management.

## Monitoring

Once the decision has been made to burn, you need to assume that despite passing the idea through all the checks you could still be wrong. Determine what criteria you can monitor from the outset that will give the earliest possible information on the direction your decision is taking you. Because the key to management of all four ecosystem processes is the soil surface, which fire can expose so ruthlessly, you generally start there. Another common indicator of early change is the types or species of plants that establish following a burn.

Apart from helping to avoid potential crises, your monitoring will help you gain a better working knowledge of the effects of fire on the land. Where a periodic burn may do little long-term damage, frequent burning, as mentioned repeatedly, can become very detrimental. Had we monitored the effects of fire on the soil surface and the incoming generation of plants during my game department days, we would surely never have set fires with such cheerful abandon. We never even considered the ecosystem processes or realized their importance. Thinking only of adult trees, grasses, and animals, we managed for species, not process or population structure, a topic covered in more detail in the next chapter.

---

Where a periodic burn may do little long-term damage, frequent burning can become very detrimental.

---

But one should manage for the whole, rather than select species, and therefore must monitor constantly the factors that reflect the health of all four ecosystem processes, such as litter cover, soil exposure, and plant spacing. Changes in these give the earliest warning of change in the water cycle through increased runoff and surface evaporation and decreased penetration. Changes in age structure of plant populations, in particular, show what is happening to community dynamics. Which plants show the greatest influx of young that survive through the seedling stage? When they mature, will that be the community you want?

You must particularly watch for any early signs indicating the need for another burn within a few years. If such develop, try to find an alternative tool to do the task, especially in more arid areas. The drier the climate, the more dramatic the effects of fire and the less frequent its use should be. But what is frequent?

For many years in Zimbabwe, the four-paddock-three-herd grazing system was the government's recommended practice. One paddock was burned each year and the animals rotated through the other three; thus each paddock was burned once every four years. In the early days of my search for better ways, I visited numerous ranches as well as the government research stations promoting this practice, to see if it offered any hope. Nowhere, including on the research stations, did I find land on which the water cycle was not deteriorating. Many published papers had attested to the healthy grassland produced by burning every four years, but all were based on the plant species present, rather than what was happening to the soil surface and reflected in the ecosystem processes.

## Conclusion

Because humans for most of our existence had only two tools—technology and fire—to manage our environment, it is not surprising that fire is so widely used and advocated. Until animal impact was recognized as a tool there was little alternative but to burn large landscapes to manage them, as the Australian aborigines did so skillfully for millennia, while causing much of the continent to desertify. Animal impact, when used properly to shape a landscape, does so without damaging it. When used following a burn, animal impact helps the exposed soil surface cover over again more quickly, thus offering a better alternative to the recommended practice of following a burn with total rest, which tends to decrease rather than increase soil cover.

Monitoring the soil surface for change will help you determine how frequent a burn should be. Where a periodic fire every twenty to fifty or more years may do good, a burn every two to five years, by exposing soil, can lead to tragedy.

Now let's move on to the last of the management guidelines, population management, which relates in some respects to many of the others.

# 37

## Population Management
### *Look to Age Structure Rather than Numbers, Communities Rather than Single Species*

*THE POPULATION MANAGEMENT GUIDELINE* bears on the tool of living organisms, but more generally on the management of community dynamics. It applies anytime we want to encourage or discourage the success of a species within a biological community. Commonly, when we say we want more corn or cattle, or fewer fruit flies and mosquitoes, or more of our team, and less of theirs, we tend only to consider the raw numbers of whatever species concerns us at the moment. The rancher asks, "What is your stocking rate?" Game enthusiasts ask, "What is the deer count?" Farmers ask, "How many bushels?" All of these questions are important, but in addition, or even ahead of them, these guidelines raise other considerations that can improve the management of populations in the context of whole communities.

One example shows both how much this broader approach can contribute to the solution of some of our most urgent problems, and how easily the best of minds fall into the old rut. Some years ago, a consulting assignment from the United Nations Food and Agriculture Organization took me to Pakistan's Baluchistan province. I had read numerous reports previously prepared by other consultants to Pakistan and the UN as well as government officials. These cited one problem that overshadowed most others: the overharvesting of desert bushes for fuel. People were scavenging an ever-expanding area surrounding their villages for desert bushes, the only fuel remaining, and didn't just lop off branches, but took roots and all. An ever-widening circle of bare ground extended around most settlements. All the reports stressed what this was doing to expand the desert, but how else could the villagers cook their food?

Most of the reports concluded that alternative stoves and fuel had to be found, but no one had an idea the villagers could afford. The report writers, however, looked only at the number of bushes available and the rate of harvesting, ignoring all other aspects of population dynamics and the maintenance of healthy biological communities. No reporting scientist apparently noticed the fact that there were no young bushes, even though new growth, lacking fuel value, did not interest wood gatherers, and billions of seeds had been produced over the life of the plants. It was like worrying about a disease killing very old people while failing to notice the people hadn't raised any children successfully in eighty years.

Without any harvesting at all, a population that does not reproduce will disappear. Bushes that produce seed that successfully establishes new bushes will provide a source of fuel that satisfies a good share of the village demand forever. The rate of consumption of mature bushes was not the problem that needed to be addressed at all. The challenge lay in determining why none of the billions of seeds produced over a great many years survived to become bushes.

We will return to this case later. The point I wish to make now, which this case highlights so well, is that there is a need for two kinds of knowledge in managing a thriving population. First, you need to be able to assess the population's age structure, and thus its health and stability. Then you need to pinpoint the cause of its condition if it is not what you desire. The population management guidelines address these questions on the basis of some rather obvious principles that modern production systems have increasingly obscured.

---

To manage a thriving population, you need to be able to assess the population's age structure, and thus its health and stability.

---

In crop farming particularly, we have come to think in terms of annual or short-term monocultures where we plant an entire population, nurture it artificially, then harvest it completely. The same logic extends to the clearcutting and reseeding of timber. We even rip up fruit orchards at a certain age and replant them because it suits our mechanized handling techniques to have everything the same age and on the same schedule. All of that, however, does violence to the natural dynamics of populations in whole communities and inevitably risks instability and failure at some point.

## Self-Regulating and Non-Self-Regulating Populations

Among animal populations there are the two fundamental types, which I referred to briefly in chapter 23: those that regulate their own numbers and those that do not. We don't yet understand how some of the self-regulating populations manage to limit their numbers, but they do, even though they have very high breeding rates and thus a potential for rapid expansion. Some of the small antelope of Africa, such as duiker or stembuck, are examples.

Self-regulating populations cause few management problems. Non-self-regulating populations, on the other hand, present a very different picture, unless the communities they inhabit are intact and complex. Most non-self-regulating populations do not appear as such, as long as they exist in complex biological communities that remain relatively stable. Predation and other forms of attrition provide limits to population growth. Among mammals and birds most of the herding, flocking, or gregarious species seem to fall in this category. Their populations often remain limited in a complex community, but can break out in problem numbers should the integrity of the community as a whole be damaged. This danger from non-self-regulating species threatens whenever we simplify any biological community.

A few species are notorious for their unrestricted growth potential in a simplified community—rabbits in Australia and quelea finches in Africa, for example. Humanity, to its sorrow, belongs to the same category. Historically we overcame so many limiting factors that our numbers exploded and continue to do so.

Since *Homo sapiens* cannot escape nature's principles we have but two options. The first is to become self-regulating through birth control and family planning, which are linked to the education and empowerment of women, who then tend to balance family size with resources (which men tend not to do). The second option is to continue as a mostly non-self-regulating population and allow massive death losses from war, accident, disease, and starvation to regulate our numbers. To date, we have most often chosen the second, which holds no promise of a bright future, as we know all too well from our past.

## Age Structure and Population Health

Figure 37-1 shows what is known as the sigmoid, or S-shaped, curve that describes the growth of almost all populations. Starting at point A with very few individuals, the population of a species increases gradually. By point B, growth accelerates as the population expands geometrically. At about point C, further growth in numbers encounters difficulties of some kind, and the rate falls off as numbers approach the biological community's capacity to sustain them. Although individuals may breed as fast as ever, the pressure for food, cover, space, and so on limit population numbers. Starvation seldom kills animals because in their weakened condition they will first succumb to accident, disease, or predation.

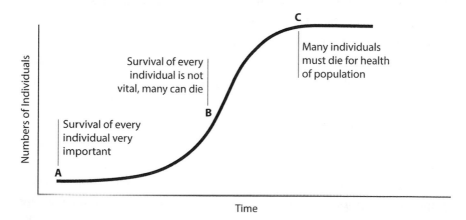

**Figure 37-1.** Virtually all populations grow according to a sigmoid, or S-shaped, curve. Numbers increase gradually between point A and B, but then accelerate geometrically until point C, when the biological community begins to lose its capacity to sustain them.

The importance of each individual changes as the population grows. At point A, the survival of each individual has a great influence on progress along the curve. Without that first plant there can be no others; without that first breeding pair, there will be no further animals. The old anecdote about a penny that doubles every day illustrates this point. If I agree to give you a penny today and then double it every day for a month you would have over ten million dollars by month's end. However, if I didn't happen to have a penny that first day and waited until the second, by the thirty-first day you would have only about five million dollars. By contrast, if I came up one penny short on the sixteenth day you would only lose about sixteen dollars. Obviously each penny, seedling, or breeding pair has tremendous impact at the outset, but less as the geometric progression advances. By point B, the loss of one individual hardly matters to the health of the whole community.

In practice, this concept is frequently overlooked. The rancher who wants to advance his annual cheatgrass range to perennial species may allow, for instance, a couple of horses free rein of the place throughout the year, thinking it unnecessary, and even inconvenient, to put them with the main cattle herd. When this happens, the first perennial grasses that try to establish are exposed to $2 \times 365 = 730$ horse-days of grazing. Rare indeed is the perennial grass that can withstand such overgrazing. The same 730 horse-days of grazing done by 365 horses over 2 days of the year would of course lead to the establishment of many perennial grasses.

By point C, the very survival of the population depends on the death of many individuals. Sometimes this occurs in ways that allow the population to remain high and relatively stable. This would be the case if the high numbers have not, in and of themselves, damaged the biological community sustaining them. When it is otherwise, a population crash can return the whole process to point A.

## Environmental Resistance

The father of game management, Aldo Leopold, called the limiting pressures on a population at point C *environmental resistance*, because they come from the entire biological community. Unfortunately, when we upset the built-in checks and balances, as we do, for instance, by removing predators; some populations explode to higher numbers that in turn exert great pressure on yet other populations and thus destabilize the whole. Predators, remember, consist of more than lions and wolves. Spraying pesticide on insect pests also kills millions of their predators, most of which are themselves insects.

At each point on the sigmoid growth curve, the population has a characteristic age structure. Figure 37-2 shows in sketch 1 how at point A (from fig. 37-1), the proportion of young is high, though numbers are low. At point B, the age structure looks more like sketch 2. The young remain numerous and numbers decline regularly through all age classes.

Sketch 2 represents a very healthy population within a biological community. It will remain healthy if kept at that level by human management or by predation that takes off individuals in a way that maintains the age structure.

Sketch 3 shows the age structure at point C, where broader environmental resistance becomes important. Because disease, starvation, and accidents affect the very young and the very old more than they do adults in their prime, the numbers dip sharply at point (A). This low reflects the high proportion of last year's young that did not survive. A herd of deer that bear young once a year, for example, might have relatively few two-year-olds. The age classes at point (B) are relatively abundant, however, because they represent individuals in their prime that can better withstand disease, malnutrition, and other stresses. By point (C) the numbers drop off again as fewer individuals reach really old age under the stress of environmental resistance. These diagrams of population age structure cover almost all situations where individuals in a population have any sort of prolonged life. Annual plant and insect populations, of course, would not follow this pattern.

Most living organisms acquire the age structure in sketch 3 when they are in balance with their biological community. Humankind, however, can get there prematurely. We are perhaps the only creature that can so damage its environment that it starts to die off before reaching its full potential. Every decade or so, in Africa in particular, millions of people starve, not because of overpopulation, but because of management that damages all four ecosystem processes, resulting in desertification.

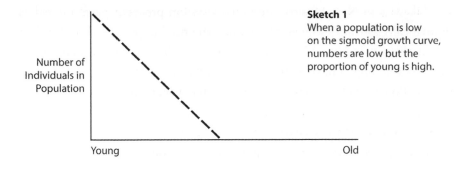

**Sketch 1**
When a population is low on the sigmoid growth curve, numbers are low but the proportion of young is high.

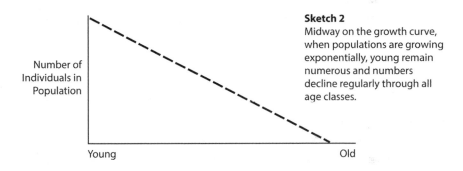

**Sketch 2**
Midway on the growth curve, when populations are growing exponentially, young remain numerous and numbers decline regularly through all age classes.

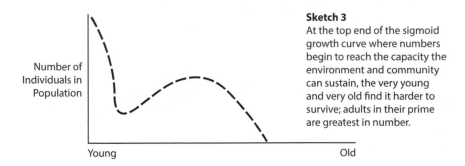

**Sketch 3**
At the top end of the sigmoid growth curve where numbers begin to reach the capacity the environment and community can sustain, the very young and very old find it harder to survive; adults in their prime are greatest in number.

**Figure 37-2.** Proportion of different age classes at various points on the sigmoid growth curve of a population.

Likewise in New Mexico, the four ecosystem processes have suffered as dramatically as I have seen anywhere, and the rural population is in fact low and declining. If New Mexico had to support the sort of population that many developing countries must on a similar area of land, its chief export would also be gruesome photographs from relief agencies.

## Age Structure versus Numbers

Because age structure reflects so precisely where on the S-curve a population lies, it provides more useful information for management purposes than numbers of individuals ever can. Knowing the size of a population seldom helps decide what to do about it, whereas the age structure often does. Accurate counts, especially of wild and mobile populations, are nearly impossible with currently available techniques, whereas random sample counts will tell a lot about age structure.

In Pakistan, for instance, I had no way to count those desert bushes. Even if I could have counted every one, how would it have helped in management? After sampling several sites, however, I can argue for the high degree of accuracy in the age structure shown by the solid line in Figure 37-3. The dotted line represents what a healthy desert bush population in this biological community should be, as people do not pull out seedlings or young plants.

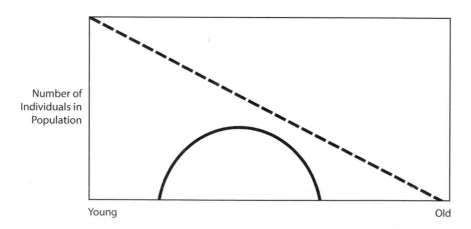

**Figure 37-3.** The solid line shows the age structure I found in the desert bush population. The dotted line shows what a healthy desert bush population would look like.

If the Pakistani villagers created their own holistic context to guide management, it would likely describe a future landscape of great diversity and health including millions of desert bushes. Given such a context for management actions, I believe the villagers would find that the correct decision would be to continue to harvest desert bushes for home cooking. Total protection of these plants is not needed. The population is nowhere near point A in figure 37-1, where every plant is terribly important to the health of the population.

The failure of people to notice and act on such readily apparent information extends far beyond the developing world. When I first visited California in 1978, I traveled through miles of oak woodlands without seeing a single seedling or young tree, but no one voiced the slightest concern. Deer and cattle populations had produced a browse line as high as adults could reach on nearly every tree and so early in the season that few fawns could possibly survive after weaning. Hunting, confined mainly to mature males, was clearly not helping to reduce pressure on the fawns. As local ranchers killed coyotes on sight, little besides annual death of most fawns by accidents or disease was keeping the deer numbers in check.

When major tree populations as well as game show no significant survival of young, bad trends already afflict all four ecosystem processes, and of course human prospects as well, but a numerical count of deer or oak trees does not reveal this, whereas an age structure sample does.

---

On rangelands, plant age structure tells us much more about the health of the biological community than numbers within key species.

---

On rangelands, plant age structure tells us much more about the health of the biological community than numbers within key species. Figure 37-4 is a close-up view of an important piece of land in New Mexico. Long rested, this relic, or pristine, site was believed to be at its optimum development when the photo was taken, and was being used by government agencies in the state as the standard against which to measure management success on similar sites as it reflects the potential they could reach. It does indeed contain the species desired by the agencies, but every plant is already dead or

**Figure 37-4.** Site used by government agencies as the yardstick against which to measure similar sites; it reflects the potential they could reach if "left to nature" (i.e., rested). "Desirable" species are present but their age structure is unhealthy—plants are dead or senile and dying. Millions of seeds have been produced but new plants have not established in many years. New Mexico.

senile after sixteen years of total rest in this very brittle environment. Despite the fact that millions of seeds were produced over those sixteen years, there is not a single young plant. When such rested sites are used as the standard for management, it renders most government statistics on success with range management suspect at best.

### *The Limitations of Game Counts*

The management of game provides many dramatic examples of the limitations of counting, and yet we persist in doing it. In game management programs, people usually sink a large part of the initial budget into a census. However, the numbers counted in the census seldom come close to the numbers actually present and provide little or no information on the health of the populations counted.

When construction began on one of the world's first mega dams, the Kariba Dam on the Zambezi River between Zimbabwe and Zambia, I was working in the game department, which was responsible for rescuing game from the islands that formed as the lake filled. The large number of islands and the logistics involved required a high level of planning and good estimates of game numbers. The perfect hindsight we gained, as we took the last animals off the dwindling bits of land, consistently revealed the utter inaccuracy of our best techniques.

---

The numbers counted in a census seldom come close to the numbers actually present and provide little or no information on the health of the populations counted.

---

I recall one island with a large, flat, fourteen-hectare (thirty-five-acre) top, grassy and easy to move over. It had large trees scattered about that made good viewing platforms and the game had browsed off all leaves below about two meters (six and a half feet), making visibility excellent. Our highly experienced crew, amounting to more than two men per hectare, counted 60 kudu, 150 impala, and numerous other animals. But there were in fact 120 kudu and over 300 impala when all were finally captured. How many people base management decisions on the counting done by one or two people over a few days over thousands of hectares of heavily vegetated land?

The mystique of aerial counting has also proven hollow in my experience. I used to allow tourist planes to fly over my own game reserve regularly when the public game reserves and national parks forbade them. If I was flying myself and happened to spot a herd of elephant or buffalo, I would radio the tour pilots and tell them where they could find the animals for their clients, but often they could not see the animals until I flew close overhead and dipped a wing. At other times, they would see herds and report to me, but I could not find them.

When I see people wasting money on aerial game counts in brushy country I like to remind them that it is so difficult for pilots to spot bright yellow rafts on blue seas that we use pigeons to do so, trained to peck when they see such life rafts. Large masses of game can render themselves practically invisible, particularly in brushy country. I remember one eight-thousand-hectare (twenty-thousand-acre) tract in Zimbabwe where I was assisting the owner to start game ranching. There were many species but the most numerous were impala. We had been using the plane to count hippo because it enabled us to look down on them in the river pools. The reedbeds were too thick to even get near them on the ground. While airborne, we decided to fly over the rest of the ranch so the owner could see his approximately five thousand impala, a figure we had established by first assessing them on the ground using standard estimating techniques. We flew the area at various altitudes, searching all the places we knew them to spend the day, without seeing a single one. The owner, in despair, concluded that, since they obviously migrated on some days to neighboring land, he could not count on a commercial harvest. Then, driving back from the airstrip along the river, to his amazement we saw hundreds standing there as usual.

Ultimately, I came to mistrust aerial counts more than any other technique, and more recent research has confirmed their unreliability.[1]

If aerial counting is hopeless, what about estimates by people who live among the game and "know the place like the back of their hand." Surely after many years in the field, a rancher or park ranger knows roughly how many deer, elk, impala, or kangaroos he has.

I once spent a full week strip-counting game on an African game ranch in the company of the owner. In this procedure, we covered roads and tracks morning and evening with several observers recording the distance traveled and the distance from the center of the track to the nearest animal sighted in each herd. Using the distance traveled and the average width of the sample strip one assumes a constant density over the whole area and computes estimated population figures, which, though better than most estimates, usually turn out low. After spending a week at this, the rancher declared that in all his years on the place he had never actually seen so much of it. When we worked out the percentage of the total land we had in fact covered, removing repeated coverage of the same land, it was only two percent of his land. What did either of us really know?

In order of importance, probably twenty other questions deserve more attention than numbers in the management of game. Besides age structure, other factors such as the sex ratio in adults; the feed, cover, and water requirements; home ranges (whether small, large, or seasonal); levels of use of feed plants; the age structure of those feed plants; and so on deserve far more attention than game counts.

### Monitoring a Harvesting Program through Age-Class Sampling

Years ago an American wildlife biologist, Archie Mossman, and I assisted the Forestry Commission in Zimbabwe to start game ranching in their forests. The project called for harvesting and marketing sable antelope and eland, Africa's largest antelope. The forest was dense and large, and eland in particular avoid people. I personally did the first survey and found them impossible to even sample. Even though their tracks and dung indicated a large population, I saw only six eland after traveling the tracks for days and lying in wait countless hours at waterholes.

In this case, we said, "Let there be X number of eland that we assume from evidence to be large enough to allow harvesting of two hundred animals." Although we had seen few from our vehicles or at water holes, professional hunters had no trouble tracking and shooting that number. Then, of course, we had a random sample, as the animals were shot on sight without regard to age, condition, or sex, as long as they appeared adult.

The age structure of such a sample can be worked out by weighing the eye lenses, which get heavier with age, or by noting tooth wear and replacement. Ranking the jaws of all two hundred eland on a scale of 1 to 10 produced the age structure shown in sketch 1 of figure 37-5. This, plus the signs of browsing on vegetation, confirmed that the X number of eland had arrived at point C in figure 37-1.

After three years of harvesting at this level, the curve changed to that shown in sketch 2 of figure 37-5. This indicated that the harvesting had begun to reduce the number of young animals dying each year and to increase the number becoming breeding adults. By steadily watching this age structure and sampling the density of tracks and the response of vegetation, we

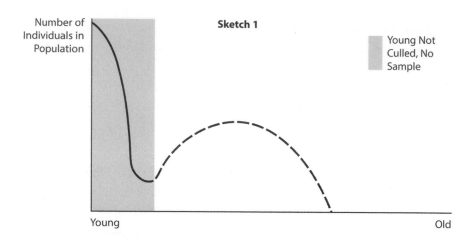

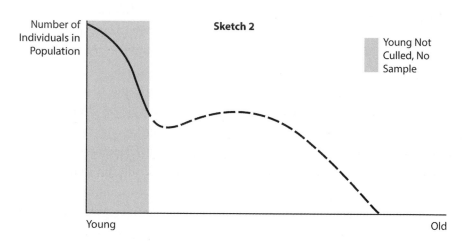

**Figure 37-5.** Age structure of eland population before (sketch 1) and after (sketch 2) well-planned harvesting operations.

gradually refined herd management without ever knowing how many animals we had.

I have dwelt perhaps overlong on this matter of numbers because it so obsesses many people. My own examples come mostly from game management, because natural cases give the purest illustrations, but the same principle applies to all plant and animal populations.

## Limiting Factors

To survive, all creatures must be able to satisfy basic needs for food, cover, and water *throughout the year*. If there is even a short period when any one of these needs cannot be met in full for high numbers of a particular species, its population will be limited. We refer to this phenomenon as a limiting factor.

When attempting to increase the numbers of a certain species you should make sure that the actions taken first address the weakest link in the species' life cycle, as the weak link context check reminds you. If you then find that numbers still don't increase, look for a limiting factor that could be keeping numbers low.

Water for crucial short periods often constitutes a limiting factor for populations that must drink routinely. They may have feed, cover, and water enough for enormous numbers, but if water lacks for several days at one point in the year, the entire population, if nonmobile, dies.

Having water present is often not enough, as it has to actually be available to the species. Many a rancher has told of placing waterpoints out on the land that can be used by birds, small animals, and game, but failed to understand that steep-sided stock troughs in sites of maximum disturbance and no cover don't serve those species and may even kill them. At low water levels, birds and other creatures frequently can clear the sides, but drown when they can't get back out. I like to ask such ranchers how they would fare if they crossed a hot desert toward the scent of water to find a thirty-eight-thousand-liter (ten-thousand-gallon) reservoir fifteen centimeters (six inches) out of reach. They soon get the idea and modify their troughs, as the rancher in Mexico did (chap. 21, fig. 21-1) by adding sloped ramps that made access easy for many species.

Many limiting factors, however, require more diligent observation than that. I had a falconer friend in Scotland with whom I have enjoyed some good days of sport on his grouse moor. Once after many hours of trudging through the heather behind the pointers, he commented that the grouse population should be higher given the quantity of food, cover, and water, the most common limiting factors. After mulling the evidence, we hit on the fact that a more productive moor nearby had graveled roadsides that might offer more grit for a bird's crop and digestion of heather.

Soon afterward he put out mounds of broken shells at several sites, but later reported that the grouse ignored them, so he abandoned the idea. Close inspection a few years later when I returned, nevertheless showed that, although the piles had not visibly diminished, some creature, presumably grouse, had methodically removed the tiniest bits of shell, leaving only larger, unusable pieces.

The limiting factor principle applies equally to plants. A good example would be the potentially very productive vleis (grassy valleys) in much of southern Africa. Moisture, soil depth, and all other conditions for a highly complex biological community exist. However, in the dry season, the low-lying vleis can experience a frost or two that limits development to communities dominated by grasses and forbs. Beyond the frost line, the biological community can develop to woodland.

Absence of a particular trace mineral is another common limiting factor in plant growth. I once worked with a rancher in Africa in open, grassy country that, for no obvious reason, lacked a good mixture of woody, deeper-rooted plants. One day while surveying the bleak countryside from the ranch house veranda, I noticed that the scraggly hedge his wife had tried to establish round the house was noticeably healthier at one point. Neither the rancher nor his wife could explain this, so we dug up the earth for evidence and discovered the copper ground of their lightning rod. As a consequence, I recommended the addition of copper to the supplementary feed given the cattle so they could begin to spread it on the land. Unfortunately, as the war heated up in Zimbabwe, I was unable to get back to the ranch and never knew the outcome.

## Dealing with Predators That Become a Problem

Typical policies toward predators reveal a deeply ingrained blindness that an understanding of community dynamics and population management might enlighten. We need much more research into the role of predators, but we at least sense how the relationship between predators and herding animals keeps entire communities in brittle environments vital. We can only guess at the number of other situations where our uninformed destruction of predators has cost us dearly.

Livestock owners the world over have tended to regard all predators as enemies. Nowhere has this aversion led to further extremes than in the United States, where ranchers and government agencies go to incredible lengths to kill predators while making little genuine effort to live with them or to protect livestock, which is far easier and more cost-effective than eradication.

I have worked with ranchers who went out of business killing predators, while not making the slightest effort to protect their stock. In one case, the rancher had access to at least half a dozen well-known and tested methods to eliminate losses to coyotes without killing a single one, yet he kept killing coyotes until he went broke. That his war on the coyote did not save him was not surprising. Typically the predators that take on humans and their livestock are particular individuals that learn to be increasingly cunning as attempts to kill them fail.

Killing coyotes does little good, particularly if you fail to get the one that is killing your stock. No matter how many you kill that haven't acquired the habit, the killer remains and becomes ever more clever, and will in time educate others. If you doubt that animals do learn destructive habits from each other, consider how quickly a cow that breaks through fences can pass on the trick.

With man-eating lions, tigers, and leopards, as well as problem hippos, elephants, bears, and others, we have long known that one must deal with the particular animal, not the whole population. We have recognized the same principle when one of our own kind becomes a murderer. Killing people at random is no response. We have to try, no matter how hard, to catch the murderer.

Some years ago in Zimbabwe, a nasty-tempered elephant had brought railroad maintenance to a virtual halt on an important section of a much-used line to a seaport. Several bulls were shot in the vicinity, but not the problem elephant. By the time I arrived the rogue had acquired a definite style, hanging around the workers' camp at night. I waited on the edge of the camp in the moonlight and when an elephant singled me out for attack I knew I had the culprit. After that, despite many elephants in the area, work resumed.

Chapter 3 mentioned a research study in which only one species of predator was removed from a community. With no other disturbance, as we saw, within one year the number of species in the community had been drastically reduced. Few simple studies have so clearly illustrated the vital stabilizing role of predators in communities. Unfortunately in many areas, particularly those now set aside as national parks in brittle environments, humans were for thousands of years a major predator keeping animals healthy, wild, and moving (rather than lingering in the same places), but today's sightseeing crowds want tame animals.

Many years ago Charles Elton, one of the earliest animal ecologists, as they were then called, described the Eltonian pyramid of numbers. In concept, it resembles the energy pyramid shown in chapter 14 (fig. 14-1). While we normally see the relationship in terms of the number of lower animals it takes to support one predator, if predation plays the crucial role we suspect it does, the pyramid also shows how many prey animals depend on a single species of predator.

Having been a rancher, I understand the frustration when a wild hunter turns on domestic stock, but it does not excuse the wholesale slaughter of predators that were innocent and play a vital part in balancing populations, including many that are agricultural pests. Certainly as I think back on the many years that I have worked with croplands, rangelands, livestock, and game populations, the healthiest situations contained high levels of predators.

Conversely, the unhealthiest situations for the land, crops, wildlife, and stock have always had a history of predator persecution. I don't believe this is coincidence. Clearly, we have much to learn and many attitudes to change before we will see intelligent and wise management of predator populations, and of our ecosystem as a whole.

## Conclusion

Humanity depends entirely upon living organisms, which are indivisible components of communities of complexity beyond our comprehension. We must, however, manage populations within those communities. The fundamental importance of the whole community can too easily be overlooked when we focus on rare, endangered, or preferred species. But in reality, the members of any one species cannot exist outside their relationship with millions of other organisms of different species. In the short term, a species might have to be favored in order to save it, but in the long run truly saving it can only rest on sustaining the biological community in which it thrives.

# PART 8

# PROCEDURES AND PROCESSES UNIQUE TO HOLISTIC MANAGEMENT

## Holistic Management Framework

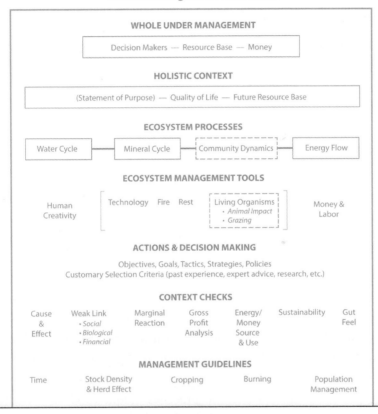

**WHOLE UNDER MANAGEMENT**

Decision Makers — Resource Base — Money

**HOLISTIC CONTEXT**

(Statement of Purpose) — Quality of Life — Future Resource Base

**ECOSYSTEM PROCESSES**

Water Cycle — Mineral Cycle — Community Dynamics — Energy Flow

**ECOSYSTEM MANAGEMENT TOOLS**

Human Creativity

Technology   Fire   Rest   Living Organisms
• *Animal Impact*
• *Grazing*

Money & Labor

**ACTIONS & DECISION MAKING**

Objectives, Goals, Tactics, Strategies, Policies
Customary Selection Criteria (past experience, expert advice, research, etc.)

**CONTEXT CHECKS**

Cause & Effect

Weak Link
• *Social*
• *Biological*
• *Financial*

Marginal Reaction

Gross Profit Analysis

Energy/Money Source & Use

Sustainability

Gut Feel

**MANAGEMENT GUIDELINES**

Time

Stock Density & Herd Effect

Cropping

Burning

Population Management

**PROCEDURES & PROCESSES**

Holistic Financial Planning

Holistic Land Planning

Holistic Planned Grazing

Holistic Policy Development

Research Orientation

**FEEDBACK LOOP**

Plan
*(Assume Wrong)*

Replan          Monitor

Control

# 38

## Introduction
### *Departing from the Conventional*

THE NEXT CHAPTERS INTRODUCE procedures and processes used in five areas where Holistic Management has enabled us to depart substantially from conventional management: annual financial planning, the layout of infrastructure on land where livestock are grazed, the management of grazing animals, developing sound policies, and orienting research to management needs.

The first three chapters are an overview of a procedure that is addressed fully in the *Holistic Management Handbook* (Third Edition).* Chapter 39, on Holistic Financial Planning, applies generally, though for reasons of focus it is written for those who are running a small or medium-sized agricultural business. Chapters 40 and 41 on land and grazing planning should prove enlightening, even if you are not managing grazing animals.

The last two chapters cover two processes unique to Holistic Management that enable us to analyze and develop policies that are socially, environmentally, and economically sound and less likely to produce unintended consequences (chap. 42), and to merge research and management more effectively (chap. 43). Both chapters provide an overview of a process that includes

---

* Savory Institute also publishes annually updated e-books covering the same material.

suggested steps, in the case of policy development, and a series of guidelines in the case of research orientation. If you hope to inform policy, or push for more enlightened policies, the overview provided in chapter 42 may be all you need. But anyone developing policy will most likely require additional training, which is available through the Savory Institute. To design a research project or to assess the value of available research isn't likely to require much more information than I provide in chapter 43.

## Holistic Financial Planning

The Holistic Financial Planning procedure is so closely allied to holistic decision making that its development over the last forty or so years has followed a similar convoluted path. Chapter 3 described my frustration as a consultant in watching a number of my ranching clients go bankrupt even though we were healing the land and improving livestock production. In the main, this was occurring because of other ventures these people had been lured into by attractive government programs and soft loans.

Their financial situation improved once we brought in a couple of consultants who worked with these clients to develop a sound financial plan that went beyond the "back of the envelope" calculations most of them had operated on for years. But this success was short lived. Income increased, but within no time at all expenses rose to match it, and by year's end they might be worse off than ever. This was especially the case when they planned expenses based on the income they *anticipated* receiving and that income didn't come in on time, or came in short. But even more worrisome was that the planning didn't account for any "externalities"—the social and environmental consequences stemming from their plans. Holistic Financial Planning grew out of these challenges.

## Holistic Land Planning and Holistic Planned Grazing

The procedures for grazing and land planning were developed in tandem in the 1960s. When we had understood the role of time in grazing and trampling, we knew that animals would have to move continually, but to do that required new thinking in the way fencing, water points, and handling facilities were laid out.

Prevailing wisdom required fencing off range sites of different types to prevent overgrazing or overtrampling of favored areas. Animals were to feed in limited numbers in specific paddocks at certain times. Now, however, we knew that limiting animal numbers did not prevent either overgrazing or overtrampling, but time did. And with animals to be moving constantly another problem arose because most of my clients' farms were in broken, hilly country where arable land lay in small pockets among a matrix of roads, tracks, rivers, grassy valleys, and woodland remnants. We had to find a new approach that would handle such complexity, both in planning the movements and in finding the most suitable fencing layout.

After several years working with hundreds of ranchers and farmers in five countries we finally developed a combination of the grazing planning approach summarized in chapter 41 and the land planning covered in chapter 40. The two go hand in glove. The land planning procedure should be used whenever large tracts of land requiring a considerable investment in infrastructure, such as fencing roads, water development, and working facilities, are involved. The old rules no longer apply when planning new developments or when modifying old ones. As you will see, the opportunities for improvement are immense and only limited by the creativity of the planners.

Small farms with only a few animals—a dairy cow or two, a few pigs or chickens—rarely require the sophisticated land planning described here. However, reading chapter 40 (plus the detail in the *Holistic Management Handbook*), may provide ideas to spur your creativity. Many more ideas for small farms can be found in permaculture design, for which there are a number of online sources for materials and training.

Holistic Financial Planning is tactical in scope. You plan once a year and implement the plan within that year. Holistic Planned Grazing is also a tactical exercise, though the planning is usually done twice a year, or in extremely dry regions over a two-year period. Holistic Land Planning, on the other hand, is strategic in scope and generally done once in a lifetime. But the plan is implemented incrementally in conjunction with Holistic Financial Planning to ensure that new infrastructure is developed when it makes money rather than costs money to do so.

## Creating Sound Policies

Once we understand the intent of a policy, and can create a generic holistic context for those affected by it, we can use the Holistic Management framework to help us analyze the policy. We do not need to wait for the result to learn that yet another effort to fight crime, halt deserts, save a community, or whatever, has failed. We can determine before application whether or not a policy is likely to succeed, and if not why. If we can do that, we can then work out how to modify the policy to increase its chances for success, as covered in chapter 42, or to develop a new policy altogether. Because development projects are designed to address problems similar to those just described, chapter 42 mentions them too, and shows how they can be modified to create successful outcomes.

## Orienting Research to Management Needs

In any number of management situations, we often find we are lacking knowledge of certain finer points—the establishment conditions for certain plants, for example—that pinpoint areas for further investigation. But how can we be sure, once the information is available, that it is relevant to our particular situation? The context checks enable us to find out. As chapter 43 explains, there is also much to be said for collaboration between land managers and researchers, starting with the design of the research to make sure it is likely to deliver the needed information and is applicable to the management reality.

# 39

## Holistic Financial Planning
### *Generating Lasting Wealth*

GOOD FINANCIAL PLANNING IS ESSENTIAL IN ANY BUSINESS that seeks to be viable and profitable, and there are many methods to choose from. Agricultural producers have an additional requirement and that is to ensure that the means to achieving viability and profit also generate ecological wealth. The wealth that ultimately sustains any nation or community is derived from green plants growing on regenerating soil, a fact that even the most sophisticated conventional financial planning methods do not take into account.

By integrating this simple-to-use, cash-based planning process into your current accounting or budgeting systems, you will be able to both increase your profit and improve your quality of life and your land's health and productivity. The full planning procedure is described in the *Holistic Management Handbook: Regenerating Your Land and Growing Your Profits* (Third Edition). In this chapter I provide an overview of the fundamentals, including the guidelines we use in the planning that generally result in a rapid increase in profitability—a three hundred percent increase on average, according to one study.[1] Even if you aren't engaged in agriculture there will be much that applies to your situation, no matter what your business or occupation.

Two key principles in Holistic Financial Planning differentiate it from other methods:

- *Plan profit before planning expenses.* Just as work expands to fill the time available, so expenses often rise to the level of anticipated income. As you will see under "Psychology of the Planning," by planning profit first, we overcome this tendency.

- *Check for context alignment.* Which actions are actually moving you in the direction you want to go—as you have indicated in your holistic context? As you will see under "Preliminary Planning," the context checks make sure we prioritize those actions that do move us in that direction while ensuring we achieve a healthy triple bottom line (financial, environmental, social).

---

### Why Another Financial Planning Process?

*I initiated the development* of Holistic Financial Planning in the late 1960s to assist farmers, many of whom disliked paperwork. They tended to simply hand figures over to their accountants, who then advised them on what and what not to do based largely on how it would affect taxation. Some farmers did receive guidance from agricultural extension officers who assisted with budgeting and cash flow planning, but far too many farmers, myself included, were still in deep financial trouble. The key principles outlined in this chapter grew out of that experience.

If you already have a good financial planning and management process you may find that you can improve your results by using some of the ideas mentioned in this chapter. If you are an agricultural producer some of the concepts presented are certain to lead to much better results. If you are a farmer or rancher who prefers spending your time being "active" on the farm or ranch and are allergic to spending time with pencil and paper, you would be wise to learn about Holistic Financial Planning. It was developed with the help of people just like you. You can get assistance, including self-teaching materials, from the Savory Institute and training through any one of the global network of Savory Hubs (http://savory.global).

## Psychology of the Planning

After years of consulting in many countries for clients of great variety in sophistication, enterprises, and economic circumstances, I was struck by what they all shared in common. Each of them finished the year in the same nail-biting suspense over their bottom line. No matter what state, country, or currency, no matter what size of business, what product, market, or price conditions, the same picture emerged consistently. Planned income: $200,000; expenses $195,000. Planned income: $10,350,000; expenses, $10,340,000. Like the unanimous elections in totalitarian countries, this defied logic. Profit margins simply could not be so uniform and proportionately small across so many widely differing situations.

Eventually it dawned on me that the problem must lie in the only common factor: human nature. Like most people, my clients were allowing their expenses to rise to meet the income they anticipated receiving. I suffered from the same weakness.

Point A in figure 39-1 shows my condition on graduating from university. The left-hand bar of the graph shows my income, from which I buy the essentials of life at a cost represented by the right-hand bar. The minuscule

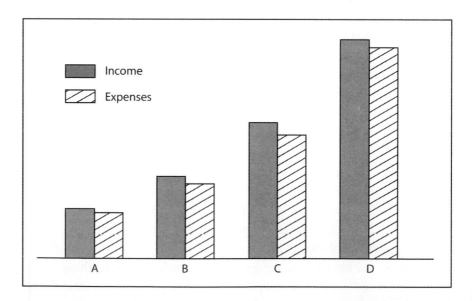

**Figure 39-1.** My personal expenses always rose to match the income I anticipated receiving.

surplus surprises me as I had thought that earning real money after years of scrounging through university would put me in clover, so I looked forward to getting a raise.

Point B shows my situation after getting the raise. I still don't have any extra money and wouldn't even say I'm living better. On the other hand, I traded my motorbike and some cash for a car so I could get around more comfortably, and I really did need that new rifle, new suit, new pair of binoculars, and so many other things.

Points C and D represent further raises, but each time the expected savings account never materializes. Most people manage life in this way until they have to make do on a meager pension. If you look back over the years at what you earned and what you saved, they probably bear little relationship. Far more important than savings were the things you bought because of peer pressure, advertising, or simply because the material culture around you made them seem necessary.

---

### The Debt Trap

*Many businesses automatically* allow their expenses to rise toward the anticipated income. When the income only comes once or twice a year, however, as it does for those engaged in seasonal production, the tendency to let expenses rise to anticipated income often leads to trouble because the money to cover expenses is commonly borrowed. The farmer can easily calculate the income she expects to receive once a crop is harvested or livestock are ready for sale. And almost always expenses will come in close to that anticipated income. When wheat, lamb, or beef prices suddenly fall, however, the farmer is unable to repay what she borrowed when the bill comes due, and the debt trap snaps shut. Servicing that debt now becomes a major expense, most of which is interest, and all of which is unproductive in that it will not generate any additional income.

### Key Principle: Plan Profit Before You Plan Expenses

To counter the tendency to let the costs of production rise to the anticipated income level, in Holistic Financial Planning you plan the profit *before* allocating money to expenses. Once you have figures for the total income you expect to receive, cut that figure by up to half and set that amount aside as your profit. You have now set a limit on the amount of money available for expenses and to which costs can rise. The principle is illustrated simply in figure 39-2.

The profit planned needs to be substantial—closer to fifty percent than ten percent of anticipated income—so that severe restraints are placed on the funds left for running the business. If the amount of profit set aside is too great, however, people will be demoralized as there is so little left to run the business. If the profit planned is too low, plenty of money is left and there is little challenge in keeping costs of production down.

This mental exercise is vital, yet most people at first blush consider it impossible to do, especially if they are struggling financially or deeply in debt.

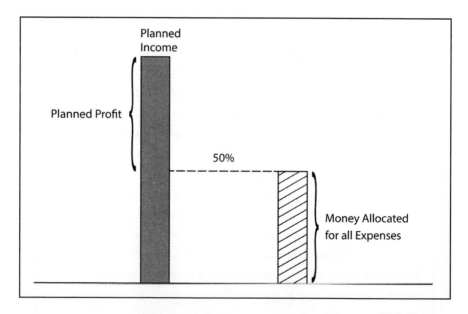

**Figure 39-2.** Plan the profit before you allocate any money to expenses. Once you have calculated the total income you expect to receive, cut that figure by up to half and set the amount aside as your profit.

They cannot see how to limit expenses even a little. Brainstorming ways to cut expenses will help, but attitude is what really makes the difference. Winston Churchill once said, "If something has to be done, and all of your experts convince you that it cannot be, then change your experts and do it!" *Do not accept can't. Do not make or accept excuses. Do it.* Cut those costs. Cut those costs. Cut those costs.

## The Planning Process

The planning process includes two parts. The first is devoted to reviewing the current year's plan, gathering information and figures for the new plan, and running decisions through the context checks. All the thinking, decision checking, and sorting of the information you compile then comes into play in the second part, when you put your plan on a spreadsheet. If you've done the first part well and researched the costs involved in the ideas you want to implement, you probably won't need more than a day or two to put your plan on paper.

### Key Principle: Check for Context Alignment

Actions you may never have questioned in the past, because the expenditures involved were so routine, may now have to be reconsidered. Use the context checks for guidance, the following in particular:

- **Gross profit analysis (chap. 28).** This check uses a technique that separates fixed costs (costs you have regardless of what or how much you produce) from the costs directly linked to production, which are variable. By separating out the fixed costs, rather than allocating a portion of them to each enterprise, you can compare many enterprises or combinations of enterprises to determine which ones bring in the most money to cover those fixed costs, or overhead, and thus contribute to profit.

- **Weak link—financial (chap. 26).** You are trying to build an operation and an environment (future resource base) that will endure, so you and following generations can sustain a profit. To do that, you want to ensure every year that your major investments of money and labor keep strengthening the weak link in the chain of production that exists at any point in time for each of the enterprises you engage in. Once you have identified the year's weak link in each enterprise, you consider all the actions you could take that would strengthen that particular weak link as soon as possible. Expenditures that address a weak link are considered *wealth generating* because they boost production, and thus profit, to a new level, though perhaps not until the following year.

- **Marginal reaction (chap. 27).** When you are considering two or more actions for addressing a weak link and every dollar matters, use this check to determine which of the alternatives gets all the dollars you can allocate to it this year. But also consider the marginal reaction those same alternatives provide in relation to the quality of life expressed in your holistic context.

- **Energy/money, source and use (chap. 29).** This check helps you avoid actions that lead to an addictive use of borrowed money with compound interest, or of fossil-fuel-derived inputs at ever-increasing costs. In avoiding both you enhance profitability on a sustained basis. And if you are a farmer or rancher it is a reminder that you want most of the internal money you invest in any action or enterprise to come from solar dollars, and not mineral dollars gained through mining soil in a nonrenewable manner.

## Preliminary Planning

As you gather the information needed for planning, you will begin to formulate a general strategy based on your analysis of your situation. Your holistic context and the context checks will help you keep a perspective on when to use money as the yardstick, when not to, and how to decide priorities. Your information gathering should provide answers to the following questions:

- *Is there a logjam, and if so, how will you address it?* You are looking for anything that might be blocking the business as a whole from making progress toward its goals. To get your answer, you need to step back from day-to-day operations and concerns to gain greater perspective, just as a logger does when he climbs a hill or tree to see which of the thousands of logs floating down the river has caused the blockage, or logjam. When he then removes the log responsible the rest flow. It's that one log you need to find so you can make progress once more. And it can be difficult to identify and deal with as it commonly involves people in key positions and is related to their attitudes, mindset, trustworthiness, and so on. Alternatively, and in the beginning especially, the logjam may be tied to how the whole being managed was defined (e.g., a key decision maker may have been left out). Or it could be tied to how the holistic context was created (e.g., it was borrowed from someone else or created superficially and has little meaning for managers and as a result it does not influence any actions). The actions you take to clear a logjam in your business receive the highest priority. And if they require money, it is allocated to them prior to any other expense.

- *Are there other factors adversely affecting the business as a whole?* In this case, you are mainly looking for things that reduce overall efficiency and productivity. They are less urgent than something that is blocking progress altogether, but still important. Give thought to potential remedies, which may or may not involve additional expense.

- *Are current enterprises profitable and have you diversified your risk?* A check of the health of current enterprises should reveal whether they are as profitable as planned. Have they enabled you to diversify the risks you could face in terms of your ability to cover the fixed costs of the business? A gross profit analysis helps you find out.

- *What is the weak link in each enterprise and how will you address it?* Once you have identified the weak link—resource conversion, product conversion, or marketing—determine what actions you can take in the coming year to address it.

## Brainstorming New Sources of Income

*No matter what enterprises you are engaged in,* you need to challenge them as you start to manage holistically and periodically thereafter. All too often businesses only add, modify, or drop enterprises when faced with hardship or necessity. Why wait until circumstances force a change? You are generally better off, and more secure, if you take a proactive attitude. Periodically brainstorming new ideas for increasing income within existing enterprises or for adding new ones will help keep you ahead of the game.

The formal brainstorming process we use is described in detail in the *Holistic Management Handbook* (Third Edition). Briefly, it should involve those who work with you, some who don't know your business well, and thus, what *can't* be done. Include children too, because their creativity and imagination are boundless. Within a very short time this group should be able to brainstorm a list of a hundred or more possibilities for new enterprises, or modifications to existing ones, ranging from the patently ridiculous to the eminently practical. The *Handbook* gives criteria for narrowing down the list to the ideas with the most promise, which you then run through the context checks to help determine the best.

*A word of warning:* Few businesses survive by chopping and changing enterprises, so be careful when adding new ones. New enterprises always involve a learning curve and can take a year or more to prove out. Before adding a new enterprise consider the following:

- *It takes time to develop skills and perfection in any enterprise.* The learning curve can be costly. As you change to the new, plan a solid overlap with the old where possible.

- *It is often easier to alter an existing product, or develop new uses for it, than to create something entirely new.* There may be other uses for your current products that you aren't promoting.

- *Managerial effectiveness is diluted by the number of enterprises one manager is responsible for.* This is especially applicable to small businesses with few staff. In taking on new enterprises, the staff is often stretched too thin, which tends to destabilize all enterprises and the business as a whole. This problem can be overcome if the management of a new enterprise is contracted to someone else.

- *There is a direct relationship between management effectiveness and the distance to what is being managed.* This is the reasoning behind the old saying that the finest fertilizer in the world is a farmer's footsteps. Obviously, the more frequent the contact with the enterprise, the greater your chances of spotting trouble early, and the more opportunities you have for finding ways to improve the enterprise.

## Putting the Plan on Paper

The planning procedure includes fourteen steps for creating your plan that I have consolidated into four for this overview: plan the income, plan the profit, plan the expenses, and assess the plan before implementing it. The bulk of your time, however, is devoted to planning the income and keeping expenses from rising to equal it, which is the key to generating profit.

- **Plan the income.** Enter the anticipated income for all enterprises onto a spreadsheet that covers the entire year in the months you plan to receive it. Assume income will come in later than you expect and will be lower than your most optimistic projections.

- **Plan the profit.** Then plan how much of your total planned income to set aside as profit. Remember, your intention in planning the profit now is to put a severe limit on all expenses, not allowing them to rise to the anticipated income. Later, after planning all expenses, including debt servicing, you may have to adjust the planned profit, but for now plan it high.

- **Plan the expenses.** The amount of money remaining after planning your profit will have to cover all your expenses. Start by allocating money to expenses involved in addressing a logjam (if there is one). Then plan any nonnegotiable expenses that you have a moral or legal obligation to pay (if there are any), just to get them out of the way. Next plan the expenses addressing a weak link, allocating as much as you feel you can because they are wealth generating. Then plan the *maintenance expenses*—those needed for running the business but not generating new income this year. These will include your own drawings, wages, and all things to do with running your business. Chances are that you will run out of money to allocate before you can cover all maintenance expenses.

- *Adjusting the plan.* Now is when the hard planning begins. Only you can decide which figures to adjust, and your success in doing so depends on your determination. Start with the maintenance expenses and seriously challenge them all. Every dollar you can shave off maintenance expenses and transfer to wealth-generating ones can make a great difference indeed. If you can't trim them enough, then the profit planned is probably too high.

- *Check the cash flow.* You have planned the income to come in later than planned and to be lower than your most optimistic projections. Expenses are likely to come earlier than planned, and perhaps higher than planned, so don't use your most optimistic projections here either. Make any final adjustments you need to ensure the best cash flow you can month by month.

- *Debt management.* If the business is currently running on borrowed money, or an overdraft arrangement with the bank, each month's interest should be calculated and given a separate column at the end of the spreadsheet. The *Holistic Management Handbook* shows the details.

- **Assess the plan.** At this point your plan represents the actual cash flow into and out of the business, including the level of borrowing, if any, and when it will peak during the year. Next, check to see that the plan will produce a profit in real terms at the end of the planned period. Have your accountant review the plan and assess your tax liability as you have not accounted for noncash income or expenses, depreciation of assets, an increase in net worth on which you could be taxed, and so on. If, after this final analysis, the plan looks good, then proceed. If not, then replan right away.

## Prioritizing Expenses

*Categorizing expenses into capital costs*—investments made to increase production—and the operating costs that maintain the business has long proved useful in financial planning, but in Holistic Financial Planning, and particularly when planning for an agricultural business, it is more helpful to categorize expenses differently.

The need for doing so dawned on me when struggling to survive on my own farm while assisting several hundred near-bankrupt farmers and ranchers to do the same in the midst of a guerilla war and world sanctions in what was then Rhodesia.

Capital investments generally do lead to higher production *at some point*, but I realized that when ruin is staring you in the face, they were like runway behind or altitude above a pilot whose engine has failed at takeoff—absolutely useless. This led me to start categorizing expenses and prioritizing them accordingly: *inescapable*—nonnegotiable expenses that you are morally or legally obligated to meet and that must be given top priority; followed by *wealth-generating* expenses that will actually increase income now or as soon as possible; and last, *maintenance* expenses, which are vital as they include salaries (apart from a year when a new person is engaged to generate yet more income), drawings, fuel, insurance, and all the many things essential to running the business, but they do not produce a single dollar.

By categorizing expenses in this way it gave us an order of priority for cutting expenses to ensure our planned profit was met. There was nothing we could do about the inescapables, so we just paid them. If we were going to get out of the fix we were in we had to concentrate on the wealth-generating expenses, which would increase income as soon as possible. We then shaved the maintenance expenses down to what we needed to operate and no more (based on the marginal reaction context check). Every dollar spent on a wealth-generating expense rather than a maintenance expense will make any business more profitable.

## Monitoring the Plan

Be prepared to modify the plan as you progress through the year. Events will rarely transpire exactly as planned, which is why you must plan in the first place; if you knew what was going to happen beforehand, there would be no need to plan. Chapter 44 describes in some detail the approach we take to monitoring and controlling the financial plan, and the *Holistic Management Handbook* elaborates further. In brief, if your monitoring indicates a deviation from your plan you make minor adjustments, controlling your plan to keep it as close as you can to your original. If these adjustments are not sufficient, or if some unforeseen tragedy strikes, it may become necessary to replan entirely from that point on. Thus *plan* is now a twenty-four-letter word: *plan-monitor-control-replan*.

## Conclusion

If you aim to be profitable Holistic Financial Planning will help ensure that you are, and in the most socially, economically, and environmentally sound way. All of us have been guilty in the past of contributing to social, economic, and environmental problems by the lifestyles we have adopted and the purchases we have made, and we can begin to make changes to rectify this. Those engaged in agricultural production carry a much bigger burden than the rest of us, however, because in making a profit they have the ability to enhance or diminish the biological capital that sustains us all. That ability has now become a responsibility that people who make a living directly from the soil or the seas have no choice but to accept.

The bill for decades of treating their businesses as industries independent of nature has come due in the form of lost or lifeless soil and water. To reflect a true profit, a successful business must also enhance the soil and water, and the life within them that fuels their production. If soil is destroyed rather than enhanced, or water polluted or depleted of life, the profits gained will not be genuine because biological capital is being consumed. On the other hand, when you enhance biological capital, you benefit not only the land but also yourself; biological capital is the one form of capital gain no government can tax, even though it is the most productive.

# 40

## Holistic Land Planning
### *Designing the Ideal Layout of Facilities for a Grazing Operation*

ONCE UPON A TIME farms were planned around the original homesite, and fields were dictated by roads and tracks or hedgerows and more recently by the machinery used to work them. Ranches were planned around homesites and handling facilities as well but with livestock ranging freely prior to the development of barbed wire fencing. Then fencing went in according to where the water was and more recently where "range sites" differed, the belief being that certain soils and plant communities needed different grazing regimes to limit damage from overgrazing or overtrampling.

Now, where livestock are involved, you have the opportunity to rethink these developments according to a whole new set of conditions and priorities. The time factor in grazing and trampling means that different soil or range types do not need to be isolated. Animals are no longer in any one place month after month and cannot do the damage they once did. You can consider new fencing layouts that minimize the number of water points required and that give you much greater flexibility in your grazing planning and more options for improving animal performance. You can anticipate the facilities you need to have in place as animal numbers and herd sizes increase, and thus avoid costly redesigns in the future.

If you are not to be tied forever to an inferior layout of facilities and infrastructure, which can have many hidden costs associated with it, it pays

429

to start afresh and plan now based on what we have learned over the last fifty years or so and guided by your holistic context. The plan you create will be for the long term (a hundred years or more) and will be implemented in steps and stages, determined largely by your annual financial planning, but also your grazing planning, which you can initiate with the infrastructure you have now. The changeover from old to new need not be costly; in fact it can usually be accomplished in ways that earn money.

After going through the planning procedure outlined in this chapter and covered in detail in the *Holistic Management Handbook*, you may find that your existing layout is the best that can be devised. In most instances, however, even minor revisions will make a big difference in efficiency and profitability. My own experience bears this out.

When I took over the last ranch I owned, I inherited a fence and water layout that greatly decreased my efficiency as manager and was very expensive to maintain. There were eighteen water points for cattle and only two that were available to wildlife, whose presence was one of the main reasons I bought the place. Along with all those water points was a complex of roads and tracks that had to be maintained as well as water pipelines, troughs, and float valves. Once I had completed the planning process outlined here, the ranch only had three water points serving cattle, and five lesser ones serving wildlife, and far less pipeline, road, and track to maintain. The annual savings in time and dollars created by the new plan was considerable.

But I did not go out and borrow money to make the changeover. I moved toward the new plan gradually, generating the capital from the land to do so and only adding a new feature when my financial planning indicated it would make money.

## The Planning Process

Holistic Land Planning involves four basic steps. The first is devoted to gathering information and preparing planning maps and can take several months to complete. In the second step you hold a brainstorming session where you create many possible layouts for the planned developments. In the third step

---

### When Fencing Is Not an Option

*If you are involved in the management* of desertifying pastoral lands where fencing is not an option—due to distances involved, wildlife movements, or social factors—and herding is in line with the holistic context of those running livestock, you will find many of the ideas in this chapter relevant. The same applies if you are managing a national park, wildlife sanctuary, or game ranch where it becomes necessary to use livestock as the main tool to regenerate land and wildlife habitat.

In both cases even though fencing is not used you will still need to divide up the land into "paddocks," demarcated by natural features, for purposes of planning the livestock moves. The Holistic Land Planning process will help you do this more effectively.

---

you create the ideal plan, using the information gathered in the first step to help determine what aspects to incorporate of the many layouts brainstormed in the second step. In the fourth step you gradually implement the plan using the Holistic Financial Planning process to determine the order of implementation.

To properly plan you will need more information than I provide in this brief overview. You can get assistance, including self-teaching materials in paper or digital format (using map planning software), from the Savory Institute, and training through the Savory Hub Network. The process described in the following steps is for those planning on paper, but it can easily be modified for digital planning.

## Step 1. Gather the Information

You will need a topographical map of suitable scale that shows the natural features of the ranch. Mark in any developments such as public roads, railways, and homes that it would be illegal or impractical to change, *but nothing more.*

You will also need tracing paper or clear transparencies that you can lay on top of it. There are basically three tasks to complete:

- *Identify and map factors affecting your plan.* These will include such things as weather patterns (prevailing winds, areas under snow), wildlife breeding and birthing areas, croplands, heavy predation areas, timber extraction routes, boundaries showing ownership, mineral rights, recreational access areas, and so on. Mark in all these factors on overlays, not on the map itself. Create one overlay that shows the essential features of the future landscape you hope to produce—where croplands will be, heavily wooded or brushy areas, or open grassland areas, and so on.

- *Determine herd and cell sizes.* Determine optimistic future stocking rates so you can determine the size and number of grazing units (or *cells*) and handling facilities. Next, project future herd sizes, so you can determine how much water needs to be available in any one place. (Note that no one today has any real idea of what the future potential stocking rate could be as the land improves, but you should be safe in assuming it will be likely to at least treble.)

- *Make a list of the infrastructure needed.* After completing the preceding tasks you will have a good idea of what to include on your list. Keep the list handy for the next step.

## Step 2. Brainstorm a Series of Plans

The aim is to create as many different layouts as possible, with the help of as many people as practical divided into planning groups. Brief them on the factors you identified in step 1, plus the infrastructure needed, and provide each group a dozen or so copies (or tracings) of the map containing only natural features and absolutely fixed features. Follow the steps for brainstorming mentioned in the last chapter to ensure you do not stifle creativity but rather maximize the flow of ideas you need to make this session a success. When the

## Radial Layouts and Grazing Cells

*Many of the ideas presented* in this chapter and the next grew out of my early work with a very large ranch of 400,000 hectares (1.25 million acres) owned by the Liebig Company in a very dry area of Zimbabwe. I had the task of figuring out how to plan a fencing layout to accommodate the grazing planning for sixty thousand breeding cows. By law the ranch had to plunge dip all cattle once a week to prevent tick-borne diseases. On average twelve thousand animals were being dipped per day, and those animals also had to drink and fill their bellies. The average walking distance to and from a dip was fourteen kilometers (nine miles), which took a toll on animal performance.

To minimize costs, I designed a layout that had fences radiating out from a central waterpoint that also included a plunge dip. Radial layouts had been condemned by nearly everyone, but I reasoned that time on the land close to water would be the same as far from water and that we were likely to see the healthiest land close to water where animal impact would be higher. The amount of land overtrampled by the gates would be so small it could be sacrificed. This proved to be correct when we tested this layout on part of the ranch, and so I moved forward with planning the entire ranch, but my first plan proved to be unworkable.

Although I used distance to water as a crucial factor, my plan would result in increased running costs year after year with all the infrastructure needed. *But water points could be moved!*

That led to the development of easy-to-make transparent planning circles whose radius represented the maximum distance cattle would have to walk to water at the center. We could now move these circles around on a topographical map to find the best place for siting each grazing unit and its water point. That made the planning much easier, and a much more creative exercise, which in turn resulted in more practical, affordable, and effective land plans.

A legacy of this early attempt at large-scale land planning is the term *cell*. On paper, the layout of the perimeter fences separating each water point looked like the top view of a honeycomb with its myriad hexagonal cells, as shown in figure 40-1, so I took to calling the area in which one herd would run by that name. Planning for livestock grazing, not fencing layout, continues to define a cell, which is any piece of land on which livestock moves are planned as a unit and recorded on a single grazing chart. Division of a cell into smaller areas (what I call paddocks) for grazing planning purposes is essential too. They can be any shape or size, but the radial shape still has many advantages in large, dry areas. You can shape paddocks to fit rough and hilly country and still have them feed into a common center.

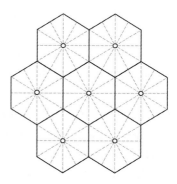

**Figure 40-1.** The first planned layout of radial fences looked like the top view of a honeycomb with its hexagonal cells, and so I called the area in which one herd would run a cell.

groups have exhausted all possibilities, have them draw up a final layout that takes into account all the existing developments the brainstorming ignored.

## Step 3. Design and Select the Ideal Plan

Over the next weeks, make overlays that incorporate the best features from the brainstormed layouts, and superimpose them on a map showing existing facilities. You should be able to come down to two or three options. Consider the following:

- Can you foresee a sequence of development that will keep prior structures in use until it makes sense to replace them?

- Review water supplies. If you run your animals in a single herd all the time, or amalgamate herds in a drought, do you have enough flow to sustain the higher numbers?

- Are the grazing units (cells) laid out in a way that will facilitate development of many divisions, or paddocks (permanent or temporary) over time?

Check each plan against the overlays you created in step 1. After doing so, you will probably draw up new plans that combine the best features of the others. Set them aside for a few days or weeks and then come back and create the final plan, checking to make sure that what you have on that plan matches the reality on the land.

## Step 4. Implement Your Plan

Once you have settled on the ideal plan for the future, you can begin the gradual process of changing over from the old to the new. Commonly the cost of the changeover is a major limiting factor in the rate of change. However, if adequate money is available, the rate of change need not be slow, but it should be sound and in line with your holistic context in every respect. In creating your annual financial plan you will be allocating money toward the desired developments in a different way. Normally you would regard most of the physical structures to be developed as capital expenditures—the capital coming from your previous earnings or more commonly from an outside source that involves repayment with compound interest. Eventually capital expenditures yield a return on the investment, although there can be some delay before this return is seen. Provided a profit can be penciled in, the capital is usually invested. This approach can be very costly.

Fortunately for most ranchers and farmers, having a lot of capital available up front is no more necessary than it is in a small start-up company run out of a home. No development—either a new one, or moving an old one—should cost you outside money unless you choose to apply it. Over the years, the land itself should be able to generate most, if not all, of the funds you will need.

Implementing your plan in this gradual way, and generating the capital to do it from the land, means that you can continue to operate at a profit year after year, and that each year other needs also continue to be met.

---

### Before You Implement Your Plan

*You don't have to have your long-term land plan finalized* before you start to plan your grazing. In fact, there are good reasons to plan your grazing using the infrastructure you currently have so that you begin increasing land and livestock productivity and the added income that brings. Almost all ranches are overcapitalized because of the belief that overgrazing results from running too many animals. As a consequence, ranchers commonly have too much money tied up in the land, buildings, wages, vehicles, and machinery, and too few animals to generate adequate income. Many try to boost income by improving individual animal performance. But no amount of improvement significantly changes the fact that *there are too few animals earning income.* Double the number of animals and the picture changes dramatically. And it does so even if individual animal performance drops, as it sometimes does.

Long ago, in Namibia, I worked with the mathematician son of a ranching client to determine that doubling cattle numbers was consistently more profitable unless individual animal performance dropped more than forty percent, *which I have never seen happen under Holistic Planned Grazing.* So, before you even begin to implement your future land plan, plan your grazing and start increasing animal numbers—through purchase or leased grazing. This is the quickest way to generate capital from the land to finance improvements.

---

## Conclusion

Many people try to avoid any long-term planning of this sort, especially when the land involved has a considerable amount of infrastructure in place. These people automatically assume that any changes will be very costly. I hope, after reading through this overview of the Holistic Land Planning process, that you now understand that the cost of the planning is very low indeed—mainly a few pencils, paper, maps, and overlays. The cost of *not* planning, however, can easily amount to hundreds of thousands, or even millions, of dollars over the years. Likewise, as you gradually implement your ideal plan, its greater efficiency and soundness will cause income to rise and running costs to fall. Few can afford not to plan afresh.

# 41

## Holistic Planned Grazing
### *Getting Animals to the Right Place, at the Right Time, and with the Right Behavior*

*When livestock are included in the whole you are managing,* their movements must be planned. If left in any one place too long, or if returned to it too soon, they will overgraze plants and compact and pulverize soils. But more than this is at stake. The traditional goal of "producing meat, milk, or fiber for profit" generally becomes a by-product of more primary purposes—regenerating soils and harvesting sunlight to create a landscape that can sustain you and future generations.

In the process of creating a landscape, you must also plan for the needs of wildlife, crops, and other uses, as well as the potential fire or drought. To harvest the maximum amount of sunlight, you have to decrease the amount of bare ground and increase the mass of plants. You must time livestock production cycles to the cycles of nature, market demands, and your own abilities. If you seek profit from livestock production, as most ranchers do, you will need to factor that in too.

Because so many factors are involved, and because they are always changing, most people throw up their hands at the idea of planning anything so complex. One can easily be swayed by those who say you can ignore all the variables—you'll do all right if you just watch the animals and the grass, or if you just keep your animals rotating through the pastures. What a relief! You hate planning anyway. But choosing to ignore the whole, and the many variables that influence it, is not the answer. In this case what you don't know *can* hurt you. And hurt you plenty.

Holistic Planned Grazing is an adaptation of a formal military planning procedure developed over hundreds of years to enable the human mind to handle many variables in a constantly changing, and often stressful, environment. The technique reduces incredible complexity step by step to absolute simplicity. It allows you to focus on the necessary details, one at a time, without losing sight of the whole or of what you hope to achieve.

Each of the factors influencing your plan—when you expect to breed and wean, when and where areas will be covered in snow or threatened by fire, when and where antelope are dropping their young, when and where ground-nesting birds are laying, when and where you will need to create herd effect, or graze or trample a crop field, and so on—are recorded on a chart. This provides a clear picture of where livestock need to be and when, and this determines how you plan their moves.

A good plan can deploy livestock to reduce or cure excessive growth of problem plants, reduce brush and remove its causes, heal a gully, maintain wildlife habitat, or decrease grasshopper breeding sites, and at the same time produce a high volume of forage and quality animal products. Since most livestock owners want to be profitable, their animals must enjoy the best possible plane of nutrition. Planning must also routinely handle unexpected fires, flash floods, droughts, poisonous plant infestations, and other catastrophes. In low and erratic rainfall environments, droughts are not an unknown; they are predictable more than half the years and thus can be planned for.

Holistic Planned Grazing requires consideration of many factors simultaneously. Yet the human brain has difficulty working on two thoughts at once, and large numbers of animals, extensive tracts of land, and long periods of time are particularly hard to conceptualize, even singly. Clearly, a methodical planning process, *on paper*, offers the best hope in most situations. However, because many parts of the world are under the influence of partially or nonliterate livestock operators, there is a greatly simplified version of this procedure that requires minimal paper but a great deal of knowledge of the land and livestock. I will touch on this again later.

In this chapter, I have limited the discussion to the *principles* involved in Holistic Planned Grazing, rather than the *details*. Study these principles first, and then refer to the *Holistic Management Handbook* for the step-by-step instructions on developing your own plan. You can get assistance, including self-teaching materials, from the Savory Institute, and training through the Hub Network.

## The Planning Approach

My academic specialty, biology, had no history of forward planning, and I was unable to find any other discipline that offered a planning process we could adapt for use in managing grazing within a holistic context. But I did find what I was looking for in a planning technique developed at the Royal Military Academy at Sandhurst, England, and adapted it for this purpose. It thus represents several hundred years of experience in fields of battle, and now nearly five decades of use in agriculture have shown the same approach to be just as effective in managing complex land, wildlife, and livestock situations.

## The Aide Memoire for Planning Grazing

Because so many factors influence any plan, you cannot tackle them all at once, but as anyone who has reassembled a complicated machine knows, putting things in a sequence that is not carefully planned has risks too. Toward the end of the job you discover a piece that won't fit without the disassembly of half the day's work, or worse, a piece that appears important but remains in the parts bucket when you're ready to clean up and go home.

The expression used at Sandhurst for a guide that prevents those problems is *aide memoire*, which is derived from the French for "memory aid." This is much more than a simple checklist, because it gives a sequence for making decisions that takes into account the effect of one decision on another. The questions raised in the aide memoire demand creative, detailed thought, and they are arranged in a specific order so that the answers build upon one another. The aide memoire for Holistic Planned Grazing has undergone half a century of adjustment and development by ranchers, farmers, foresters, wildlifers, and pastoralists, and others using livestock to sustain themselves or regenerate the land they manage. The challenges experienced by thousands of people on several continents have influenced the development of the aide.

A main benefit of planning according to such a tested procedure is peace of mind. Using the aide, you can truly relax in the most alarming situations, have confidence in the process, and concentrate fully on one step at a time without worrying about something that might come first or get left out. You will find that the overall plan that emerges has covered every imaginable detail

---

### Planning in Emergencies

*I got a call one night* from a client after a fire had raged through about half of his ranch. It had struck worst in the areas where he had water for stock but spared some areas where water points had already dried up. He wanted advice on how much to destock.

When he met me at the local airstrip he was eager to take me off to see the burned area and the cattle as soon as possible, but I asked to see his grazing plan. "No," he said. How could that help, if I hadn't seen the fire damage or his livestock? He got extremely hot at the suggestion that one more look at blackened ground and idle cattle would waste my time and his money, so perhaps he'd rather muddle through by himself.

He did not of course have a plan. The charts we had developed together had disappeared, but by luck his wife retrieved the aide memoire, reeking of tomcat, from the bottom of her son's toy chest. I got out a fresh chart and insisted, against vigorous protest, that we now plan step by step. Gradually, however, his protests weakened as a picture of his ranch began to emerge on the planning chart he'd been taught how to use but hadn't thought important. His enthusiasm really began to mount after we laid out all the problems and began to plot actual cattle moves.

It was a very sheepish man who finally accompanied me out onto the land to confirm some final judgments. Without any input from me, other than the knowledge that the longer he put off planning, the more hours I would add to my bill, he had proved to himself on paper (followed by a couple of final field checks), that he could carry his whole herd through without risk or leased pasture. In a couple of hours of planning, which brought clarity to the confused picture in his mind, he had saved many thousands of dollars.

in a logical fashion and represents the best plan possible for the present. This ability to concentrate completely and confidently on one point at a time bears fruit, particularly in emergencies, when a tendency to panic and lose focus can prove disastrous.

## The Planning Chart

As you cover each step in the aide memoire, you record the details on the planning chart, the principles of which are shown in figure 41-1.

The body of the chart is divided into seven-month sections for the sake of size but across the chart smaller divisions account for time down to the day. The rows represent paddocks, or grazing divisions (whether fenced, or unfenced if herding). At the base of the chart and on the left and right sides are space and format for various planning figures and calculations, including the plant recovery periods you will use and the grazing periods derived from them. In the central body of the chart covering both days and areas of land, all problems and needs can be shown in color-coded marks. An orange line across paddock 3 in April might show poisonous plant danger. A brown line across paddock 4 in certain months might note lack of water. A red line across some months in any paddock might indicate animals should not be there over that period because antelope are fawning. Then within the context of all these factors, you plan the livestock moves.

## When Do You Plan?

Grazing planning can be done at any time, but you commonly plan twice a year: at the beginning of the season when most growth occurs (*the growing season plan*) and again at the end when growth has greatly slowed or stopped

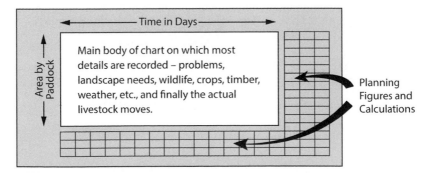

**Figure 41-1.** The layout, in principle, of the grazing planning chart.

### Using the Chart to Plan Backward

*The first inclination of anyone* planning livestock moves is to plan them forward, with animals moving from one paddock to the next over the days or months ahead, but Holistic Planned Grazing is different. Animal moves can be plotted both backward and forward. Picture the planning chart at this point with every single problem, issue, danger, or limitation color coded in every paddock, and vertical color-coded lines showing the months in which livestock are breeding, weaning, and so on. If for instance, these lines show you that cows are dry and pregnant you know they can be moved anywhere easily because they are most hardy at that stage, and if they drop in weight they can calve more easily. Later, as you approach breeding time, you can make sure they come to the bulls on a rising plane of nutrition to conceive more easily. When the lines show you are calving, lambing, or kidding, you will want to make sure livestock moves are always to adjacent paddocks, to minimize mismothering. Armed with this clear picture, from all points of view—livestock, land, weather, wildlife, crops, or any other activities on the land—you can plan the animal moves backward in places, forward in others. If, for instance, you want to wean in the paddocks near headquarters, plan to have animals there at weaning time, and then figure out where they will have to come from to get there—planning backward. Because you note all moves as penciled lines on the chart, you can think, erase, and adjust. And when you are happy with all the plotted moves your plan is complete.

Think of the chart as depicting a minefield with all the mines (problems and issues) clearly marked so that you can easily step the animals through it over many months, getting them to the right place, at the right time, for the right reason, and with the right behavior (so that they regenerate soils and keep plants healthy).

(*the nongrowing season plan*). The aide memoire has been separated into two to reflect the modifications needed in planning for each season.

Make the growing season plan shortly before the onset of main growth. In this plan your aim is to grow as much forage as possible, and you do not have to plan to a specific date. The plan remains *open-ended*, because you don't know when growth will slow or end or exactly how much forage will grow before that date.

Make the nongrowing season plan toward the end of the growing season, when forage reserves available for the nongrowth period will be known. In this plan, you ration out the forage over the months ahead to a theoretical end point, making this a *closed* plan. That end point should be a month or more after your most pessimistic estimate of when new growth could occur. The additional "month or more" becomes your *drought reserve*.

Figure 41-2 illustrates the open-ended and closed plan principle, showing how the time reserve for drought fits in. In some regions where rainfall is

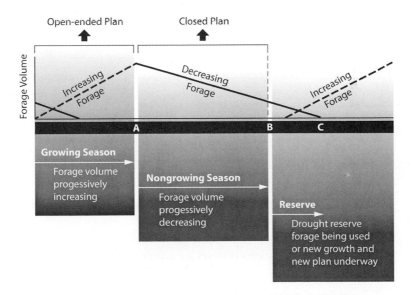

**Figure 41-2.** The growing season plan is open-ended because you don't know when growth will end or exactly how much forage will grow before that date. The nongrowing season plan goes into effect once growth stops (point A). It is a closed plan in that you are rationing out a known amount of forage over a specific period of time (from A to B), which should extend a month or more later than you estimate new growth could occur. This additional time (from B to C) is the drought reserve.

## Planning a Drought Reserve in Days Rather Than Area

*Droughts, especially in brittle environments,* are so common that the aide memoire for the nongrowing season assumes that the next growing season will arrive late and has you plan accordingly.

Traditionally ranchers have planned for drought by reserving areas where forage can accumulate. But there is a high cost in taking this approach. Consider a ranch with one cell of twenty paddocks, of which two are withdrawn for a drought reserve, and what would follow:

- Because you plan from a fixed recovery period you would have to hold animals longer in the eighteen paddocks remaining to ensure plants are given adequate recovery time from grazing and trampling.

- Holding animals longer in every paddock automatically depresses the production on all plants grazed in all paddocks. A few ounces less grass produced on each square meter adds up to many tons of forage lost on the ranch overall.

- Holding animals longer in all paddocks also depresses livestock condition because the plane of nutrition is lower.

- If the growing season arrives on time, or even early, you have two paddocks that have lost a season of animal impact and grazing, leaving overrested forage that will likely depress livestock condition when grazed.

- Should a fire occur in the two reserved paddocks the forage would be lost.

For all of these reasons, area reserves are a very costly form of drought insurance.

Reserving a month or more of forage spread out over the entire ranch—the time reserve shown in figure 41-2—makes much more sense. It reflects the fact that we also measure drought in days—days until the rain comes, days until growth starts. Your banker doesn't ask, "How many hectares do you have in reserve?" but simply, "How long can you hold out?" No production is lost on the land or from livestock, rainfall is made more effective over the entire ranch, there is no risk of fire wiping out any reserved area, and if drought does not materialize no areas have accumulated an excess of aged plants leading to reduced animal performance.

In the event of a midseason drought, when the rain is very poor or ceases altogether, there are still measures you can take to save the situation well before the end of the year. Immediately replan your grazing using a closed plan that carries you through the next growing season and drought reserve. In doing so you will know if it can carry all the animals you have now, and if not, by how much you have to reduce numbers. A slight reduction now leaves many more days of grazing for the animals that remain. Thus the earlier you reduce numbers the fewer you have to cut.

very low and unreliable, there may be a major overlap of the drought reserve in the closed plan and the start of the next open-ended plan due to the necessity of making drought reserves extend to as long as a year or more.

## The Planning Procedure

The planning generally takes a person two to three hours twice a year to complete. Implementing the plan is of course a continuous process, and it will always require adjusting. The first twelve or thirteen steps (the growing season aide memoire has one more) are the actual planning steps, and they center on these questions:

- What sort of landscape are you trying to create?
- How much total forage will the cell, or cells (grazing units), have to supply in the current planning period?
- How much forage will an average hectare/acre of land have to supply?
- How long will standing forage at the end of the growing season last in a nutritious state (including reserves for a late start to the growing season, drought, fires, wildlife, and so on)?

- How long will animals spend in each paddock, and when will they return (the vital recovery period grazed plants require)?
- Where and when will you need to concentrate animals most to maintain healthy grassland, reduce weeds or woody vegetation, or prevent soil erosion.

The next-to-last step in both the growing and the nongrowing season aide memoires provides guidelines for implementation and monitoring of the plan, and how to make the necessary adjustments as you move through the season. The last step provides guidelines for keeping a record of what happens, which you will use to improve future plans.

## Monitoring and Controlling the Plan

As you implement your plan be on the lookout for any assumptions you have made that prove faulty—and there will always be some. You may, for instance, have misjudged the quality of a paddock and find you have to move animals out of it before they have grazed their allotted number of days. If you move out a day early in only one paddock, you should be okay. But if this happens in several paddocks, particularly in the growing season, you need to take action. This is because you determined the grazing periods to use based on a predetermined recovery period (or range of recovery periods) that you do not want to violate.

Every time animals move out of a paddock a day early means a day of recovery has been lost in *every* paddock, including the one they were just in. If this happens over several paddocks, all paddocks have lost several days of recovery. Time to act and control your plan! This you do by going out to see which paddocks can take a few more days of grazing in the period ahead than you had originally planned for them. Each day animals are held longer in a paddock now adds a day of recovery to every paddock, and you should be able to get back to providing the recovery time every paddock needs.

Given good planning and control, replanning only becomes necessary in the event of some major disturbance outside your control, such as a fire or mid–growing season drought. When replanning does become necessary you simply get out the aide memoire and start afresh.

## Holistic Planned Grazing with Herding

Using herding rather than fencing to control livestock movements is of course an old practice, but one few ranchers use today, citing the cost of labor as the main reason. Nonetheless, there are three key advantages that I hope will encourage you to change your mind if you have ruled out the idea of herding:

- Herded animals provide consistently higher animal impact.

- Herding provides you with hundreds of virtual "paddocks" that if fenced would come at such great economic and environmental cost, particularly for wildlife, that they would not be feasible.

- Because herders can control precisely where animals graze within a paddock, you only need to create a few virtual paddocks to ensure adequate recovery periods. Because there are so few paddocks, grazing periods for each paddock will be rather long, but herders will graze the animals for no more than three days on any section of land within a paddock, so grazing periods from the plants' point of view remain brief and recovery periods adequate.

Over the past decade, the first Savory Hub, the Africa Centre for Holistic Management, in Zimbabwe, has learned much about herding livestock to regenerate large tracts of land and wildlife habitat using predator-friendly practices in the midst of abundant predators (lions, leopards, African wild dogs, etc.). The aim was to be able to share what they were learning with the agropastoralist communities surrounding the Centre's learning site, which they started to do in 2006.

What we have developed and learned to date about Holistic Planned Grazing with herding is captured in the materials now available through the Savory Institute and the training programs offered through Savory Hubs, including the Africa Centre. I provide highlights here of the key modifications to the grazing planning that need to occur when working with whole communities that herd their livestock and plan their grazing with virtual paddocks.

## The Context for Planning

The context we based our mobilization, materials, and training on in southern Africa is widely shared among Africa's pastoral regions and will be similar wherever pastoral populations still exist:

- The land involved is under communal ownership, or tenure, rather than privately held.
- The environment is brittle.
- Communities farm crops as well as livestock (agropastoralism).
- Livestock are herded on foot, which requires mastery of low-stress herding and handling skills that are sometimes limited in agropastoral communities.
- Livestock are placed in a movable enclosure at night for protection from predators and thieves.
- No or very little fencing is used to demarcate paddocks.

## Land Planning

In this context, we now refer to land *and social* planning for grazing because of the agreements that need to be in place at the start, such as those between people in the community who are and are not participating, between herders and livestock owners, and around water point maintenance. Some infrastructure is still required and needs to be planned for: bulk water storage if water has to be pumped; movable livestock enclosures, if animals need to be penned at night; and, in some cases, handling and dipping facilities.

A map identifying paddock boundaries is essential in all cases. Fortunately, because animals are herded, relatively few (five to ten) paddocks are needed. There is a need for access to water from every paddock, which often involves demarcation of sacrificial corridors that livestock move through to avoid overgrazing and overtrampling wider areas through repeated movement. The map and boundaries can be drawn by hand on butcher paper, as shown in figure 41-3, or whatever else is available. In very dry areas where the grazing unit or "cell" would likely cover an extensive area, hand-drawn maps work less well than satellite maps available from the Internet.

**Figure 41-3.** Map showing virtual paddock boundaries created by a community for planning their grazing. Zimbabwe.

## *The Grazing Planning Procedure*

The aide memoire for communal lands grazing planning includes twelve steps, one of which is applicable to the nongrowing season only. Both the aide and the planning chart have been reduced to the bare essentials and most of the arithmetic eliminated. This becomes possible because there are so many virtual subdivisions within paddocks that the recovery period will always be long enough to ensure plants are not overgrazed nor soils overtrampled. Even though many pastoral communities have low literacy rates, there are generally at least a few members who are literate (and numerate) and can lead the planning process. These people are often part of a grazing committee, which is responsible for planning and monitoring implementation. Once the plan is complete the planners share it with the whole community, or at least all members impacted by the plan.

## Using Movable Livestock Enclosures to Create Impact

*Although the aim* in developing movable livestock enclosures was to prevent predation and theft, once they began operating successfully other benefits emerged.

We could regenerate severely damaged sites simply by placing the enclosure on them. Longstanding, hard-capped, and crusted soils would sprout grasses in the following growing season, as shown in color plate 19 We could also place the enclosure over gullies, or seriously eroding sites, and see them begin to recover too in the following season. This cost us no money or extra effort as the enclosure had to be moved somewhere every few days.

We quickly realized when working with our neighbors in the nearby communal lands that the enclosure could be moved across crop fields and not only fertilize them but prepare the soil too, eliminating the need for plowing. Yields of the staple crop, maize, increased dramatically. Figure 41-4 shows the difference between maize plants grown on the animal-impacted area (right), compared to those outside it (on the left). Not only did yields double and even quadruple in some cases, labor was reduced as women no longer had to carry manure to the fields prior to planting.

**Figure 41-4.** The dense, healthy maize plants mark where the enclosure was placed on this field. Zimbabwe.

### Herding to the Grazing Plan

The planners generally include the herders who help decide the moves through the paddocks over the months planned. The herders, using their knowledge of the land, then decide where and when to graze *within* each paddock, where to water, and where to place the movable nighttime enclosure. While grazing each day they follow two simple rules:

- Keep the herd reasonably concentrated.

- Do not graze the herd in the same area (virtual subdivision) for more than about three days.

Any slight overlap in the areas grazed within a paddock is not a concern because few plants would be overgrazed, and animal impact is so high.

In this manner plants *anywhere* in the grazing unit, or cell, will not be grazed for more than about three days followed by adequate recovery time, which has been guaranteed by the planning.

## Conclusion

From the very first series of lectures I gave in the United States in the late 1970s, and even earlier in Africa, I have stressed that the Holistic Planned Grazing procedure underpins all my work in grazing situations and all my claims for success. Yet the vast majority of research projects conducted, allegedly to study my methods, have ignored the heart of the matter—this planning process and the continual monitoring allied to it. Instead researchers have set up countless trials or on-farm studies of rotational grazing systems of one form or another, but paid no attention at all to the planning procedure.

Many of the rotational grazing systems are derived from the work of André Voisin, who first discovered the link between overgrazing and time. He developed *rational* (meaning well-thought-out) grazing in response to this discovery. He would probably turn in his grave to see that so many have converted it to rotational grazing when he spoke out so vehemently on its dangers. Table 41-1 attempts to show where rotational grazing, rational grazing, and Holistic Planned Grazing differ.

**Table 41-1.** Rotational, rational, and Holistic Planned Grazing—how they differ

| | Rotational Grazing | Rational Grazing | Holistic Planned Grazing |
|---|---|---|---|
| Grazing periods are based on: | Number of paddocks and desired rest period | Recovery periods needed during fast and slow growth | Recovery periods needed during fast and slow growth |
| Adjustments to grazing periods based on: | Height of grazed plants in paddock | Daily growth rate of plants | Daily growth rate of plants, livestock performance, and/or wildlife needs |
| Stocking rate based on: | Estimated dry matter intake and/or rainfall received | Animals days per hectare/acre (ADH/ADA) | ADH/ADA available for the nongrowing season plus a "time reserve" for drought, and *effectiveness of water cycle* rather than rainfall record |
| Animal nutritional needs addressed by: | Estimated dry matter intake and daily monitoring of animals | ADH/ADA estimates and daily monitoring of animals | ADH/ADA estimates, daily monitoring of animals, and allocating best paddocks for critical times, then planning backward from those critical periods |
| Use of herd effect for land restoration | Not planned | Not planned | Incorporated into plan—essential in brittle environments |
| Wildlife and other users/uses | Not planned for | Not planned for | Incorporated into plan so livestock can be used to enhance |
| Drought planned for by: | Reserving grazing areas | Reserving time (days of grazing) spread over all paddocks | Reserving time in all paddocks, and ADH/ADA estimates at end of growing season in a closed plan |
| Performance in brittle environments (most of the world) | Breaks down in brittle environments | Breaks down in brittle environments | Does not break down in any environment |
| Performance in less brittle environments | Good short term, but likely to break down long term | Good short and long term | Good short and long term |
| Fire prevention: | Not planned | Not planned | Routinely planned |
| Management decisions based on: | Gaining high production (forage, meat/milk, fiber) from the land | Gaining high production (forage, meat/milk, fiber) from the land | A holistic context and actions aligned with it that are socially, environmentally, and economically sound |

Holistic Planned Grazing was developed nearly fifty years ago. Since that time, it has undergone continuous refinement based on the experiences of thousands of people on several continents—an effort that continues to be co-ordinated by the Savory Institute. This procedure offers the simplest way we have found for managing the complexity that exists when livestock share the land with wildlife, crops, and other uses. This procedure will lead to the best possible plan in the most difficult and seemingly hopeless situations. Even when the rains have failed to come at all, and even through times of crisis, including war, this planning procedure has never failed me. Nor do I believe it will ever fail you.

# 42

## Holistic Policy Development
### *Creating Sound Policies*

P*OLICIES ARE CREATED* in an attempt either to solve a problem or to prevent a problem from occurring by prescribing a course of action to follow. In either case, a policy developed within a holistic context generally fares better because the course of action will be assessed in terms of the desired outcome, rather than in terms of the problem itself.

Policies designed to prevent problems are used in a variety of situations. In your home, you might have a policy that limits the number of hours your children watch television (to prevent them watching too much of it). In the workplace, you might have a policy for handling customer complaints (to avoid confusion), a policy covering safety precautions when working in a dangerous area (to avoid accidents), and so on. Because these sorts of policies are fairly straightforward and easy to form, I will not dwell on them.

Policies designed to solve problems are more common than they should be, because all too often they fail to tackle the underlying cause of the problem and merely work to suppress symptoms. When enforced, they may in fact alleviate some of these symptoms, but because the cause hasn't been addressed, the symptoms (and the problem) persist. Another policy may then be developed but it too fares no better for the same reason. The many policies created to combat crime are an example. When more and better-equipped police officers don't solve the crime problem, we build more prisons, and when that doesn't solve it, we enforce stiffer penalties on repeat offenders, and so on. Each of these policies might help reduce crime, but because none of them address the underlying cause, which is likely to be related to complex social or economic factors, the problem never goes away.

When policies aim to solve a problem, the holistic context will help put the problem in perspective, but a thorough analysis of the policy using the Holistic Management framework is also needed. This should start with a diagnosis of the problem's root cause. Then each of the actions prescribed in the policy needs to be passed through the seven context checks. If none of those actions address the root cause of the problem (cause and effect check), the policy may require fairly drastic revision. If an action fails any of the remaining checks, slight modifications may be all that are required to make the policy economically, socially, and environmentally sound and in line with the holistic context.

Nowhere is such an analysis more needed than when forming policies that attempt to solve natural resource management problems—the focus of this chapter. In policy analysis courses I facilitated in the United States in the 1980s for university educators and researchers, government officials, and advisers—those who most often provide technical advice to politicians—I had the participants analyze the following policies actually implemented at that time.

1. With the aim of rescuing a vanishing breed of trout in a wilderness area, a predatory trout was being poisoned.

2. To stave off the day when a large and vital dam inevitably fills with silt, its wall was being raised. There was some controversy on whether to raise it a few feet at a time so as not to immediately destroy eagle nesting sites, or all at once (the cheaper course).

3. To help rid rangelands of noxious plants, livestock numbers were being reduced and the plants poisoned. At the same time ranchers were being encouraged, with cost-sharing programs, to invest in more fencing and water points.

4. To destroy grasshoppers that threatened crop and forage yields, an aerial spraying program was being implemented.

5. To reduce brush encroachment (mesquite trees), a liquid herbicide was being applied.

6. In order to heal the land after a severe drought, extension agents were advising a prolonged rest period.

Every policy except the last prescribed actions involving the tool of technology and in nearly all situations the context checks would show it to be counterproductive. All the proposed solutions addressed a symptom, and they failed the cause and effect check. None passed the sustainability check either. Where profitability and ecosystem stability were factors, the weak link and marginal reaction checks would have eliminated most, if not all, of them.

After the brief exposure to holistic policy analysis given in the courses, all the participants—numbering close to two thousand, and including some of the very people who had masterminded these policies—came to similar conclusions. All agreed that public funds had been wasted on these policies because none of them could succeed in the long run. All agreed that the sample was not biased. Such policies were typical, they said, both in the United States and in countries the United States was assisting. One class of thirty-five, after some discussion, actually stated as a group that *they could now see that unsound resource management was universal in the United States.*

The only point on which they were not in agreement was in determining who actually produced the policies. But when asked to think about where we might start to rectify such a situation, most participants arrived at a similar breakdown. About seventy-five percent believed the policies ultimately emanate from our educational system and the professional advisers and consultants it produces. About twenty-five percent felt that, though this accounts for most cases, vested interests (financial, professional, lobbyist groups) also influence a certain proportion of policies. Today, because so many educational institutions receive corporate funding that influences both curriculum and research, these percentages might differ significantly.

Even assuming that some participants might have publicly accepted the majority opinion while not entirely agreeing, the consensus indicated a real need to rethink our approach to developing resource management policies.

But that did not happen in the 1980s, or even in the two decades after when I also trained senior foresters in India and senior civil servants in Lesotho (southern Africa). I believe this lack of movement is tied to the fact that those who attended my courses worked within institutions, which reflect the general views of the society in which they are established. If new ideas emerge that run contrary to societal beliefs, institutions not only cannot adopt them, they also tend to lead the opposition to them.

Institutions do not adopt the new idea until a significant shift occurs in how a society views the new idea, which, paradoxically, individuals within those institutions help bring about. We were simply too early in the 1980s, but the time is now ripe for a shift to occur.

## Analyzing an Existing Policy

I could take any one of the six policies in the preceding list to illustrate the process in detail, but I'll use the grasshopper-spraying policy because it is large in scope and representative of many others. Emergency appropriations in the millions of dollars to spray malathion or some other pesticide on offending grasshoppers are not unusual. In this case, thirty-five million dollars was appropriated to address the grasshopper problem in the state of Wyoming.

### *The Cause of the Problem*

When diagnosing the cause of a problem involving an increase or decrease in the numbers of a particular organism, focus your attention on community dynamics. As you will remember (from chap. 13), populations cannot build in numbers unless an environment has been created that enables them to flourish. To figure out what that ideal environment is, you need to know something of the basic biology of the species. What stage in the grasshopper's life cycle is its weakest, and what conditions does it require to survive at that point? In grasshoppers, the weakest point is at the egg or nymph stage, so we look for the conditions that would promote survival of the eggs and nymphs. It turns out that the grasshopper species that tend to become pests prefer to lay their eggs in bare ground, which must remain warm and dry for the nymphs to hatch. Bare ground has steadily increased in Wyoming, as it has in most western states, and thus so have egg-laying sites.

---

### The Parallels with Development Projects

*All that I write about* policy applies equally to the development projects launched in so many countries that are designed to address a problem, such as poverty, hunger, land degradation, or recurring droughts and/or floods, and yet all too often the context is the problem, and the problem addressed is really symptomatic of a deeper problem. While the project may be successful in the short term success is seldom sustained.

The Millennium Development Goals, developed in 2000, were meant once and for all to address this lack of success. Fifteen years later, having only partially succeeded in reaching those goals, the United Nations launched new *Sustainable* Development Goals. One thing both sets of goals have in common was that each goal addresses a symptom of a deeper issue, which was not addressed in the first set of goals and does not appear to feature in the new set. That could change. Although I refer mainly to policy in this chapter, do remember that in analyzing why a development project is succeeding or failing, or in designing a new one from scratch, you can use the same process described in this chapter.

---

In most years, the majority of eggs laid in the bare ground, or the nymphs that emerge from them, won't survive because temperatures and rainfall are too high or too low. But when the right temperature coincides with the right amount of rain, millions of eggs will hatch, grow into nymphs, and eventually become grasshoppers. The cause of the problem is most likely to be the production of vast areas of bare soil between plants providing ideal breeding sites.

### *The Whole*

In this case, all of the people in the state of Wyoming are affected, but most directly the farmers, ranchers, and others whose livelihood has been damaged by the loss of forage and crops. The land within the state, and the biological communities it encompasses, serves as the resource base. The money available for implementing this policy is thirty-five million dollars, gleaned from state and federal treasuries.

## *The Holistic Context*

We can probably surmise that the people in the state of Wyoming would like to have stable families, prosperity, good health, physical and financial security, and so on. To produce all this, they are bound to require a healthy land base in which water and mineral cycles are effective, energy flow is high, and communities are rich in biological diversity—all of which would reduce the likelihood of grasshopper outbreaks.

Even such a simple statement as this is enough to enable us to use the context checks to discover if the course of action outlined in the policy is likely to produce desirable results.

## *The Actions Proposed*

The policy only prescribes one action—the use of a poisonous spray (the *tool* of technology) that kills grasshoppers.

## *The Context Checks*

The spraying clearly fails the cause and effect check because it does nothing to treat the bare ground—the underlying cause of the outbreak.

We might also predict a negative verdict on the sustainability check, particularly when we gauge the likely effects of spraying on community dynamics.

Wyoming probably hosts more than two hundred organisms that prey on grasshoppers in their various stages and help limit outbreaks,[1] but spraying does not offer the slightest possibility of benefiting any of these predators. It is likely to kill off high proportions of many and thus promote the hatching and survival of more grasshoppers. Spraying that also decimates a broad range of predators could lead to further outbreaks because prey populations generally recover before the predators. The next time ideal conditions prevail, an abundant egg supply and reduced predators will likely produce another outbreak. Clearly this policy fails the sustainability check.

Even without further checking, we can definitely call this policy unsound. The spraying program, apart from not addressing the cause of the problem, adversely affects community dynamics by reducing the diversity within biological communities. People face increased pesticide pollution danger. And a country that is struggling to balance its budget wastes thirty-five million dollars. Obviously we should seek other solutions.

Addressing the underlying cause of the grasshopper outbreaks would figure largely in the revised policy we would create using the Holistic Management framework to guide us.

### Revising the Policy

Given the generic holistic context already described, consider how you as a politician might approach the grasshopper problem. You would base your revised policy on actions that address the underlying cause of the problem, taking into account what you have learned from your analysis of the original. Then, given the possibility of a thirty-five-million-dollar appropriation, you might propose allocating it in the following way:

1. To soften the immediate impact of the grasshoppers you might use some of the money to directly compensate people who truly suffered damage to their livelihood. Suppose this costs twenty million dollars.

2. Of the remaining fifteen million dollars, five million could go to training farmers, ranchers, refuge and park managers, and so on, in management practices that would promote the generic holistic context and address the cause of the outbreak. Increased ground cover would reduce egg-laying sites. Increased ground cover would enhance the water cycle as well, which in turn would enhance the population of fungi and microorganisms that destroy grasshopper eggs. Increased biodiversity would help keep predation levels high. All of this would be in line with the holistic context and provide a long-term solution. More species of grasshopper might inhabit the increasingly complex communities that develop, but in lower numbers characterized by smaller, if any, outbreaks.

3. You could return ten million dollars to the treasury to help balance the budget and reduce taxation. If such a commonsense solution proved hard to sell, one million dollars might go toward a public awareness campaign.

From my own political experience, I believe it possible to sell such a program to voters far more easily than the poisoning that will add to taxation, endanger human health, and ultimately lead to more grasshoppers.

## Developing a Sound Policy

The six suggested steps for developing a new policy involve the same thinking involved in analyzing one, but the process in moving through those steps is also critical. Below is a summary of the steps and of a suggested process. It is followed by a real-life case of a policy I worked on with a group of Zimbabwean lawmakers that indicates how flexible the process needs to be.

### Step 1. Identify the Underlying Cause of the Problem

Put together a policy task force made up of people with broad experience and education, and possibly one or two with special expertise on the problem, to work through all six steps. It may take the group a few days to determine the underlying cause of the problem and they may come up with several causes, one or two of which may later be ruled out.

### Step 2. Loosely Define the Whole the Policy Encompasses

The task force completes this step in a brief session that also includes the next step. What people are affected, what resources are involved, and what money is available?

### Step 3. Create a Generic Holistic Context

The group develops a generic holistic context they believe those affected by the problem would agree to.

### Step 4. Draft an Outline of the Policy

*It is essential that once the first three steps are completed, discussion ceases.* Now task force members work independently to create their own outline of a policy, which they will then share with the larger group. This rule is essential to follow because discussion in a group will automatically gravitate to detail before there is big-picture clarity on a policy that will solve the problem and be acceptable to those affected by the policy. It is also essential because task force members with expertise in the problem being addressed usually dominate discussion, which discourages expression of new ideas and the deeper reflection needed for that big picture clarity. In my experience to date I have found that when people move the discussion to detail without first having established an overarching strategy, it is all but impossible to return their thinking to the big picture. In moving through this exercise, bear the following points in mind:

- *Time allotted for this individual work is important;* too brief and people don't have time to think deeply enough; too long and they lose focus and their ability to concentrate and think deeply.
- *Keep the focus at the strategic level;* think in broad principles, and then broad actions.
- *Make each individual solely responsible for the policy* so that the buck stops with him or her.

Then have each person think out the broad outline of a policy that would

1. address the underlying cause (or causes) of the problem;
2. be aligned with the holistic context, passing whatever context checks apply;
3. address the concerns of all those affected by the policy, and how to ensure their support of the policy; and
4. indicate whether the full policy could be developed using the generic holistic context or whether a real one is required and, if the latter, how its creation would be accomplished (this is commonly how point 3 is resolved).

When the group reconvenes, each individual, including those with expertise in the subject, presents his or her draft policy outline while the others listen for ideas that meld with their own. Then the group as a whole pools

their ideas together and agrees on a first draft of a policy in very rough outline, including how it could be developed so that public support and acceptance are assured.

### Step 5. Develop the Policy

When the policy is a simple one, limited in scope, the task force can be
responsible for adding the detail needed to finalize the policy. This is likely
to involve months of work by staff members and others, including advisers
with specialized knowledge of the problem or the field(s) it represents. If it
is a more complicated policy that requires citizen involvement in its development, including the creation of a real holistic context, the task force will
create a plan for bringing that about in a way that ensures full participation.

Every action proposed in the eventual policy will need to be run though
the context checks to make sure the policy is aligned with the generic holistic
context (or the context created by the people most affected by the policy) and
is socially, environmentally, and economically sound.

### Step 6. Determine the Criteria to Monitor to Ensure the Policy's Implementation Is Successful

If the policy is sound, then obviously the problem will be resolved or will
never materialize. But what if the task force was slightly off the mark in identifying the cause of the problem, or what if its members missed the mark altogether? And what if the modifications they've made in getting to a final draft
fall short in achieving what they are supposed to achieve? Task force members
must identify *what* can be monitored for the earliest sign that the policy is
working. If applicable, they should also identify how they will know when a
policy has lived out its purpose and is no longer needed.

## A Land and Agriculture Policy for Zimbabwe

In 2012 I ran a policy workshop for thirty-five members of Parliament who
had come together in a government of national unity to ease a very sizeable
rift between two opposing parties. Zimbabwe's agriculture had collapsed following implementation of some disastrous policies by the previous government that had resulted in experienced farmers losing their land, idled and

derelict farms, the displacement of hundreds of thousands of farm workers, plus hundreds of thousands more as farm-related businesses also collapsed. Once considered a regional breadbasket, Zimbabwe was now importing much of its food.

I explained the holistic policy-making process to the group. But knowing I intended to have them tackle the land distribution and agriculture policy—the most contentious in the nation—I did not have them analyze the cause. To do so could only have resulted in violent disagreement between members of the two political parties. We started instead with defining the whole and creating a national holistic context. The whole included every citizen, all of whom were affected, as was the land base of the whole country. The money to implement a new policy would have to be generated from taxation or borrowed.

When we created this generic national holistic context expressing how all Zimbabweans wanted their lives to be and tied that to the state of the environment as it would have to be far into the future to sustain that way of life, the atmosphere changed from fairly hostile to one of palpable goodwill. Suddenly it was simply thirty-five Zimbabweans working together as people rather than as politicians.

At this point I reminded them of a speech made by their president three decades previously in which he stated that Zimbabwe had no greater problem than its rising population and deteriorating environment and that as politicians they did not know what to do. They could only rely on their professional advisers, but when things went wrong the politicians always got the blame, not their advisers. I pointed out that we had no advisers present, and at this point suggested they tackle the land distribution and agriculture policy to see how one might be developed holistically. To this they readily agreed in an atmosphere that had already soared above partisan politics. We proceeded without anyone raising a political point or taking a position. There were of course no lobbyists or pressure groups present, but had there been they would have appeared foolish had they tried to push their cause in that atmosphere.

We were now at step 4 and ready to draft the outline of a policy, based on the holistic context, that would not only enable agriculture to thrive, but would also amicably redistribute land from those now owning ten or so unproductive farms to those who had the skills to farm but were sitting idle, and could put hundreds of thousands of people back to work. We still hadn't discussed the underlying cause of the problem, which we all knew but couldn't talk about. Yet I realized there was still a way to address the cause. When they came back as a group from reflecting individually and shared their different ideas, I posed a few big-picture what and how questions.

With our national holistic context in mind, we arrived at the first obvious question: what is the role of government in this case? The answer was obvious: government's role was to create the policies that would help revitalize agriculture and distribute land equitably. And how could that best be done? After quite a lot of back and forth discussion it came down to two broad principles:

1. Use policy to reward farmer creativity and productivity.
2. Use policy to encourage farming practices that were in line with the holistic context and to discourage practices that were not.

Our task now was to come up with possible actions, without going into detail, that would create a workable policy based on maximizing farmer creativity and production while rewarding or penalizing farmer behavior based on whether or not it was aligned with the national holistic context. After a few hours we had a solid list of possible actions that were aligned with the holistic context and likely to lead to greatly increased agricultural production, while simultaneously addressing the problem of inequitable land distribution. These actions included the following:

- No income tax to be paid by bona fide farmers on income from agricultural products
- All bona fide farmers to pay an annual land tax based on farm size and agroecological region (rainfall and soil types well known in Zimbabwe)
- Land tax to be reduced each year based on the number of families living and working on a farm (not only to get families back on the land but to encourage the use of labor over machinery in a country producing neither machinery nor oil)

- Taxes from people living and working on farms to be gathered by the farmers and remitted to government as evidence of the number of people working on the farm
- Land tax to be reduced based on the number of different products produced and marketed from a farm
- Land tax to be reduced based on the purity of water flowing off the farm at the lowest points

All participants then discussed the likelihood of such a policy succeeding and were agreed that it would result in a massive increase in crop diversity and food production, idle farms coming up for sale, many thousands of people back on the land and working, regenerating soils, clean and nutritious food, and the government's tax base increasing dramatically at very low cost.

The group's final task was to decide whether the full policy could be developed using the generic holistic context we had created. All agreed that, because this policy was fundamental and national buy-in so important, it would require a real holistic context developed by Zimbabweans.

In discussing how to go about this they suggested two major policy retreats for leaders of all sectors of society, not just agriculture but also industry, local and national government, churches, women's organizations, chiefs representing the communal lands, and more. The aim of the first retreat would be to develop a national holistic context to guide the final policy draft, which the leaders would then take back to their constituents for approval. The aim of the second retreat would be to incorporate any changes the leaders brought back and then to ratify the national holistic context. The leaders would take this final version back to their constituents and ask them for their ideas on how to create a thriving agriculture and equitable land distribution. The leaders would then feed these ideas back to the task force to address in the final policy draft.

Developing the final policy (step 5) would involve months of detailed work by a policy task force that would need to ensure that each of the actions finally proposed was passed through the context checks. The criteria to monitor to ensure the policy's success (step 6) would round out the document. The task force would then release the policy to the public and present it to Parliament for consideration, debate, and ratification.

We ended the workshop at this point, and there were two immediate responses from this group of thirty-five lawmakers. The first was a mixture of anger and sadness that they had come this far in their careers and only now learned about this process. The second was an appeal to repeat the workshop at the cabinet level, which, regrettably, we were not able to do as the unity government was dissolved shortly afterward. Few of us, however, will forget the experience and what we learned from it. And no one has given up on the idea.

I described this rather unusual case in some detail because not only does it show how much latitude you have in modifying the process or the order of the steps, but also the power of the holistic context in setting the scene when the parties involved are in conflict.

## Conclusion

The steps and suggested process provided in this chapter can serve as a guide to analyzing or developing any policy that aims to solve a problem, no matter how large or small in scope or importance. Though this chapter has focused on resource management policies, I hope you can see that policies of many kinds would benefit just as much from holistic analysis and development, as the example of one U.S. town illustrates (see box).

We will always have problems that need to be addressed through sound policy (and sound development projects), but what we should aim for is to have those problems largely be due to natural causes. Currently, far too many problems are due to our own making and largely, I believe, because of "reductionist" management that attempts to bypass the social, environmental, and economic complexities that do indeed exist. But reductionist policy making is also responsible, even when sophisticated interdisciplinary teams lead the effort.

Such teams invariably begin by identifying the problem the policy is intended to address. Aware that their policy will have economic, social, and environmental consequences, the team includes experts in these areas who contribute their ideas on how best to solve the problem, but from each expert's point of view (remember those colored balls in chap. 3 and color plate 3?). Some ideas conflict and there are disagreements, which the group does its best to resolve, and ultimately the most persuasive view prevails. But it doesn't

**One Town's Experience with Holistic Policy Making**

*Longtime Holistic Management educator* Joel Benson *also serves as mayor of Buena Vista, a high-country Colorado town, population 2,700, with a predominantly youngish outdoors-loving population. Joel and his board of trustees have the distinction of being the first town to incorporate Holistic Management into its governance. Joel reflects on the experience here.*

Buy-in and a critical level of dedication from staff as well as perseverance from consistent, thoughtful board members has brought us through many bumps to a point where the board and staff have incorporated the Holistic Management framework into both governance and policy development. We have a formally adopted holistic context guiding our actions, although we're still learning, and each trustee has a laminated copy of it, as well as a copy of the checking questions for easy referral.

During a recent strategic planning meeting we discussed how difficult decision making had been in the past, much of it because we were not able to uncover root causes of issues and had no common direction. We went through a series of difficult years because we had thought a perceived problem with staff turnover had to do with a variety of things that were not in fact the issue. Once we honed in on a common direction, and began governing via clearer policy, the public was more engaged and town "problems" became easier to resolve. Of course, we still have ample opportunity to be clearer in developing and implementing our policies, but our community now has a more cohesive board of trustees providing sound leadership to our staff.

end there. In too many cases, lobbyists for special interest groups are allowed input, often late in the process, which can result in so much haggling, trade-offs, and editing that the final policy lacks common sense.

Alternatively, were the team to be expanded to include people with a broader view—generalists rather than experts—and this larger team then started by creating a holistic context, the results would be quite different. The focus would move beyond the disciplines of the experts to the people most affected by the policy—*including those the lobbyists represent.* Then, when the team subjects each of the proposed actions to the context checks, the best ideas for solving the problem would float to the top and inform the policy.

Society relies on policies that are by necessity formed by institutions—be they government, corporations, large nongovernmental organizations, or international agencies—most of which have yet to consider using holistic policy analysis or development as presented here. It is my hope that this edition of *Holistic Management* will begin to change that.

# 43

## Holistic Research Orientation
### *Orienting Research to Management Needs*

*OCCASIONALLY WE DISCOVER A GAP* in our management knowledge that research could help bridge. If such research is to usefully inform our management, we have to ensure that it is relevant to our situation and in line with our holistic context. This is a different approach from that of recent decades in which reductionist research has often driven management—to the detriment of that management. With the development of Holistic Management, research can more effectively *support* management while no longer driving it.

Conventional scientific research tends to be reductionist in that it seeks to reduce phenomena to a simpler form for study by controlling as many external variables as possible. It does this to show that one factor and not another contributes to an observed outcome. Management, on the other hand, must deal with complex relationships and innumerable variables, none of which can be ignored or removed without adverse consequences. Increasingly, researchers endeavor to design studies that include more variables, but the essential difference between management and research remains: management accounts for constantly fluctuating influential variables, while research attempts to remove their influence and observe a particular phenomenon without any outside influence (which is not actually possible in research that occurs in real landscapes). The training required to gain expertise in research—even increasingly less reductionist research—is not the same training that is required for management, which needs to be holistic. However, these two types of endeavor can still complement each other, and in fact need to. This is especially true in agriculture, the focus of this chapter.

## Closing Knowledge Gaps

In seeking to address the root cause of a problem, we often make use of information derived from reductionist research. In diagnosing the cause of an insect outbreak, for instance, we rely on such research to help us determine the weakest point in the insect's life cycle so we can deal with the insect at its most vulnerable stage. A search of the literature frequently turns up no useful information, however, even with much studied species, simply because researchers have not yet seen the need to relate their studies to a particular management situation. Clearly, the greater the collaboration between managers and researchers in real-life situations the more relevant the research will be to the manager, and the more likely to produce management success, a subject we'll return to.

### How Can We Be Sure the Research Is Applicable in Our Context?

Knowing that a certain research finding comes from reductionist methodologies simply warns us to observe and monitor carefully as we apply it within the whole we are managing. Always be cautious of any findings promoted as "best management practices," as they may not be best for *you*. Farms and ranches vary in brittleness, the families managing them are constantly changing (through births, marriages, divorce, deaths, etc.), as are the economies in which they operate and the weather patterns that influence production. Each whole under management is not only unique, but uniquely different every year. Thus, only when those best management practices are checked to ensure they are in line with your holistic context can you know if they are appropriate.

#### Pass the Ideas through the Context Checks

Will the ideas you have gleaned from the research address an underlying problem or merely suppress symptoms? Will they address the weakest link in the life cycle of the organism you are trying to encourage or discourage?

Might their use create a social weak link between you and those whose support you need? Dollar for dollar would one idea provide a greater return toward your goal than another? Could the ideas lead to an addictive use of energy or money? Will they lead you toward or away from the future resource base described in your holistic context?

Consider the case for noxious plant management guided by research focused on measures that attack adult plants or seed dispersal mechanisms. From the weak link (biological) check we know that we need to address the most vulnerable point in the life cycle of any plant, and that point is at germination and establishment. The right conditions for both must coincide—if millions of seeds germinate but do not find ideal establishment conditions, they will die. Managers generally fare much better by tackling this problem through soil management practices that favor the germination and establishment of the plants they want to increase, or that discourage the germination and establishment of the plants they want to decrease.

### Where Are You on the Brittleness Scale Relative to the Research?

If research on plants, animals, or soils has been conducted in brittle environments be cautious in applying it. Far too often the land in question will have been subjected to the damaging effects of partial rest, as chapter 20 explained, and this will compromise the results of any research. This concept of partial rest is new and generally unrecognized as a critical influence by those conducting the research. But even total rest, or nondisturbance, is still viewed by many as beneficial in these environments, even though long-rested areas degrade, which will influence research results.

In the more humid, nonbrittle environments this problem does not arise. Resting an area—either partially or totally—to conserve it is a powerful way to restore biodiversity, although the restoration occurs over time, and the various populations of species in communities will change.

In the United States a fair amount of data has been collected on species and their interrelationships in areas set aside as bio reserves. The Sevilleta Wildlife Refuge in New Mexico, mentioned in chapter 20, is one of these. In this case, the environment is high on the brittleness scale and receives little rainfall. It thus requires some form of disturbance to vegetation and soil surfaces to make whatever rain falls effective, and to maintain plant vigor, complexity of species, and healthy soils. But the refuge, having limited hoofed species—a few pronghorn antelope mainly—has been under the influence of partial rest for many years.

Despite the obvious loss of biodiversity occurring (see color plate 8), there are no published papers that ascribe the deterioration to partial or total rest. But there are, fortunately, an increasing number of papers showing that when you do the opposite—that is, periodically disturb the soil and trample down dying vegetation under some form of planned grazing—the land improves.[1]

When the aim of the research is to provide guidelines for better management, clearly any research conducted on land that is losing much of the available rainfall to runoff and evaporation, and is thus desertifying, should not be taken at face value. The same research done on land where rainfall has been rendered more effective and soil, plant, and animal life is increasing would likely yield different results and conclusions.

## Beyond Reductionist Research

Often in diagnosing the cause of a problem we find that we do not know enough about the underlying cause we've identified to address it effectively, particularly when the underlying cause is tied to heretofore unknown or recently introduced fungi, bacteria, and viruses: these are areas for urgent research. To date, management as well as research on such problems has tended to be reductionist, but that is slowly changing.

It is highly unlikely that simple organisms, such as bacteria, live by rules outside the evolutionary process and ecosystem functioning generally. When we create ideal conditions for an organism, or where we introduce one without biological restraint on its population as it enters a new environment, an invading organism can flourish to problem levels—as rabbits and prickly pear did in Australia, or as syphilis did in Renaissance Europe, and measles and small pox in North America. It is likely that diseases exist *because an environment exists that supports them.* Our task is to find what *we* have done to produce that environment. In our passion to cure with technological solutions, rather than *prevent* these diseases, we often overlook this basic question.

One group that *is* increasingly asking this question is the scientists studying the complex life that exists in soil communities and what happens to soil, plant, and animal health when management practices destroy that life. Early in my game department career the great ecologist Sir Frank Fraser-Darling left me as a parting gift his personal copy of Sir John Russell's *Soil Conditions and Plant Growth*, telling me that if I really wanted to save elephants I should focus on the soil. By that time (the 1950s) this focus was rapidly disappearing, and over the next decades was lost altogether. Two other instances support the idea of looking to what *we* have done to create an environment more conducive to disease outbreaks: the Great Plague and foot-and-mouth disease.

### The Great Plague

Today, we know that the bubonic plague that devastated Europe's population in medieval times was derived from bacteria spread by fleas living on rats. The plague reached epidemic proportions when overcrowded urban slums produced the right conditions for a massive increase in the rat population, which in turn provided the right environment for a massive increase in fleas. Attempts to tackle the disease head-on by seeking a cure for the individuals afflicted by the plague were unsuccessful. It wasn't until the slum conditions were rectified, and thus the environment made less conducive to rats and fleas, that the disease was contained.

### Foot-and-Mouth Disease

This viral disease, which mainly affects cattle, is indirectly responsible for the destruction of game in Africa on a massive scale. American and European importers of African beef insist that the meat must come from disease-free areas because the virus, although generally not fatal in Africa, apparently takes on a more virulent form when it reaches northern climes. Veterinarians believe the virus is carried by wildlife, buffalo in particular, and spread to cattle through close contact. To prevent any contact they fence livestock areas to exclude game. They sometimes shoot out the game and vaccinate livestock as well. In tackling the problem as though the disease acted independently of its environment, today's veterinarians have failed to address the larger questions, the most obvious one being, What kind of environment is conducive to the proliferation of the virus?

In India during the 1930s, British researcher Sir Albert Howard demonstrated repeatedly that cattle running on healthy soils and maintaining a healthy diet could actually rub noses with infected animals while drinking from the same trough and not contract the disease. My own experience backs this up. The outbreaks that occurred in Zimbabwe up until 1964, when I left the game department, always showed a far greater correlation with nutritionally stressed livestock, certain soil types, and a deteriorating environment, than they ever did with game populations. In fact in our areas of greatest buffalo-to-cattle contact, outbreaks were almost unknown. When one did occur in a small cattle herd in the middle of a large game reserve, the veterinarians jumped on it as proof that buffalo were spreading the virus. When we investigated further, we found that immediately before the outbreak a veterinary officer from a foot-and-mouth-infected area in nearby Zambia had paid a visit.

Whenever an outbreak occurred, a cordon was placed around the area, which meant that all roads leading out of it had checkpoints where shoes and tires were sprayed. This control of human movement effectively controlled the spread of the disease. Many species of game, including buffalo and vultures, still moved freely back and forth, which indicated that game were not the main agent in spreading the disease. In fact, it would be hard to imagine a better experiment to clear game of the accusation that they spread the disease. Despite this, the foot-and-mouth cordon remains the official method for containing spread of the virus, which apparently satisfies European import requirements.

Clearly we need to research the relationship between health and infection—not only health of the cattle but of the entire biological community, starting with the soils. Much like the urban slums of medieval times contributed to outbreaks of plague, I believe livestock living in "slum conditions" on degraded land are more susceptible to foot-and-mouth and many other diseases than those living in healthy communities on regenerating soils.

## Partnering on Agricultural Research

To develop the truly regenerative agriculture we require to keep our planet livable, agricultural research cannot be divorced from the reality on the land, as so much of it continues to be. Partnerships between researchers and farmers or ranchers managing holistically can help overcome this disconnection, with research findings enhancing and supporting management and no longer driving it. I see two major areas that would benefit from such partnerships:

*research to fill knowledge gaps identified by managers* that provides information relevant to their management context; and *whole systems research*, to assess changes on holistically managed versus conventionally managed farms and ranches to inform policy or public opinion.

Because we use the term *system* so loosely—it can refer to a complex biological, social, or economic system, or to a prescriptive management system—confusion reigns, and some definitions are in order:

- **Whole systems research.** The aim is to use appropriate research designs so the system can be assessed in terms of the whole it encompasses. Those researching the causes of the Great Plague and foot-and-mouth disease outbreaks had to look well beyond the actual organisms causing the disease to understand why each disease spread so readily, which was due to a combination of environmental, social, and economic factors. In assessing a whole farm or ranch to compare Holistic Management practices, researchers are basically documenting what exists—socially, economically, ecologically—and then comparing that to what existed at the start of Holistic Management or to what exists on properties not managed holistically. This documentation does not provide information on *how* to manage (as this is specific to the holistic context and location of each farm or ranch), but it can be used to inform policy and shift public opinion.

- **Management systems research.** Management systems are only appropriate when what is being managed is largely predictable and simple, such as an accounting system or an inventory-control system. The aim of research in this case is to provide information on how to improve the management system. Management systems, no matter how adaptable or flexible, by definition fail when used to manage what is unpredictable and complex, which is why researching various cropping or grazing systems makes no sense. Far too many researchers have published papers critical of Holistic Planned Grazing but rather studied grazing systems (often referred to as rotational grazing systems) that bear a resemblance only in that livestock are continually moved.

## Bringing Land Managers and Researchers Together

A true partnership between the people on the ground managing holistically and the researchers supporting their efforts needs to start with mutual respect. But since the time of Descartes, and the beginning of modern science, society has so elevated the status of the academic researcher and so lowered that of the land manager that generally the researcher speaks with more authority on management today than the person actually managing the farm from day to day and producing food. And this is so even though farmers and pastoralists were the ones who discovered which plants and animals could be domesticated, and then bred thousands of varieties from them several millennia before scientists existed. The grains, roots, vegetables, fruits, poultry, pigs, sheep, goats, cattle, horses, and camels resulting from their breeding efforts made civilization possible. Farmers and ranchers have also managed to keep biologically based agriculture alive against an onslaught of research promoting chemically based crop production and factory-style livestock production over the last five decades.

Fortunately, there is a growing subculture within academia in which researchers see the value of including people on the ground in jointly designed research projects that are informed by traditional knowledge. When researchers and the people on the ground are equal in stature as well, the research and its outcomes can only benefit.

## A Real-Life Case for Partnership

The winter thorn tree (*Faidherbia albida*) has long been highly valued throughout most of Africa. Unlike other species, it produces leaves and drops seedpods in the dry, nongrowing season when shade and fodder are in short supply. The World Agroforestry Centre (best known by its original acronym, ICRAF) refers to it as a keystone species in "evergreen agriculture." As a nitrogen-fixer it can be planted in crop fields to fertilize them, and because it has no leaves in the growing season, it won't shade out valuable crops. In late winter it drops its pods, which provide a nutrient-dense livestock feed at a time when forage is lacking. For these reasons, ICRAF hopes to bring the winter thorn back to areas where it no longer occurs or is dying out. But their studies have not revealed a practical way to do this, as they have found the trees

### Tracking and the Origins of Science

*Louis Liebenberg in The Art of Tracking: The Origin of Science* goes back even further than the original farmers to make a case for reconsidering the value of the contributions coming from the people on the ground. Early hominid trackers mainly used their sense of sight when tracking their prey, whereas natural predators mainly track by scent and sound. This difference in the senses used has a profound effect. Humans then as now had to observe the slightest signs of disturbance—on the vegetation and the dew or spider webs clinging to it, and on the soil surface, noting among many other clues whether insect tracks appear over or underneath those of the prey, and much more. Then they had to quickly synthesize all this information to deduce the direction traveled, the time of the disturbance, and the state of mind of their quarry: the buffalo went that way just after dawn and is wounded in the left foreleg.

"One of the more obvious ways in which the modern scientist differs from the tracker," says Liebenberg, "is that the scientist has access to much more knowledge by means of documentation." Even though documented knowledge is open to criticism, he says, it is impossible for the scientist to critically appraise it all and he or she has to accept the validity of the information based to a certain degree on an act of faith in others. "This," he says, "has the inherent danger that well-established knowledge may become dogmatic, which may result in irrational beliefs becoming entrenched in science."[2] Such entrenched beliefs were, I believe, a major reason the key insights covered in the early chapters of this book have taken so long to find acceptance in the scientific community. On the other hand, farmers had already observed some of them: the Scottish shepherds who talked of the golden hooves of the sheep, the Navajo stockmen who talked about the plants coming up only where animals had been.

no longer establish on their own. What the ICRAF researchers have learned is how long the seed can be stored and remain viable, what insecticide the seed should be treated with prior to planting, that these trees are best propagated by potting the seeds rather than direct sowing, and that seeds should be soaked in a sixty-six percent sulfuric acid solution or mechanically scarified to encourage germination. Through such methods ICRAF is promoting the planting of winter thorn trees in the "great green wall" that stretches across Africa south of the Sahara to stop the desert's spread.

Not knowing of ICRAF's work, but wanting ourselves to encourage the establishment of winter thorn trees at the Africa Centre for Holistic Management's learning site in Zimbabwe, we did our own informal study of what it would take to get them to establish. These trees had evolved over millions of years with wildlife dispersing the seeds after consuming the tons of protein-rich pods produced by each tree. And up until the last century or so the trees had established themselves without the assistance of insecticides and acid baths. So what was *so* different now that they were not establishing naturally?

As it turns out our problem wasn't a lack of germination. We found thousands of seedlings germinating wherever our herd of cattle and goats had concentrated after feeding on the fallen pods, telling us that the gut of the animal was as good as an acid bath in promoting germination. But they weren't surviving long enough to establish and grow into healthy trees. This was obviously the weakest point in the life cycle of the winter thorn and what we had to address. When we did so by protecting the seedlings from browsing bushbuck, kudu and other animals, we found they did establish successfully, even on the tops of rocky hills where the ICRAF research said they would not grow. As I write, we are building information into the grazing plan to see how much grass cover is needed to protect the establishing trees and for how long.

This method has proved to be a more practical and affordable way to establish new trees, even though we still don't know how they established over millennia in the presence of many more browsing animals. But we can deduce how, based on what we've observed: there appears to be a strong correlation between groves of older trees that established on temporary islands that lacked browsers over the critical establishment period, and with past human settlement, suggesting their presence kept browsing species away during critical periods.

An ICRAF researcher paid us a visit several years ago and found our efforts to establish winter thorns exciting. But what a difference it would have made if we had been able to collaborate on this research—sharing what we were observing in practice and demonstrating through our management, and ICRAF documenting the results far better than we—being managers first—ever could.

## Conclusion

Researchers have long resisted a close involvement with those who would benefit from their research; often for fear that such involvement would compromise the researchers' objectivity. But research funded by corporations with a vested interest in the results has become increasingly common, and this fear is sadly now a qualified one. In the United States, state and federal governments once funded the bulk of academic research, but as the cost of such research has escalated and state and federal coffers dried up, scientists had to look elsewhere for funding, and corporations stepped up. A second challenge is that no matter who funds the research, the journal editors who publish it and the peers who review it want to see research designed and conducted along traditional lines. Those striving to do whole systems research, or to create research partnerships between traditional land managers and scientists, often find funding doors closed and tenure tracks in jeopardy.

Nonetheless, the trend toward consumer-centered research is well worth cultivating if research is truly to benefit management, and essential if it is to benefit agriculture. The researchers' lack of involvement in the practical realities of management results in conclusions that are out of touch with reality. University and government extension services tasked with promoting such findings, often as best management practices, may actually do harm. Research will become increasingly relevant, I believe, as researchers gain an appreciation of the holistic context that drives the management of the resources they study.

The next chapter addresses another aspect of Holistic Management that will influence collaborations between researchers and managers and has been referred to throughout this book: proactive management through monitoring—or monitoring to make happen what you want to happen.

# PART 9

# COMPLETING THE FEEDBACK LOOP

# Holistic Management Framework

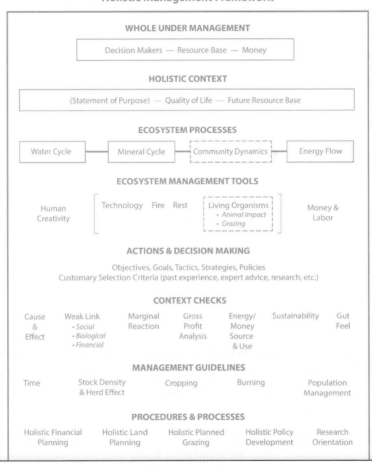

**WHOLE UNDER MANAGEMENT**

Decision Makers — Resource Base — Money

**HOLISTIC CONTEXT**

(Statement of Purpose) — Quality of Life — Future Resource Base

**ECOSYSTEM PROCESSES**

Water Cycle — Mineral Cycle — Community Dynamics — Energy Flow

**ECOSYSTEM MANAGEMENT TOOLS**

Human Creativity

Technology   Fire   Rest   Living Organisms
  • Animal Impact
  • Grazing

Money & Labor

**ACTIONS & DECISION MAKING**

Objectives, Goals, Tactics, Strategies, Policies
Customary Selection Criteria (past experience, expert advice, research, etc.)

**CONTEXT CHECKS**

| Cause & Effect | Weak Link • Social • Biological • Financial | Marginal Reaction | Gross Profit Analysis | Energy/ Money Source & Use | Sustainability | Gut Feel |

**MANAGEMENT GUIDELINES**

Time   Stock Density & Herd Effect   Cropping   Burning   Population Management

**PROCEDURES & PROCESSES**

Holistic Financial Planning   Holistic Land Planning   Holistic Planned Grazing   Holistic Policy Development   Research Orientation

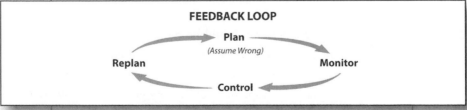

**FEEDBACK LOOP**

Plan
*(Assume Wrong)*

Replan                    Monitor

Control

# 44

## Monitoring and Controlling Your Plans to Keep Management Proactive

*ONCE A PLAN IS MADE,* monitoring becomes essential because, even though the actions involved have passed all the context checks, events rarely unfold as planned. You monitor to look for deviations from the plan for the purpose of correcting, or controlling, them. Although most people would define *monitoring* in the same way, in far too many situations we merely monitor "to see what happens."

Nowhere is this more apparent than in agriculture. The tendency there has been to seek and apply the best grazing system or the best cropping system, or the best of various management practices, and proceed confidently until things go wrong. When things do go wrong, another management system or another practice is tried until things go wrong again. In the long run, things never do go right because there never can be one management system or practice that applies everywhere when you are managing living systems and natural processes. There are always too many variables operating in such situations.

In *any* situation we manage, we should be monitoring in order to make happen what we want to happen—to bring about desired changes in line with our holistic context. The word *plan* becomes a twenty-four-letter word: *plan-monitor-control-replan.* All hope of reaching any goal or objective without great deviation or waste depends on this process: Once a *plan* is made, it is then *monitored*. If results begin to deviate from what was planned, then *control* is instituted, and the deviation is brought back to plan. Sometimes events go beyond our control, or we fail to control the deviation, and there is a need to *replan*. The better we control the plan, the less we need to replan.

---

The better we control the plan, the less we need to replan.

---

A simple analogy illustrates the process. Let's say my objective is to visit a friend who lives at the end of a winding road. My *plan* is to use my car to get there. I begin to drive, but have no hope of achieving my objective unless I do more than just drive. I will leave the road at the first bend unless I *monitor* the road well ahead. Then I have to *control* by turning the steering wheel in order to stay on the road. Now if all goes well and there are no earthquakes or flash floods, or my car does not break down and my monitoring and control are adequate, I will get to the house.

In real life things seldom go so smoothly. I may not monitor well because I am watching the scenery and my mind wanders. I may not control well because when I stop my daydreaming and with a sudden fright realize I'm going off the road, I turn back too sharply and go off the other side. If anything does cause me to break down, hit the ditch at the roadside, or whatever, I have to *replan* to ensure that I still reach my objective. In replanning I will always be working with a changed set of resources—my car has broken down and I am now on foot, for example. I now plan to walk to the house on the hill, but I still have to monitor the road ahead and turn to keep on it, and so on.

In the Holistic Management framework, we depict this process as a loop because it is a continuous effort. Throughout the process you should be seeking indicators of change and responding to the feedback you receive, constantly adjusting your actions to stay on track. Ideally, you will shorten the loop by controlling deviations so quickly and effectively that there is no need to replan.

Close attention to the feedback loop is what makes management proactive rather than reactive or adaptive. On several occasions I have been asked, "What will happen to the deer, or the weeds, or the finances, or the quality of life on my place if I manage holistically?" My routine reply is: "I don't know. What do you want to happen?" What you want to happen should be planned through management actions that are in line with your holistic context, and ensured through completing the feedback loop. If you want the deer to increase, then you will create the habitat that favors them. If you want the weeds to decrease, then you will apply the tools in a way that discourages their establishment. If you want to improve your finances, you will plan your profit and control your expenses. If you want to improve your quality of life, you will build your capacity to enhance your well-being. It becomes *your* responsibility to achieve what you want in your context. When your monitoring shows that no change has occurred where change was planned, or if any change occurs that is adverse to plan, *take action immediately.* If control is quick, a simple adjustment may be all you need to keep on track.

---

When your monitoring shows that no change has occurred where change was planned, or if any change occurs that is adverse to plan, take action immediately.

---

We have developed specific procedures (one simple, one more detailed) for monitoring ecosystem processes on land under Holistic Planned Grazing (details available in the *Holistic Management Handbook*), and a quick and practical early warning monitoring technique that will confirm whether or not a new practice is taking you in the direction you want to go.

## Monitoring Management of Ecosystem Processes

Because of nature's complexity we must take the attitude that much of what we do as land managers may lead to unanticipated effects. Anytime you are instituting a new practice, *assume you are wrong*, even though the decision, or decisions, involved have passed all the relevant checks. We refer to this practice as early warning monitoring, which is simple and quick to do. Monitoring ecological change over your land as a whole is a more involved process, but it is still important to do so that when changes you expect to see do not occur, you revise your plans to get back on track.

## *Early Warning Monitoring*

Anytime you try something new you should assume you are wrong and monitor proactively, making adjustments quickly if your monitoring shows adjustments are needed. Start with a clear idea of what you want to achieve and then ask yourself, "If I am wrong, at what point could I get the earliest possible indication?" That is the point you need to monitor in the simplest way you can devise. This information is for your own use. It is not for a research project, and it need not cost you much in time or money to gather. It is merely to guide your management—and to get you where you want to go more quickly.

---

### Identifying Early-Warning Indicators

*As you apply any one of the management tools*—technology, fire, rest, and living organisms, including animal impact and grazing—for the first time, you need to determine what to monitor that will give the earliest possible sign of change.

Monitoring changes in plant or animal species, a common practice, is a measurement that comes too late, indicating considerable change has already occurred that may not have been in the right direction. You want to detect changes well before that. *The earliest changes are most likely to occur at or near the soil surface.* They could show up as a change in spacing between plants, or the soil surface between plants (is it exposed, capped, covered in litter?), soil density, soil aeration or organic content, insect activity, seedling success, quality of water runoff, and a host of other things.

Depending upon what your monitoring indicates, you will either continue to apply the tools as you have been or you will need to make adjustments (the control). Obviously, if all is going as anticipated when the particular tool was selected, no control is necessary. If not, you will have to diagnose the cause of what is going wrong (see chap. 25) and quickly modify what you are doing.

---

Let me give a couple of examples to illustrate just how simple this early-warning monitoring can be.

In chapter 41 I mentioned the overnight livestock enclosure developed at the Africa Centre for Holistic Management in Zimbabwe to deter predators. We had long known that an excess of dung, urine, and trampling leads to bare ground and a lack of plant growth, and five hundred animals inside a confined space each night provide dung, urine, and trampling in good measure. But our thinking was that if we moved the enclosure every week or so there wouldn't be any damage; the land should in fact improve.

Assuming we were wrong, what could we monitor? The nature of the soil surface (is bare ground increasing?) and plant spacing (is it getting wider?) were the most obvious measures. How could we do that? Because there were no plants at all on the worst sites, plant spacing wouldn't tell us much. So we opted for using a camera to record what the soil surface looked like before and after the enclosure was placed on it. This provided the evidence we needed to quickly establish that the practice was safe. It subsequently led to greatly improved plant cover and biodiversity even on hard-capped soil surfaces with a cryptogrammic crust, which we captured in fixed-point photos. Over five hundred sites have since been treated with the movable enclosure, and no two have responded exactly the same because of soil types, weather patterns, and time of year treated. But all, without exception, have improved and continue to do so.

My second example is a theoretical one, but useful in that it shows how early-warning monitoring can help get things moving when controversy blocks progress, and people doubt that your management can sort it out.

In some of America's western states, desert tortoises are dying out and the standard prescription for restoring their populations is to remove livestock and rest the land, even though it is already overresting and large areas are crusted over with algae and lichen. In diagnosing the underlying cause of the problem, we find that tortoises are probably disappearing due to habitat change. Most of the grasses they formerly depended on for feed, moisture, and cover (for newly hatched tortoises vulnerable to predation) have disappeared. So you decide to bring back those grasses, and to do so using livestock because the former herds of wild grazers have all but disappeared too.

The Tortoise Defense League is extremely skeptical of your plan. They contend that livestock, cattle in particular, have trampled many tortoises to death and are thus part of the problem. But you are persuasive, and they agree to give your plan a try. Although Holistic Planned Grazing isn't a new practice for you, it is for the Tortoise Defense League, and you propose a test site on which you apply herd effect, to help reassure them. You start by assuming you are wrong, and then together with the group identify the indicators you will monitor to give the earliest warning of adverse change. This will help build the group's confidence and the continued support you will need to save the tortoises in the wild. Once more the nature of the soil surface becomes your prime indicator—bare and capped (or crusted) soils should decrease and litter cover, grasses, and other plants should increase. The easiest, most inexpensive way to document changes to the soil surface would again be through simple fixed-point photos. You would take the photos before treatment with herd effect, immediately after treatment, and then later as the growing season progresses. Did the herd break up the crusting thoroughly? If not, new grass plants will have difficulty establishing and soil erosion could increase. If thoroughly broken did the site erode or did some plants begin to establish? If plants did begin to establish you have early confirmation that you are probably on the right path.

### Routine Monitoring of Ecosystem Processes

Holistic Planned Grazing calls for daily monitoring of the plan's implementation, and the growing and nongrowing season aide memoires include guidelines for this. Periodically, it also pays to carry out an assessment of the soil surface and the life upon it.

If you are a rancher or farmer with limited funds and little passion or time for more involved monitoring there is a basic procedure you can use to get the information you need to keep your management on track. It involves taking photos of fixed plots and making notes of what you observe on them each year, which shouldn't take more than a day of your time. Even though your focus is on a longer time frame, control remains the most vital part of the monitoring feedback loop and essential for keeping your management proactive.

This is also the case for the more comprehensive monitoring that provides more "quantitative data," which many prefer to use to inform their management, particularly when monitoring change on public lands or in a collaborative grazing effort. The *Holistic Management Handbook* includes details on a comprehensive monitoring procedure that has worked well in such cases and takes a few days of work once a year.

There are many other comprehensive monitoring procedures more attuned to specific needs, such as compliance with standards related to marketing livestock products, or ecosystem services such as carbon storage. These include some developed by the Savory Institute. If you use these procedures make sure they focus on the nature of the soil surface, which tells so much about the health of the ecosystem processes. It is often better to have the monitoring done by a third party, which has several advantages. Seeing the metrics of change for the better confirmed by an outsider is a great morale booster and motivator for managers. It is also useful for convincing skeptics who still think of Holistic Management as some form of management system and want evidence "that it works." And of course, robust, statistically significant data help greatly in shifting public opinion as well as institutional mindsets.

### Monitoring Technology in the Environment

New technologies will continue to be developed that have the potential to affect the environment profoundly. While these technologies are often monitored for their effects on site, or in specific applications, their effects elsewhere and much later seldom are, although awareness of the need to do so is increasing. The classic case of chlorofluorocarbons (CFCs) used in refrigeration systems, fire extinguishers, and cleaning solvents and their effects on stratospheric ozone is one of the better examples that have alerted the public to just how much later, and just how far from the site of application, the effects can become manifest.

In the case of substances, such as CFCs, that can persist in the environment, the problem of deciding what, where, and when to monitor is enormous. To compound the problem, although a substance may on its own be proved safe, in combination with other substances in the environment it can become lethal.

How do you determine which criteria to monitor for the earliest warning that the cumulative effects of any new technology are damaging, when we have so little understanding of nature's complexity? I do not have answers, but I do know that most of us now see the need for answers.

## Proactive Monitoring and Control

*Proactive monitoring with rapid control is of course what makes* management *proactive. When done well (returning to our car analogy), you will be looking far ahead so that controlling the plan requires only small adjustments to keep the car on the road. No amount of adaptive, or reactive, driving is likely to do anything but keep you bouncing from one side of the road to the other and possibly never reaching your destination. As the following case illustrates, there is a vast difference between proactive monitoring with quick control, and reactive monitoring, which basically bypasses control and tends to go straight to replanning, making management reactive (or adaptive) too.*

In the early 1980s the U.S. Forest Service initiated a pilot project in Arizona in which Holistic Management was to be practiced on land leased by a local rancher. The necessary holistic context was created, and animal impact and grazing were the main management tools the rancher would use to move toward the future landscape described in that context.

Over the next eight years, the rancher operated more or less according to plan, and the Forest Service methodically monitored the results, ably assisted by the local university. They rightly focused on soil capping, soil erosion, and plant spacing, all of which would provide early warnings of any adverse changes. Each year the monitoring showed that, as planned, soil capping was decreasing. But it also showed that soil erosion was increasing when it should have been decreasing, and plant spacing was widening when it should have been narrowing. Something was clearly wrong, *yet no control was implemented.* Everyone involved was waiting to "see if it (Holistic Management) worked," instead of responding to what the monitoring was showing. Analysis of the adverse direction would indicate that, although overgrazing of plants had ended, partial rest had not. A simple and timely adjustment in animal numbers and behavior in the second or third years would have changed everything. Nothing was done, however, and when the project wound up at the end of the eight-year period, it was considered a failure.

When humans see a need, and put their minds to it, amazing things can happen. Prior to introducing new products, we would be wise to think out the path the products (or their by-products) will take from the earliest stages of production to their ultimate resting place. (The international life cycle assessment standards—ISO 14040—are a step in this direction.) How a product affects *any* life forms encountered along that path could provide the earliest warning of changes that would adversely affect the four ecosystem processes.

This sort of monitoring will require considerable expertise and sophisticated technology, both of which are potentially within our reach. However, in the case of unnatural substances that persist for many years once released in the environment, we may have little alternative but to ban their production altogether if containment cannot be reasonably guaranteed.

## Monitoring and Controlling a Holistic Financial Plan

Those managing any well-run business will appreciate the importance of monitoring and controlling a financial plan. In Holistic Financial Planning (chap. 39) income, expense, and inventory figures are monitored monthly to ensure that actual figures match with what was planned. This should be done *within the first ten days of the following month*. No excuses accepted because the faster the control is done and adjustments made, the greater the success by year's end and the less the likelihood of a need to replan. Unlike the monitoring of ecosystem processes, you do not *assume* your plan is wrong. You *know* it is. No financial plan ever unfolds exactly as planned.

If you use the services of an outside accountant, you can't afford to wait for her to supply the figures you need each month. You know how much money you have spent and how much you have received and can record the amounts as the money comes in or as you commit to each expense. You don't need to nail down every cent. Monitoring quickly and on time matters more than perfect accuracy. If survival prospects look critical, then monitor actual figures *daily*, and project them to month's end (or even year's end) daily, to see where they stand relative to your plan.

In developing a sugar farm from virgin bush in Africa, I came to the point of daily monitoring, and it saved me. Short of capital and heavily in debt, I was pushed to the edge of desperation by a rapid fall in prices. Planning a way out took ages. My main crop, sugarcane, would not return any income until eighteen months from the date it was planted, and I faced ruin.

Eventually I managed to plan survival around a series of crops that would mature at different stages and sustain cash flow, but all aspects of the plan simply *had* to work. Even one month of over-budget expenses on any crop would have put me out of business. Only daily monitoring and projecting trends ahead allowed me to take corrective control action in time.

Control is the hardest part of financial planning, yet many farmers and ranchers trivialize it. The key principle is to control deviations within each income or expense category and not simply allow deviations between them to balance out. You will have income and expense categories that either run down columns or across rows on your spreadsheet, depending on the computer software you use, that show the figures for twelve months. Many people note any differences in income and expense projections each month and control their plan based on overall figures. If, for example, income from category X is higher than planned in a given month and is able to cover greater-than-planned expenses in several other categories, we consider all to be well. But this leads to sloppy management and often frequent *replanning*. Treat each income or expense category separately and control within each category and you will keep control of your plan.

Although you might have spent considerable effort creating the best possible plan, when your monitoring turns up a few adverse figures you have no option but to do better than your best. You cannot afford to simply ignore the signs and grind on in hopes that fate will err in your favor somewhere down the road. If you are serious about making a profit, knuckle down and do that seemingly impossible control.

## Monitoring Quality of Life and Human Behavior

While we concentrate on measuring growth in terms of an economy, we rarely measure growth in terms of quality of life, which is even more important.

No more so than in agriculture, where the need to enhance quality of life is imperative, given the alarming fact that suicide is the leading external cause of death among food and fiber producers throughout the world. But how do we best monitor change in quality of life? Procedures and methods do exist, but very few within the context of a whole system. As a new generation of scientists puts their minds to measuring quality of life within whole systems many more practical and creative methods are likely to emerge. In the meantime, the Savory Institute continues to release new tools for monitoring quality of life as they become available.

In day-to-day management it is important to continually monitor the following:

- Your planned actions for creating the quality of life your holistic context describes
- The quality of your relationships with the people in your resource base whose loyalty, respect, and support you need

In each case, the earlier you detect signs of change the easier it is to remain on track. For example, if your holistic context talks of the desire for more quality time with family, and one of the things you do to create that is plan at least one activity each month, did you put the dates in your calendar? (It probably won't happen if you didn't.) Monitoring the quality of your relationships often comes down to monitoring behavior. The earliest signs of trouble may be reflected in a change in attitude (e.g., loss of enthusiasm or interest, lack of trust) and in body language (e.g., avoiding eye contact, a submissive or combative stance). When things are going well, the opposite would, of course, apply. In such a case you would never assume your expectations of the person's behavior are wrong; you would want to assume they are right for the obvious reason that positive (and higher) expectations lead to enhanced performance.

## Conclusion

A plan, no matter how sound, serves little purpose unless its implementation is monitored and deviations are controlled. Otherwise, even assuming no lapses at all in management, unpredictable events sooner or later render the best plan irrelevant or even destructive. Some will ask, "Then why plan in the first place?" We must plan, monitor, control, and replan, simply because it is the only sensible way we can make happen what we want to happen.

# PART 10

# CONCLUSION

# 45

## A Commonsense Revolution
## to Restore Our Environment

*OUR LOVELY PLANET NOW STAGGERS* under massive human impact, and fast-rising populations that, if unchecked, can only lead to catastrophe. Some people remain optimistic that technology in one form or another will alleviate the damage we have done and enable us to continue on our present course. Others are deeply pessimistic that anything can save us, and it's easy to see why, given the increasingly alarming reports on the rate at which our climate is changing.

Because we are accelerating toward catastrophe, minor improvements that only slow the rate of acceleration are ultimately meaningless. Slowing down won't prevent you driving your car over a cliff, only delay the time of the crash; *you have to change direction altogether*. That's what we must do now to avoid a future none of us wants, and it will require nothing short of a revolution in four areas:

- **A revolution in management**. Successfully manage what is complex—people, organizations, natural resources, and natural systems—by managing holistically.
- **A revolution in agriculture**. Decarbonize the atmosphere, reverse desertification, and provide food for life by regenerating soils through sound cropping practices and properly managed livestock.

- **A revolution in policy making.** Create policies that are ecologically, economically, and socially sound by developing policies holistically.

- **A revolution within our organizations and institutions.** Overcome institutional inertia and opposition to *holistic* management by creating awareness and shifting public opinion.

## Revolutionizing Management

Addressing the complexity inherent in anything we manage is our greatest challenge, and one we have to meet if we are to regenerate the world's soils so that we reverse desertification, address climate change, and successfully feed nine billion or more people. The Holistic Management framework gives us a way to do this for all forms of organization.

Almost all of the problems overwhelming us now are, I believe, a consequence of reductionist management, where the context is reduced to a need, desire, or problem, when we, in fact, live in a holistic world and must manage within a holistic context. The problems that continue to worsen are not the complicated ones technology adeptly addresses, they are the complex social and biological problems that management can resolve—as long as that management is holistic.

Yet many scientists, and indeed most of society, still seek salvation in technology. Such misplaced confidence is based largely on our inability to recognize the difference between what is complicated and what is complex and why that difference matters. As chapter 3 explained, all that we *make* using technology is complicated, but predictable, in that it does what we design it to do. All that we *manage* involves people, organizations, and nature, and they are complex, in that they are self-organizing and thus unpredictable. The success of our technological achievements in addressing complicated problems has blinded us to the fact that technology is of little use in resolving problems that are complex. Although technological "solutions" may enjoy short-term success, the likelihood of their producing unintended consequences is too high a risk to take. Some still seek to solve the climate crisis by using technology to geo-engineer solutions almost certain to spawn dangerous effects, so we haven't learned our lesson yet.

To resolve such mega problems as desertification and climate change we have to tackle their underlying cause, which I have tried to show convincingly in this book is tied to our failure to manage what is complex. Livestock and fossil fuels, which are merely resources to us, do not cause desertification and climate change; *we do* in our management of them. It is our management of livestock that determines whether they damage land and pollute the atmosphere, or whether they restore damaged land to health and decarbonize the atmosphere. It is how we manage fossil resources that determines whether they are burned up as fuel and pollute the atmosphere or used sparingly for centuries in the manufacture of essential products. Reductionist management has given us the former, holistic management promises the latter.

The revolution in management is the first that needs to occur, as it is the catalyst for the other three. While we have all the money and manpower in the world to bring it about, the one luxury we do not have is time.

## Revolutionizing Agriculture

Modern agriculture is a major contributor to desertification and climate change and is causing us to lose our ability to feed a growing population. Almost all of the world's vast grasslands are desertifying, leading to increased hunger, poverty, violence, and millions of environmental refugees.

Any time soil is exposed through plowing, burning, or improperly managed livestock, soil organisms die, releasing carbon dioxide into the atmosphere. In fact, as chapter 1 pointed out, carbon dioxide emissions from the destruction of agricultural soils and the burning of croplands and grasslands nearly equal emissions from fossil fuels each year.

Healthy, *living* soils can store carbon for millennia in the form of organic matter. And grassland soils, with the help of the grazing animals that evolved with them, can store the greatest amounts, due to their sheer size. This is key because once we reduce the carbon dioxide coming from agriculture and fossil-fuel emissions, there will still be many billions of tons of carbon dioxide in the atmosphere that need to be drawn down to Earth and safely stored (as carbon) to keep our climate livable.

I've mentioned farmers and ranchers throughout this book who are demonstrating what is possible on both croplands and grasslands using livestock properly managed to assist in regenerating their soils and in turn increase productivity and profitability. But their numbers aren't growing fast enough, and we can't afford to wait. So let the revolution begin.

## Revolutionizing Policy Making

If those revolutionizing management and agriculture aren't supported by sound policies, their revolutions will falter. Policies need to reward outcomes that are economically, environmentally, and socially sound, and discourage those that are not. Policy makers now have a tool in the Holistic Management framework to help them figure out which is which. A policy formed within a holistic context will always fare better than a policy formed reactively to address a singular problem. Holistic policy making focuses on the future resource base and what people want versus problem-centric policy making that continually treats the symptom. The context checks help assure that the policy does tackle the underlying cause of the problem, and doesn't just suppress symptoms, which will persist if the cause is not addressed.

Far too many of the problems policies seek to resolve are of our own making, having been brought about by reductionist management, and reductionist policies too. Policies developed holistically will help discourage this and in turn eliminate many of the problems that require more and more policies and more and more resources to address. I believe it won't be long before a young and incredulous generation asks how policy makers did not see this long ago. Such questions drive revolutions.

## Revolutionizing Our Organizations and Institutions

When new ideas emerge that run contrary to what a society believes, institutions will not adopt them and commonly lead the opposition to them, which is one of the "wicked" problems associated with human organizations. Thus, until a significant shift occurs in how society views *holistic* management, institutions will remain at best passive, even though individuals within them are not. This presents a special challenge because it is organizations and institutions that create the policies that in turn impact agriculture and management.

While there are many problems within large institutions that may need solving, this revolution needs to focus on organizations and institutions of every size that are unable to adopt new insights ahead of society because they reflect the views of the society in which they were established. Thus, while revolutionizing our institutions is a high priority, we have to do so by making the public aware of the new insights and shifting public opinion first. Then institutions can shift too.

How great that shift in public opinion needs to be is hard to determine but I believe it is not as large as we might think for a tipping point to be reached. When a few million people are using social media to state that management needs to be holistic or we're not going to make it, institutions will begin to change. But it will take more than social networking, as powerful an influence for change as it has proven to be, to reach that tipping point.

Books have long played a role in spurring societal movements, and my hope is that this one will too, even though modern attention spans are decreasing. But the most promising catalyst for change is the growing number of grassroots networks affiliated across countries and continents. The Savory Hub Network is one of them, and it in turn is linked with food and nutrition networks, regenerative farming networks, climate change activist networks, and many more. Three years ago, our goal was to establish one hundred locally led and managed hubs around the world, influencing one billion hectares (two and a half billion acres) with Holistic Management by 2025. We are set to exceed those numbers, and by a substantial amount, largely because the hubs can create awareness and provide training and coaching in Holistic Management in the cultural, economic, and environmental context of their regions. And they can share what they learn as well as unite around issues of concern with the broader network. Neighbors telling neighbors and friends sharing with friends across a local network give power to the larger network, which becomes a movement and produces the tipping point. And a revolution.

## In Addition . . .

Alongside the deterioration of land, water, and our atmosphere, the explosion of the human population that has paralleled the degradation of our resources will also have to be tackled. We cannot manage resources holistically according to the natural laws of our ecosystem if we continue to act as if those laws do not apply to humanity as well.

Various cultures and religions favor large families for reasons deeply rooted in historical conditions in which security in old age, or survival of a race, depended on many children. Now, however, uncontrolled population growth threatens our survival, and we have other means to provide old age security.

The institutionalized religions in modern society that encourage high numbers of children do not reflect the present state of the world. The sages who founded them spoke out of the conditions of the times in which they lived, but in all cases their universal message was compassion. I find it difficult to believe that, were they preaching today, they would suggest that we continue to produce high numbers of children knowing that by doing so we ensure poverty, violence, social breakdown, even genocide, and ultimately threaten our survival as a species.

I myself am not a conventionally religious man. I do however feel infinitely small and powerless in the presence of the wonders of Nature and our universe. Such marvelous creation did not occur by chance. There is a power that is greater than all humankind, and out of deference to it we should respect each other and the ecosystem that sustains us all. This means controlling our population and respecting the diversity of cultures, tribes, nations, and spiritual beliefs as a great gift to all humankind, and the same duty includes the companion task of halting the deterioration of life on Earth.

Fortunately we do not have to wait for an era of world peace and collaboration to make a start because Holistic Management leads to conflict resolution, and the Holistic Management framework will function for any people in any place, regardless of religion, system of government, economic base, or climate.

A great many of our conflicts arise directly out of the deterioration of resources and ignorance about the tools we have to manage them, but even such a materialistic explanation has a philosophical aspect. I once heard of a Navajo medicine man who, in mediating a grazing dispute between two families, said, "You are neighbors whether you want to be or not, because the land itself unites you. It links you both as you walk on it today, and you will both lie in it together when you die. Then the plants that grow in the soil you become will infect your children with either your hatred or affection as you can choose now. If you bless your land, it will return the blessing and your present argument will become insignificant."

To the prayers, songs, and practical gestures the medicine man had in mind I would add Holistic Management. In case after case we have already witnessed what it can contribute to conflict resolution, and this role will surely become ever greater. The conflict often begins to resolve itself as antagonists brought together to create a holistic context discover that what they have in common is far greater than their differences. That is why we have to start this revolution by revolutionizing management. Our efforts to save our planet and ourselves hinge on this one transformation occurring. The great innovator Buckminster Fuller talked of the trim tab—that one small thing that if moved would automatically move the rest that needs moving. That is the best description of Holistic Management I can think of. When we acknowledge what is complex and manage it holistically, the rest will move too.

Earlier in my life, the magnitude of problems without solutions depressed me utterly. Now, at last, I see the possibility of wonderful times ahead as we both enjoy the fruits of technology and learn to live within our ecosystem's rules. Had I another stock of years to replace those I've spent already, I could not imagine a more exciting time to live them than now.

# NOTES

## Chapter 1

1. Norman Myers, J. Kent, and K. Smith, *The New Atlas of Planet Management*, rev. ed. (Oakland: University of California Press, 2005), 39.

2. Sara J. Scheer and S. Sthapit, *Mitigating Climate Change through Food and Land Use*, Worldwatch Report no. 179 (Washington, DC: Worldwatch Institute, 2009), 50.

3. Mark Z. Jacobson, "Effects of Biomass Burning on Climate, Accounting for Heat and Moisture Fluxes, Black and Brown Carbon, and Cloud Absorption Effects," *Journal of Geophysical Research: Atmospheres* 119 (2014): 8980–9002, doi:10.1002/2014JD021861.

4. National Geographic Society, "Ocean Acidification: Carbon Dioxide Is Putting Shelled Animals at Risk," accessed October 10, 2015, http://ocean.nationalgeographic.com/ocean /critical-issues-ocean-acidification/.

5. USDA Agricultural Research Service, "Glomalin Is Key to Locking Up Soil Carbon," *Science Daily*, June 16, 2008, 5, accessed October 10, 2015, http://www.ars.usda.gov/is /pr/2008/080617.htm.

   Robin White, Siobhan Murray, and Mark Rohweder, *Pilot Analysis of Global Ecosystems: Grassland Ecosystems* (Washington, DC: World Resources Institute, 2000), 51.

   Contribution of Working Group I to the Fourth Assessment Report of the Intergovernmental Panel on Climate Change, *Climate Change 2007: The Physical Science Basis* (New York: Cambridge University Press, 2007).

6. J. Hansen et al., "Target Atmospheric $CO_2$: Where Should Humanity Aim?, *Open Atmospheric Science Journal* 2 (2008): 217–31.

7. W. R. Teague et al., "The Role of Ruminants in Reducing Agriculture's Carbon Footprint in North America," *Journal of Soil and Water Conservation* 71 (2016): 156–64.

## Chapter 3

1. Jan C. Smuts, *Holism and Evolution* (Westport, CT: Greenwood Press, 1973), 336.

2. Jan C. Smuts, *Jan Christian Smuts* (London: Cassell & Company Ltd., 1952), 290.

3. Robert T. Paine, "Food Web Complexity and Species Diversity," *American Naturalist* 910 (1966): 65–75.

4. Zev Naveh and Arthur Lieberman, *Landscape Ecology: Theory and Application* (New York: Springer-Verlag, 1983), 56.

5. John Ralston Saul, *Voltaire's Bastards: The Dictatorship of Reason in the West* (New York: Vintage, 1993), 8.

## Chapter 4

1. Allan Savory, "The Utilisation of Wildlife on Rhodesian Marginal Lands and Its Relationship to Humans, Domestic Stock and Land Deterioration," Proceedings of the First Congress of the Associated Scientific Societies of Rhodesia Symposium on Drought and Development, Bulawayo, 1966, 118–28.

2. Christopher Sandom et al., "Global Late Quaternary Megafauna Extinctions Linked to Humans, Not Climate Change," *Proceedings of the Royal Society of London B: Biological Sciences* 281.1787 (2014), accessed October 14, 2015, doi:10.1098/rspb.2013.3254.

## Chapter 5

1. A. B. Harker, "Sheep Husbandry," in *Intensive Livestock Farming,* ed. W. P. Blount. (London: William Heinemann, 1968), 64. According to Harker:

"The 'Golden Hoof' of sheep is a description which has perpetuated through the centuries of British history and is indicative of the wealth which has resulted from sheep. Many areas owe their existence to the development of the marketing and weaving of wool. The actual hooves of sheep have also been praised for contribution to land and crops in compressing soil in such a way unequalled by mechanical means. Such is the reputation of sheep, which has done much to maintain its vast population in Britain today."

2. George Blueeyes was interviewed in 1978. His interview appears in:

Sam Bingham, Janet Bingham, C. Arthur, and Rock Point School, *Between Sacred Mountains: Navajo Stories and Lessons from the Land*, Suntracks Series (Tucson: University of Arizona Press, 1984), 177.

3. Researcher Sam J. McNaughton (Biological Research Labs, Syracuse University, New York) has conducted a number of field studies in East Africa documenting the relationship between herding grazers and the plants they feed on:

Sam J. McNaughton, "Grazing as an Optimization Process: Grass-Ungulate Relationships in the Serengeti," *American Naturalist* 5 (1979): 691–703.

Sam J. McNaughton, M. B. Coughenour, and L. L. Wallace, "Interactive Processes in Grassland Ecosystems," in *Grasses and Grasslands: Systematics and Ecology*, ed. J. R. Estes, R. J. Tyrl, and J. N. Brunken (Norman: University of Oklahoma Press, 1982), 167–93.

Sam J. McNaughton, "Grazing Lawns: Animals in Herds, Plant Form, and Coevolution," *American Naturalist* 6 (1984): 863–83.

Sam J. McNaughton, F. F. Banyikwa, and M. M. McNaughton, "Promotion of the Cycling of Diet-Enhancing Nutrients by African Grazers," *Science* 278 (1997): 1798–1800.

## Chapter 6

1. George Mossop, *Running the Gauntlet: Some Recollections of Adventure,* 2nd ed. (Pietermaritzburg, South Africa: G.C. Burton, 1990), 7–8.

2. The findings of John Acocks are reported in his paper, "Nonselective Grazing as a Means of Veld Reclamation," *Proceedings Grassland Association of South Africa* 1 (1966): 33–39.

3. William J. Ripple and Robert L. Beschta, "Trophic Cascades in Yellowstone: The First 15 Years after Wolf Reintroduction," *Biological Conservation* (2011), doi:10.1016/j.biocon .2011.11.005, accessed September 15, 2015.

In a later paper, Ripple and colleagues do not see any link between the Yellowstone example and what might be possible using livestock managed to mimic the behaviors of wild ruminants under threat of predation. In fact, they believe that reductions in livestock numbers, because of the greenhouse gases emitted in modern meat production, would help mitigate climate change and yield important social and environmental cobenefits. See:

William J. Ripple et al., "Ruminants, Climate Change and Climate Policy," *Nature Climate Change* 4 (2014), 2–5, doi:10.1038/nclimate2081, accessed November 14, 2015.

## Chapter 11

1. P. A. Yeomans, *Water for Every Farm*: *Using the Keyline Plan* (Katoomba, Australia: Second Back Row Press Pty Ltd., 1981).

## Chapter 13

1. Roderick MacDonald, "Extraction of Microorganisms from the Soil," *Biological Agriculture and Horticulture* 3 (1986): 361–65.

2. Chris Maser, *The Redesigned Forest* (San Pedro, CA: R & E Miles, 1988), 24–38.

3. David Tilman, David Wedin, and Johannes Knops, "Productivity and Sustainability Influenced by Biodiversity in Grassland Ecosystems," *Nature* 379 (1996): 718–20.

Forest Isbell et al., "Biodiversity Increases the Resistance of Ecosystem Productivity to Climate Extremes," *Nature* 2015, doi:10.1038/nature15374.

4. Nyle C. Brady and Ray R. Weil, *The Nature and Properties of Soils*, 11th ed. (New York: Prentice Hall, 1996), 333.

5. André Voisin, *Grass Productivity* (Covelo: Island Press, 1988), 45.

6. Douglas H. Chadwick, "What Is a Prairie?" *Audubon* (November–December 1995), 36.

## Chapter 16

1. Bank for International Settlements, *Triennial Central Bank Survey 2013, Foreign Exchange Turnover in April 2013: Preliminary Global Results,* September 2013, Basel, Switzerland, accessed December 7, 2015, https://www.bis.org/publ/rpfx13fx.pdf.

2. David Pimentel and Michael Burgess, "Soil Erosion Threatens Food Production," *Agriculture* 3 (2013): 443–63, doi:10.3390/agriculture3030443.

3. Norman Myers, Jennifer Kent, and Katy Smith, eds., *The New Atlas of Planet Management* (Oakland: University of California Press, 2005), 304 pp.

## Chapter 17

1. Brené Brown, *Daring Greatly: How the Courage to Be Vulnerable Transforms the Way We Live, Love, Parent, and Lead.* New York: Penguin Group (USA), 2012.

2. Kirstin McGuire, "Joyful Work: Re-imagining Engagement, Creativity and Performance," *Integral Leadership Review* (August–November 2013), accessed February 1, 2016, http://integralleadershipreview.com/10553-8152013-joyful-work-re-imaging-engagement-creativity-and-performance/.

## Chapter 18

1. Big Think Editors, "#5: Stephen Hawking's Warning: Abandon Earth—Or Face Extinction," BigThink.com, August 9, 2010, accessed February 4, 2016, http://bigthink.com/dangerous-ideas/5-stephen-hawkings-warning-abandon-earth-or-face-extinction.

2. Arthur Grube et al., "Pesticides Industry Sales and Usage: 2006 and 2007 Market Estimates," U.S. Environmental Protection Agency, Biological and Economic Analysis Division, Office of Pesticide Programs, February 2011, https://www.epa.gov/sites/production/files/2015-10/documents/market_estimates2007.pdf.

3. Max Kutner, "Death on the Farm: Farmers Are a Dying Breed, in Part Because They're Killing Themselves in Record Numbers," *Newsweek*, Europe Edition, April 18, 2014, 23.

## Chapter 19

1. Tim Flannery, *The Future Eaters* (Chatswood, Australia: Reed Books, 1994).

2. Jean Michel Brustet et al., "Characterization of Active Fires in West African Savannas by Analysis of Satellite Data: Landsat Thematic Mapper," in *Global Biomass Burning*, ed. Joel S. Levine (Cambridge, MA: MIT Press, 1992), 53–60.

3. J. Fishman et al., "Identification of Widespread Pollution in the Southern Hemisphere Deduced from Satellite Analysis," *Science* 252 (1991): 1693–96.

4. Stein Mano and Meinrat O. Andreae, "Emission of Methyl Bromide from Biomass Burning," *Science* 263 (1994): 1255–56.

Molly O'Meara, "The Next Hurdle in Ozone Repair, *World Watch* 9, no. 6 (1996): 8.

5. R. K. Pachauri and A. Reisinger, eds., *Fourth Assessment Report of the Intergovernmental Panel on Climate Change* (Geneva: IPCC, 2007), 36.

## Chapter 20

1. David Sheridan, *Desertification of the United States* (Washington, DC: Council on Environmental Quality, 1981), 21.

## Chapter 21

1. Masanobu Fukuoka, *The One-Straw Revolution: An Introduction to Natural Farming* (Emmaus, PA: Rodale Press, 1978), 181 pp.

2. Paul R. Ehrlich, *The Machinery of Nature: The Living World around Us and How It Works* (New York: Simon & Schuster, 1986), 162.

3. Rachael Feltman, "Invasive Asian Carp Could Overtake Lake Erie," *Washington Post*, January 5, 2016.

   Hongyan Zhang et al., "Forecasting the Impacts of Silver and Bighead Carp on the Lake Erie Food Web," *Transactions of the American Fisheries Society* 145, no. 1 (2015): 136, doi: 10.1080/00028487.2015.1069211.

4. NASA Earth Observatory, "Spring Dust Storm Smothers Beijing, March 19, 2002," accessed November 15, 2015, http://earthobservatory.nasa.gov/IOTD/view.php?id= 2285.

5. Fred Pearce, "Wall of Trees Keeps Deserts at Bay in China," *New Scientist*, December 3, 2014, no. 2999, accessed November 22, 2015, https://www.newscientist.com/article /mg22429994.900-great-wall-of-trees-keeps-chinas-deserts-at-bay/.

6. Figures provided in an on-site meeting with the director of Barari Forest Management (the company managing the forests), Abu Dhabi, February 3, 2014.

## Chapter 22

1. International Atomic Energy Agency, "Belching Ruminants, a Minor Player in Atmospheric Methane," 2008, accessed December 3, 2015, http://www-naweb.iaea.org/nafa /aph/stories/2010-methane-ruminant-livestock.html.

## Chapter 23

1. Y. Baskin, *Under Ground: How Creatures of Mud and Dirt Shape Our World* (Washington, DC: Island Press, 2005).

   E. Pucheta et al., "Below-Ground Biomass and Productivity of a Grazed Site and a Neighbouring Ungrazed Exclosure in a Grassland in Central Argentina," *Austral Ecology* 29 (2004): 201–08.

## Chapter 25

1. *Agenda 21*, Chapter 12: "Report on the Plan of Action to Combat Desertification," U.N. Conference on Environment and Development, Rio de Janeiro, 1992.

2. United Nations, "Sustainable Development Goals: 17 Goals to Transform Our World (2015)," accessed January 25, 2016, http://www.un.org/sustainabledevelopment /biodiversity/.

## Chapter 28

1. Information obtained from personal correspondence with David Wallace.

## Chapter 29

1. Silvio Marcacci, "US Solar Energy Grew an Astounding 418% from 2010–2014," Clean Technica, April 24, 2014, accessed February 14, 2016, http://cleantechnica .com/2014/04/24/us-solar-energy-capacity-grew-an-astounding-418-from-2010-2014/.

2. Daniel Cusick, "Solar Power Enjoys Unprecedented Boom in the U.S." *Scientific American*, March 10, 2015, accessed February 14, 2016, http://www.scientificamerican.com /article/solar-power-sees-unprecedented-boom-in-u-s/.

3. L. Hunter Lovins and Amory B. Lovins, "How Not to Parachute More Cats: The Hidden Links between Energy and Security," paper prepared for the Center for a Postmodern World conference, Toward a Postmodern Presidency: Vision for a Planet in Crisis, held at the University of California–Santa Barbara, June 30–July 4, 1989; available from Rocky Mountain Institute, 1739 Snowmass Creek Road, Snowmass, CO 81564.

4. Alliance Commission on National Energy Efficiency Policy, *The History of Energy Productivity* (Washington, DC: Alliance to Save Energy, 2013), 44 pp.

## Chapter 30

1. David Williams, "Plastic to Outweigh Fish in Oceans by 2050, Study Warns," January 19, 2016, accessed February 22, 2016, http://phys.org/news/2016-01-plastic-outweigh-fish -oceans.html.

2. Joseph W. Thornton, Michael McCally, and Jane Houlihan, "Biomonitoring of Industrial Pollutants: Health and Policy Implications of the Chemical Body Burden, *Public Health Reports* 117 (2002): 315–23.

3. ISO Survey of Certifications, 2015, accessed February 2, 2016, http://www.iso.org/iso /home/standards/management-standards/iso14000.htm.

## Chapter 34

1. D. M. Gammon and B. R. Roberts, "Aspects of Defoliation during Short Duration Grazing of the Matopos Sandveld of Zimbabwe," *Zimbabwe Journal of Agricultural Research* 18 (1980): 29.

## Chapter 35

1. André Voisin, *Better Grassland Sward* (London: Crosby Lockwood & Son Ltd., 1960), 95–124.

2. Aldo Leopold, *Game Management* (Madison: University of Wisconsin Press, 1986), 124–36.

3. W. C. Lowdermilk, *Conquest of the Land through 7,000 Years,* Agriculture Information Bulletin no. 99, U.S. Department of Agriculture, Soil Conservation Service, issued 1953 and slightly revised in 1975.

4. Clive Ponting, *A Green History of the World: The Environment and the Collapse of Great Civilizations* (New York: St. Martin's Press, 1991), 81.

5. Ibid., 292.

## Chapter 37

1. H. Jachmann, "Comparison of Aerial Counts with Ground Counts for Large African Herbivores," *Journal of Applied Ecology* 39 (2002): 841–52.

## Chapter 39

1. Deborah H. Stinner, Benjamin R. Stinner, and Edward Martsolf, "Biodiversity as an Organizing Principle in Agroecosystem Management: Case Studies of Holistic Resource Management Practitioners in the USA," *Agriculture Ecosystems and Environment* 62 (1997): 199–213.

## Chapter 42

1. J. A. Onsanger, "An Ecological Basis for Prudent Control of Grasshoppers in the Western United States," *Proceedings of the Third Triennial Meeting, Pan American Acrididae Society,* 5–10 July, 1981 (Pan American Acrididae Society, 1985), 98.

## Chapter 43

1. A sampling of papers include the following:

B. G. Ferguson et al., "Sustainability of Holistic and Conventional Cattle Ranching in the Seasonally Dry Tropics of Chiapas, Mexico," *Agricultural Systems* 120 (2013): 38–48.

R. Alfaro-Arguello et al., "Steps toward Sustainable Ranching: An Energy Evaluation of Conventional and Holistic Management in Chiapas, Mexico," *Agricultural Systems* 103 (2010): 639–46.

T. Wang et al., "GHG Mitigation Potential of Different Grazing Strategies in the United States Southern Great Plains," *Sustainability* 7 (2015): 13500–521.

W. R. Teague et al., "Grazing Management Impacts on Vegetation, Soil Biota and Soil Chemical, Physical and Hydrological Properties in Tall Grass Prairie," *Agriculture, Ecosystems and Environment* 141 (2011): 310–22.

Keith T. Weber and Shannon Horst, "Desertification and Livestock Grazing: The Roles of Sedentarization, Mobility and Rest," *Research, Policy and Practice* 1 (2011): 19.

2. Louis Liebenberg, *The Art of Tracking: The Origin of Science* (Cape Town: New Africa Books, 2012), 174; also available as a free download at http://www.cybertracker.org /downloads/tracking/The-Art-of-Tracking-The-Origin-of-Science-Louis-Liebenberg.pdf.

# GLOSSARY

**Animal days per hectare (ADH) or acre (ADA).** A term used to express simply the volume of forage taken from an area in a specified time. It can relate to one grazing in a paddock or several, in that more grazings than one can be added to give a total ADH or ADA figure. The figure is arrived at by a simple calculation as follows:

$$\frac{\text{Animal Numbers} \times \text{Days of Grazing}}{\text{Area of Land in Hectares or Acres}} = \text{ADH or ADA}$$

**Animal-dependent grasses.** Perennial grass plants that grow in bunched form with growth points close to the ground. Ungrazed, they tend to die prematurely due to old oxidizing material.

**Animal impact.** The direct physical influences animals have on the land other than grazing—trampling, digging, dunging, urinating, salivating, rubbing, and so forth.

**Biodiversity.** The diversity of plant and animal species—and of their genetic material and the age structure of their populations.

**Biomass.** The mass, or volume, of life—plants, animals, and microorganisms.

**Brittleness scale.** All terrestrial environments, regardless of total rainfall, fall somewhere along a continuum from nonbrittle to very brittle. For simplicity, we refer to this continuum as a 10-point scale—1 being nonbrittle and 10 being very brittle. Entirely nonbrittle environments are characterized by (1) good distribution of humidity throughout the year and (2) a high rate of biological decay in dead plant material, which is most rapid close to the soil surface. Very brittle environments are characterized by (1) poor distribution of humidity throughout the year and (2) chemical (oxidation) and physical (weathering) breakdown of dead plant material in the absence of herding ungulates.

In nonbrittle environments, dead grass leaves and stems are soft, moist, and crumple when squeezed in your hand. In brittle environments, they are brittle, dry, and shatter when squeezed in your hand.

**Capping, immature.** A soil surface that has sealed with the last rainfall and on which there is as yet no visible sign of successional movement.

**Capping, mature.** An exposed soil surface on which succession has proceeded to the level of an algae-, lichen-, and/or moss-dominated community and has stalled at that level. In more brittle environments, if not adequately disturbed, such communities can remain in this state for centuries provided the soil is level enough to inhibit erosion by water. Some people refer to this as soil crusting or cryptogamic soil.

**Community dynamics**. The ever-changing dynamics within a biological community. This process is ongoing due to the constant interplay of species, changing composition, and changing microenvironment.

**Crumb structure**. A soil that has good crumb structure is made up largely of aggregates or crumbs of soil particles held together when wet or dry with a glue provided by decomposing organic matter and soil life forms. The space around each crumb provides room for water and air and this in turn promotes plant growth.

**Desertification.** A process that starts with an increase in bare soil between grass plants, which leads to reduced rainfall effectiveness and thus a gradual reduction in plant life. Symptoms include increased incidence of flood and drought, with no major change in rainfall, and numerous social problems tied to the loss of land productivity and water sources: poverty, social breakdown, dislocation, and violence.

**Forbs.** Tap-rooted herbaceous plants often referred to as *weeds*.

**Grazing, frequent**. Grazing that takes place with short intervals between the actual grazings on the plant. With most plants, frequent grazing is not harmful as long as the defoliation is light.

**Grazing, low-density**. (Sometimes referred to as *patch* or *selective grazing*.) This refers to the grazing of certain areas while others nearby are left ungrazed and on which plants become old, stale, and moribund. Normally it is caused by stock grazing at too low a stock density, too small a herd, or a combination of these, or with too short a time in a paddock. Once it has started, even by only one grazing, it tends to get progressively worse, as the nutritional contrast between regrowth on grazed areas and old material on ungrazed areas increases with time.

**Grazing, planned**. (A common abbreviation for *Holistic Planned Grazing*.) The planning of livestock grazing in line with a holistic context that caters simultaneously for many variables: animal behavior, breeding, performance, wildlife needs, other land uses (such as cropping), weather, plant growth rates, poisonous plants, dormant periods, droughts, and so forth.

**Grazing, rotational**. Grazing in which animals are rotated through a series of paddocks, generally on some flexible basis, but without planning that caters for the many variables inherent in the situation.

**Grazing, severe**. Grazing that removes a high proportion of the plant's leaf in either the growing or the nongrowing season.

**Grazing, strip**. The grazing of animals in narrow strips of land, generally behind a frequently moved electric fence. In some cases, different areas are strip-grazed within a paddock.

**Grazing, ultra-high density**. The grazing of livestock in such a manner that they are at extremely high stock densities continuously. Generally, these densities are achieved by enclosing livestock in small areas with the use of movable fencing.

**Grazing cell**. (Also referred to as *grazing unit*.) An area of land planned for grazing management purposes, normally as one unit for one herd to ensure appropriate timing of grazing/trampling and recovery periods. It is generally divided into smaller units of land (paddocks, or grazing divisions) by fencing in some form, or marks on the ground or vegetation that herders can observe. A grazing cell will normally contain livestock year-round, or at least for a prolonged period of time.

**Herd effect**. The impact on soils and vegetation produced by a herd of bunched animals. Herd effect is not to be confused with stock density, as they are different, although often linked. You can have high herd effect with very low stock density (e.g., the bison of old that ran in very large herds at very low stock density, as the whole of North America was the paddock.) You can have high stock density with no herd effect, such as when two or three animals are placed in a half-hectare (one-acre) paddock. At ultra-high densities, the behavior of livestock will change adequately to provide herd effect.

*Note:* Herd effect is the result of a change in animal behavior and usually has to be brought about by some actual management action—stimulating the behavior change with the use of an attractant, crowding animals to ultra-high density, or, on large tracts of land, herding animals—the larger the herd, the better—and keeping them bunched through much of the day.

**Overgrazing.** Land cannot be overgrazed, as commonly stated, only plants can. Overgrazing occurs when a plant bitten severely *in the growing season* gets bitten severely again while using energy it has taken from its crown, stem bases, or roots to reestablish new leaves. Generally, this results in the weakening or eventual death of the plant.

**Overrested plant**. A bunched perennial grass plant that has been rested so long that accumulating dead oxidizing material prevents light from reaching growth points at the plant's base and hampers new growth, weakening or eventually killing the plant. Overrest occurs mainly in the more brittle environments, where, in the absence of large herbivores, most old material breaks down in sunlight through oxidation and weathering rather than biological decay.

**Paddock.** (Also referred to as a *grazing division*.) A small division of land within a grazing cell in which stock are grazed for short periods of time—hours to days. Paddocks can be fenced or demarcated by natural features known to herders.

**Rest, partial**. Takes place when grazing animals are on the land but widespread with little bunching behavior and herd effect. Even when plant overgrazing is stopped with Holistic Planned Grazing, partial rest remains a problem on large fenced ranches (as opposed to where livestock are herded), where it can cause land improvement to stagnate if not detected and acted upon.

**Rest, total**. Prolonged nondisturbance of soils and plant/animal communities. A lack of any physical disturbance and/or fire.

**Rest-tolerant grasses**. Perennial grasses that are able to thrive under rest in brittle environments. Commonly, such plants have some growth points, or buds, well above ground (detectable by branching stems) where unfiltered sunlight can reach them; or they are short in stature or sparse-leafed, enabling unfiltered light to reach their ground-level growth points. In the past, such grasses tended to be found in steep gorges and other sites large grazing animals did not frequent, but today, where overgrazing is believed to be linked to animal numbers and thus numbers have been reduced, these grasses can dominate large areas of land.

**Stock density**. The number of animals run on a subunit (paddock) of land at a given moment of time. This could be from a few minutes to several days. Usually expressed as the number of animals (of any size or age) run on one hectare or acre. Some express this as total weight of animals on a unit of land.

**Stocking rate**. The number of animals run on a unit of land usually expressed in the number of hectares or acres required to run one full-grown animal throughout the year or part thereof.

**Succession**. An important aspect of community dynamics, succession describes the process, or stages, through which biological communities develop increasing complexity. As simple communities become ever more diverse and complex, succession is said to be advancing. When complex communities are reduced to greater simplicity and less diversity, succession is set back. If the factors that set it back are removed, succession will advance once again.

# ABOUT SAVORY GLOBAL

*Savory Global facilitates large-scale restoration* of the world's grasslands through Holistic Management. Savory's goal is to regenerate one billion hectares of grasslands through our global network to mitigate climate change and enhance water and food security. Savory Global includes:

- The **Savory Institute** equips the global network with innovate tools and curricula, informs policy, engages commercial markets, increases public awareness, and conducts relevant research on ecological, social, and financial outcomes. Additionally, the Institute develops and incubates these programs through its experiential learning site, West Bijou Ranch, in Colorado, USA.

- The **Savory Global Network**, comprised of Savory Hubs, Accredited Professionals, and Regenerating Members, advocates, trains, and supports land managers to regenerate land in their own context through Holistic Management.

- The **Land to Market** program, deployed by the global network is the world's first outcome-based verified **regenerative sourcing** solution. The program connects conscientious brands, retailers, and consumers directly to producers who are regenerating their land. The more farmers and ranchers the hubs train, support, and verify, the more grasslands regenerated, and the greater the environmental, economic, and social impact globally.

 http//:savory.global

# INDEX